Sustainable Food System in the European Union

Sustainable Food System in the European Union

Editors

Mariarosaria Lombardi
Vera Amicarelli
Erica Varese

MDPI • Basel • Beijing • Wuhan • Barcelona • Belgrade • Manchester • Tokyo • Cluj • Tianjin

Editors

Mariarosaria Lombardi
University of Foggia
Foggia, Italy

Vera Amicarelli
University of Bari Aldo Moro
Bari, Italy

Erica Varese
University of Turin
Turin, Italy

Editorial Office
MDPI
St. Alban-Anlage 66
4052 Basel, Switzerland

This is a reprint of articles from the Special Issue published online in the open access journal *Sustainability* (ISSN 2071-1050) (available at: https://www.mdpi.com/journal/sustainability/special_issues/Sustainable_Food_System_EU).

For citation purposes, cite each article independently as indicated on the article page online and as indicated below:

LastName, A.A.; LastName, B.B.; LastName, C.C. Article Title. *Journal Name* **Year**, *Volume Number*, Page Range.

ISBN 978-3-0365-8318-1 (Hbk)
ISBN 978-3-0365-8319-8 (PDF)

Cover image courtesy of Mariarosaria Lombardi

Contents

About the Editors

Mariarosaria Lombardi

Mariarosaria Lombardi is an Associate Professor Ph.D. in Commodity Sciences at the Department of Economics, University of Foggia (Italy). She has been a Rector's delegate for Coordination of Planning and Accreditation Activities since May 2023 and a departmental delegate for the Third Mission (relationship with the territorial stakeholders) since June 2017. She is currently engaged in two university courses: "Innovation and Sustainable Technologies" and "Economics and Technologies of Energy Sources." She is the author of 80 papers published in refereed journals and academic volumes. She deals with the innovations and socio-economic and environmental sustainability assessment of supply chains, using mainly Carbon Footprint and Material Flow Analysis methodology, focusing mostly on agro-energy and agro-food systems in relation to climate change. Particular attention is also paid to social innovation as an alternative approach to improving the sustainability of the agro-food sector. She has been part of different national and international research groups' projects, as coordinator of some Erasmus Intensive Program projects on bioenergy and leader of a WP for the European FP7 project, STAR*AgroEnergy, coordinated by the University of Foggia. Currently, she is a member of a Ph.D. Program and co-advisor of a Ph.D. project at the Department of Economics of the University of Foggia. She is a reviewer for several international scientific journals and guest editor for some open access journals, including Sustainability, where she is also a member of the relative editorial board (Food section). In 2019, she received the Future Food Institute and FAO certification as a "Climate Shaper." Since 2015, she has been a member of Vazapp, a social innovation initiative that promotes relations among farmers to develop a more sustainable agro-food system. She is a member of the Italian Commodity Science Academy and the International Society of Commodity Science and Technology.

Vera Amicarelli

Vera Amicarelli is an Associate Professor Ph.D. in Commodity Science at the Department of Economics, Management, and Business Law at Aldo Moro University of Bari, Italy. In undergraduate and graduate courses, she teaches Industrial Ecology, Quality Environment and Sustainability, and Resource and Waste Management. She is the author of more than 80 papers published in refereed journals and academic volumes. Her current research interests focus on Material Flow Analysis, Environmental Indicators such as Water and Carbon Footprint, Food and Textile Waste Minimization, and Circular Economy. Her main academic activities are related to the Erasmus+ exchange program; she is the department delegate and coordinator for several agreements with foreign universities. Additionally, she is a member of the Ph.D. Professor Committee and advisor of Ph.D. projects. She is a member of the Italian Commodity Science Academy (AISME), IGWT (International Society of Commodity Science and Technology), and the Italian Circular Economy Stakeholders Platform (ICESP), and is included in the CREA (Council for Agricultural Research and Analysis of the Agricultural Economy) expert register. She is a reviewer for several international scientific journals, and since 2014, she has been a guest editor for various scientific journals, including the International Journal of Sustainable Economy (InderScience), Sustainability (MPDI), and British Food Journal (EMERALD). She has been and is a member of national and international research groups' projects.

Erica Varese

Erica Varese earned a Ph.D. in "Culture and Enterprise—Entrepreneurial Strategies in the Age of Globalization" in 2005. She is currently an Associate Professor in Commodity Science at the Department of Management "Valter Cantino," University of Torino (Italy), where she teaches undergraduate, graduate, and master courses in Italian and English. Her main research areas include food-related aspects (such as Food Businesses, Labeling, Quality Assurance and Certification Schemes, Food Waste Management, and Industrial Tourism); Consumer Science, Protection and Perception; International Trade and Customs; and Circular Economy. She has been an Erasmus visiting professor at Canterbury Christ Church University (Canterbury, UK). At the Université Nice Sophia Antipolis (Nice, FR), where she is a standing invitee, she has attended teaching and research activities. She has published articles in many refereed journals, contributed to chapters and books, and presented papers at conferences on a global basis. She is a member of the International Society of Commodity Science and Technology (IGWT), of the Accademia Italiana di Scienze Merceologiche (Italian Academy of Commodity Science), and Full Academician of the Accademia di Agricoltura di Torino (Agricultural Academy of Torino, IT). She is a reviewer for several international scientific journals, has been a guest editor for the scientific journal Sustainability (MPDI), and has been and is a member of national and international research groups' projects.

Preface to "Sustainable Food System in the European Union"

Among the efforts to tackle climate and environmental challenges, the European Green Deal (EGD) plays an important role in proposing a holistic approach in which all European actions and policies contribute to the objectives of the Green Deal itself. The EGD plans new, sustainable, and inclusive growth strategies to enhance the economy, improve people's health and quality of life, protect nature, and ensure no one is left behind. The Farm to Fork (F2F) Strategy is one of the EGD strategies focused on sustainable food systems and the complex links between healthy people, healthy societies, and a healthy planet. The F2F strategy, while emphasizing the awareness that food systems remain one of the key drivers of climate change and environmental degradation, highlights the need for a transition towards a green agri-food system.

The main goals of the F2F strategy are to ensure sufficient, affordable, and nutritious food within planetary limits; to guarantee sustainable food production through a substantial reduction in pesticides, antimicrobials, and fertilizers use, and an increase in organic farming; to promote more sustainable food consumption and healthy diets; to minimize food loss and waste; to fight food fraud; and to improve animal welfare. The F2F strategy's aim can be summarized as a shift to a sustainable food system that brings environmental, health, and social benefits, offers economic gains, and ensures sustainable growth.

This Special Issue of Sustainability has contributed to this field by collecting high-quality studies and research related to the complexity of food systems to measure the progress in achieving F2F and EGD strategies' goals.

The Guest Editors are glad that this Special Issue has attracted the attention of several researchers who shared their results by submitting original research articles, case studies, reviews, critical perspectives, and viewpoint articles. The Special Issue's final results provide an updated picture of the effectiveness of the F2F strategy and any corrective actions needed.

Mariarosaria Lombardi, Vera Amicarelli, and Erica Varese

Editors

Article

Consumer Choice for Milk and Dairy in Romania: Does Income Really Have an Influence?

Diana Maria Ilie [1], Georgiana-Raluca Lădaru [2], Maria Claudia Diaconeasa [2,*] and Mirela Stoian [2]

[1] The Agricultural Economics Office, Research Institute for Agriculture Economy and Rural Development, 010961 Bucharest, Romania; necula.diana@iceadr.ro

[2] The Department of Agrifood and Environmental Economics, Faculty of Agrifood and Environmental Economics, The Bucharest University of Economic Studies, 010961 Bucharest, Romania; raluca.ladaru@eam.ase.ro (G.-R.L.); mirela.stoian@eam.ase.ro (M.S.)

* Correspondence: maria.diaconeasa@eam.ase.ro; Tel.: +40-760-312-002

Abstract: Milk and dairy are basic food products and their importance in healthy human development is well known. However, this does not mean that the consumers' requests for these products are not evolving and fitting into the new context of sustainable development. By conducting a quantitative analysis on 847 answers regarding milk and dairy consumption offered by Romanian consumers, the objective of this study is to reveal what are the main factors of influence for respondents when choosing a milk or dairy product, and to see if these factors are evolving towards including sustainability-related aspects. The results point out that while price and store availability are still present as choice criteria, new aspects that might be related to a sustainable behavior, such as ecologic certification, country of origin or traditional products, are considered by the respondents when purchasing milk and dairy. However, this depends on the level of income; higher incomes allow respondents to consider new criteria.

Keywords: consumer behavior; milk and dairy choice; sustainable choice; influencing factors; income influence

Citation: Ilie, D.M.; Lădaru, G.-R.; Diaconeasa, M.C.; Stoian, M. Consumer Choice for Milk and Dairy in Romania: Does Income Really Have an Influence? *Sustainability* **2021**, *13*, 12204. https://doi.org/10.3390/su132112204

Academic Editors: Mariarosaria Lombardi and Giuseppe Todde

Received: 22 September 2021
Accepted: 3 November 2021
Published: 5 November 2021

1. Introduction

The debate around milk and dairy consumption has become more important along with the increase of nutritional information [1,2], the consumers' need for ensuring balanced and healthy diets for themselves and their children [3], but also due to the possible environmental impact of animal farms [4], and even possible health risks determined by this type of products such as allergies or intolerance [5]. Increasingly, how the choices made by consumers affect the development of the planet, meaning sustainable choices [1,6], including food products, are getting to be more present in the regular choice patterns [7].

The international funds and grants for agriculture always aim a significant percentage of their support at farms for milk production as this product is considered a basic one [8]. Yet, a slight change in the agricultural policy and support schemes, such as the lift of milk quotas in the European Union (EU), has major impact for the producers, affecting them differently based on the market size and farm size, determining important progress for Danish farmers and the incapacity of being competitive for Greek farmers, therefore bringing major changes for the local markets [9].

The milk and dairy market potential of Romania, as a member of the EU since 2007, serves as a particular case for this study considering on one hand the tradition of consuming milk and dairy from a variety of species (cow, sheep, goat, buffalo and even donkey), the country being part of the Balkan region [10], and on the other hand considering the constant negative trade balance for milk and dairy, meaning that the local products are insufficient for satisfying the consumers' needs [11]. Additionally, the GDP per capita for the EU countries in 2020 placed Romania as the last but one among the 27 member states. The GDP per capita in Romania was 8810 euro, compared to the EU27 average of 26,380 euro [12]. Meanwhile, the harmonized index of consumer prices in 2020 for the milk and dairy category shows a higher increase for Romania than for the EU 27 average, meaning that the prices for this type of product has raised faster than in the EU [13]. Moreover, Romania is in the last place considering the disposable income reported to the consumption expenditure in the EU, meaning that the people spend much of their income on satisfying their basic needs, such as providing food [14]. Therefore, the influence of income on the consumption choices that consumers make should be a key factor to look into for Romania and is considered as the main aspect of investigation for this study.

Several studies focus on determining the aspects that influence the consumer choice for specific food products, both positively and negatively. Therefore, they investigate consumer behavior [15]. Beginning with the obvious factors such as price or availability, which have been observed by marketers to have a high influence on the purchasing and consumption behavior, the new socio-economic and environmental context presents itself with new factors that change this behavior, such as the willingness to pay for more sustainable products [1]. In the case of milk and dairy, Nam et al. [16] has observed such a shift regarding the consumers' willingness to pay for mountain dairy produced in sustainable farms.

Understanding the factors that influence the consumer choice for milk and dairy, as important nutritional providers, and determining if there are any tendencies towards sustainable choices serve as the purposes of this study.

The paper should be of interest both to local and international producers in the milk industry, as they should be aware of the consumer expectations and purchasing power so to adjust their offer accordingly, but also to policy makers in documentation for future food policies intended at supporting and educating production and consumption in the milk and dairy sector.

2. Theoretical Background

2.1. Importance of Sustainable Food

Since the Brundtland Report [17], which proposed sustainable development as a solution for improving the quality of the environment and society in the long term, along with economic development, and also until the Sustainable Development Goals [18], the ways of production in domains such as agriculture [19,20], construction [21,22], industry [23] (including the dairy industry [24]), and, more recently, consumption of different products such as food [1,25] and fashion [26], or services such as tourism [27], have been questioned and solutions for making them more sustainable have been proposed.

The case of sustainable dairy is a sensitive one. On one hand, milk and dairy, along with meat and eggs, represent a prime source of superior protein, known for thousands of years, so it is natural to observe increasing trends in the consumption of these products while countries register economic and social development [16,28]; this also being the case of EU Central and Eastern countries, where GDP values increased compared to the EU average [29]. On the other hand, the intensive dairy farming industry is recognized by the high environmental impact and contribution to global warming, acidification, energy consumption and land occupation [30,31], which makes it unsustainable. Therefore, the alternative may reside in traditional farms with a small production of traditional products [32] or mountain products [30], which are increasing on the consumers preference list [16,33].

Regarding the notion of sustainable food, the FAO [34] envisions it as food that is nutritious and accessible for everyone, while natural resources are managed to support the current and future human needs. Otherwise, there are different accepted characteristics that can make a food product recognized as sustainable, such as plant-based [35] or insect-based [36], with a less meat-based composition [37], seasonal food [38,39], locally grown and produced food [39], and organic food [40]. Additionally, there are a series of accepted barriers to consuming sustainable food. For example, cultural barriers as the reluctance to consume cultured meat or insects [39], financial barriers [41], or even habituality barriers [42].

2.2. Importance of Milk and Dairy in Diets

Around the world, milk and dairy have been known as food sources for a long time. Milk is acknowledged as a complete food, composed by all nutrient categories. Moreover, other dairy products such as yogurts are included in the category of functional foods, meaning products that are beneficial for the health and wellbeing of the consumer [2,43].

Several authors mention the importance of milk and dairy consumption especially for pregnant women, children, adolescents, and older people, due to the increased composition of mineral salts and vitamins, responsible for the proper development and maintenance of bones and muscles [3,44,45]. Adding on this, Givens [3] mentions that threats of increased cardiovascular disease due to milk and dairy consumption are disproved by clinical studies, while the correlation between yogurt consumption and type II diabetes needs to be further studied.

Regarding the regular consumption of milk and dairy, studies have determined an average of 2–3 servings per day, depending on the availability of these products and their presence in the culture of a country [46], being a regular presence in an extended part of the globe [47].

Since this type of product may contribute to ensuring food security through the nutritional values and its widespread, the level of income should not be a factor of influence in milk consumption. Nevertheless, studies show that lower incomes lead to poorer choices in milk quality [48,49]. Moreover, other research points out that the lower income groups have a higher sensitivity than medium and high-income groups to income and price fluctuations when choosing dairy products [50]. In addition, the income inequalities significantly influence the quality of life of people in developing countries, including their possibility of spending on high quality food products [51].

Demographic factors, such as gender, are known to influence the choice of diet. Women pay more attention to low fat diets and healthy diets than men [52]. Even more, there are studies claiming that men are less willing to pay for higher quality in food products [53] and read the labels superficially [54]. Nevertheless, the consumption of milk in men and women should not differ as it has lifelong benefits [55].

2.3. Sources of Milk and Dairy

While in general terms, milk refers to the product of the cows, they are not the only type of animal which produces edible milk. Park [43] observes that the general tendency is to skip the importance and nutritional value of milk coming from other animals, especially since cows have adapted so well in farms all over the world. Hoowever, the milk and dairy coming from other types of animals such as buffalos (mozzarella), sheep and goats (yogurt and chesses) or even donkey (milk) puts renowned specialties on the market. For example, the Italian mozzarella is a certified product made especially from buffalo milk, which offers it a superior taste and texture [56]. Zicarelli [57] shows that buffalo milk has a higher nutrient content and a lower cholesterol level than cow's milk. However, the farming of such animals is more difficult, needing more water and space, and therefore being less suited for large farms [56]. The case of sheep and goats is also special, as the extensive methods of farming specific to the Mediterranean or Balkan region offer the dairy products particular sensorial qualities and place them among the traditional products sought especially by locals. However, whether they will be able to adapt to the standardized market of the developed countries or they will remain a hard-to-get traditional product is still not known [10,58]. Donkey milk is more known as a treatment for diverse types of affections, such as milk intolerance in infants, having a chemical composition remarkably close to human milk [59,60]. Moreover, using it in the treatment of lung disease, including lung cancer, has raised the interest of scholars [61]. Depending on the local culture and natural fauna, there are other species of animal that provide sources of milk, which are less known or understood at the general level, such as camels, mares, or reindeers [43,62]. In addition, the innovation vector [63] has not jumped over the dairy sector. Research and development have presented alternative plant-based results for milk, such as soy milk, rice milk or almond milk [64].

Nevertheless, due to its high availability and recognizable taste, cow's milk is expected to be the preferred source for consumers [65].

2.4. Factors Influencing the Consumer Choice of Milk and Dairy

Determining the factors that trigger or suppress the purchasing and consumption decision for several types of products has been of interest to researchers and marketers for a long time [1,66]. The universality of these products has attracted attention from researchers in various geographical regions. For example, in Kosovo, a study [67] revealed that the factors that have a significant influence on the choice of dairy are consumer gender, trust in the products, perceived quality, origin, and price of the purchased product. The Slovak consumers consider that price, taste, and quality of the local dairy are strengths, being perceived as healthy, while the imported products excel in packaging and variety [68]. Other researchers [69] show that Chinese consumers are significantly influenced by the country-of-origin of milk and dairy products, trusting them more than the local products, while the preference for a specific country is guided by consumer familiarity and experience with the products, ethnocentrism, and animosity, and even some cultural value differences. For the Italian consumers, the low price and high availability in the supermarkets of cows' milk are main reasons for consumption, while the health benefits of the donkey milk are seen as superior, but the difficulty of finding it in the supermarkets proves to be a significant barrier [70].

The sensory properties of milk and dairy products, such as color, smell, taste, fat quantity or density are powerful indicators for consumers in choosing a particular product [2]. Others focus on factors that may be related to a sustainable choice, such as origin of the products, determining a preference for local and mountain products [13] or the certification of Good Agricultural Practice, in the case of Japanese consumers [71]. Other authors observe that basic factors such as availability of products, price and packaging significantly influence the consumers in making a choice for milk and dairy products [70,72].

Some newer factors indicate that not all consumers are open to trying organic products, but there are some for which ethical aspects and green consumerism are motives for purchasing organic products [40]. Other authors point out that in higher-income countries, green purchases have the role of bringing people closer to the environment. Therefore, sustainable food choice is becoming more pressing especially in these countries [73]. However, other studies [74] claim that emerging economies have a higher willingness to pay for environmentally certified food produce. Roman et al. [75] find that for people who give a higher importance to natural foods, the willingness to eat ecological or organic food increases, while other studies point out that consumers are willing to pay more for sustainable food products or food with sustainable characteristics. For example, Gao et al. [33] claim that the willingness to pay for sustainable dairy is 40% higher than for regular dairy in the case of Chinese consumers. Other authors [76] claim that Spanish consumers are willing to pay more for locally grown almonds, as opposed to long traveled almonds. Adding on this, other studies point out that some European consumers are willing to pay more for locally captured fish, due to the trustful standards and effective communication regarding the standards [77]. Other aspects considered by consumers as worthy of paying more are innovative packaging solutions in the case of milk and dairy [78], or the provenance from small farms, that actually diminishes the need for organic certification [79]. Due to these previous studies, we consider that the willingness to pay more for milk and dairy with sustainability related aspects from the Romanian consumers is of further interest. In this case, the hypotheses of the current research were based on the previous studies on the influencing factors of consumption of milk and dairy.

Given previous studies [51–54,68], we consider that the correlation between gender and the choice criteria for milk and dairy should be further investigated, and we expect that some considerable differences between respondent gender groups would be revealed. Hence, hypothesis one was formulated.

Hypothesis 1 (H1). *There is a significant correlation between gender and the considered choice criteria for milk and dairy.*

Considering earlier information [12–14,48–51], we expect income to be significantly correlated with the availability of ecological products [16,33,40,75], given the fact that higher income groups would afford purchasing products with higher price [55,70,72]. Additionally, a significant correlation is expected between income and perceived quality of the products [67,68,77], given the fact that people expect to have the highest benefits from their purchases; and between income and traditional products, such as local products, especially coming from small producers [16,33,68,76,79]. The testing for these criteria has been considered through the willingness to pay for products that are certificated, traditional and have a high perceived quality, but also the declared expenditure for them is at least a medium one per week. Therefore, the second hypothesis for this study is the following:

Hypothesis 2 (H2). *There is a significant correlation between the income of the respondents and the environmental-related criteria (availability of ecological products, perceived quality, or traditional products), expressed through their willingness to pay more for these products (H2a) and by having at least a medium weekly expenditure for them (H2b).*

Because several authors mention the country of origin in their research [67,69,76], this characteristic complying with both expectations for lower price [67,68,70] and contribution to supporting local production [76,79], we considered the correlation between income level and the country of origin in the third hypothesis.

Hypothesis 3 (H3). *There is a significant correlation between the income of the respondents and the country of origin for milk and dairy.*

Store availability is mentioned constantly in previous studies [70–72], meaning that the consumers would buy what is available if they come to the store with the purpose of buying milk, even if it may not satisfy their ethical or environmental expectations; therefore we expect income and store availability to be strongly correlated, as opposed to a lower correlation between possible health recommendations or long term health benefits of milk consumption and income [68,70].

Hypothesis 4 (H4). *There is a lower correlation between the health recommendations of milk and dairy consumption and income of the respondents than between large retail store availability of these products and income of the respondents.*

3. Materials and Methods

Considering that Romania has a negative trade balance regarding milk and dairy products, as it may be seen in Figure 1, especially regarding cheese and curd, but also for raw milk, as a total for all species that are traded, it presents a particular case for studying the factors that guide Romanian consumers in their choice of purchase and consumption for milk and dairy. Since the import of milk and dairy is at a high rate, how consumers take into consideration the country of origin for these products and their appreciation for the local produced ones is of interest and will be shown later in the study. Additionally, the low level of income and GDP per capita, compared to the steep increase in the harmonized index of consumer prices, earlier presented [12–14], support the choice of the case study considered in this research.

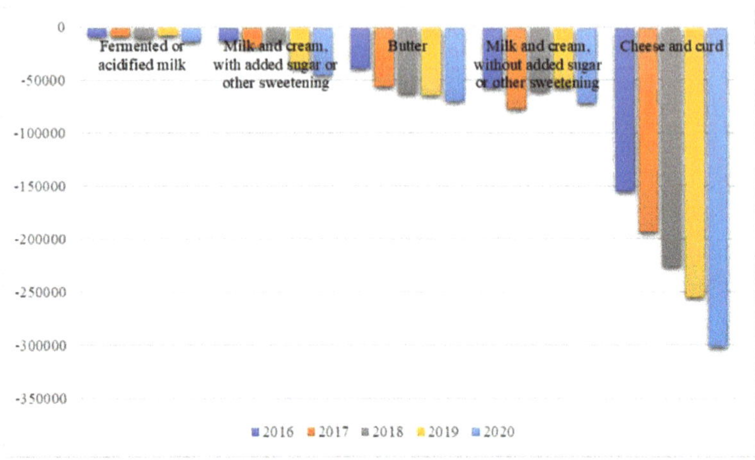

Figure 1. Romanian trade balance for milk and dairy between 2016–2020 (thousand US Dollars). Source: International Trade Center [11].

Regarding the method of gathering the information, the quantitative survey was considered, using the structured investigation technique through a self-administered questionnaire [80]. Convenience sampling using the "snowball" method was used as a sampling method [81].

In order to determine the sample size, the Taro Yamane method was used [82], according to which:

$$n = N / (1 + N * e^2), \quad (1)$$

where n is the sample size, N is the total population size, and e is the accepted error.

For a total population of 17,592,625 people over the age of 15 years old [83], the size of the determined sample is 847 people, using an error of 3.43%.

The questionnaire contained 29 questions, of which 26 were closed questions and three were open questions. Once developed, the questionnaire was tested on twenty people to gather feedback on understanding the questions and thus improve the quality of the research. Then it was released for the general public.

The variables used in the study of the milk and dairy choice are:

- dependent variables: consumer preferences for milk and dairy products; willingness to pay for sustainability characteristics (ecological and traditional products).
- independent variables: gender, age, and income.

The data were interpreted using a quantitative analysis software SPSS [84] and the semantical differential scale [85] in order to capture and present the main characteristics of the respondents. The answers to the open questions regarding suggestions from the respondents for the milk and dairy producers were interpreted using a map generating software, KH Coder, based on frequency and correlations of the words in the open answers [86].

4. Results and Discussion

From the total number of participants in this study (847 persons), 96.5% declare themselves as consumers of milk and dairy and 3.5% declare they do not consume these types of products. In order to see the structure of the respondents, in Table 1, the frequency of consumption by gender, age groups and income groups is presented.

Table 1. Milk and dairy consumption frequency.

		Frequency of Consumption (%)			
	Distribution (Total %)	Daily	2–3 Times/Week	Weekly	Occasional
Males	26.3	28.17	35.68	12.68	23.47
Females	73.7	34.05	28.60	15.87	21.49
<20 y.o.*	7.7	27.69	36.92	9.23	2.08
20–29 y.o.	58.4	30.64	33.19	16.81	19.36
30–39 y.o.	14.3	38.66	21.01	11.76	28.57
40–49 y.o.	12.4	37.50	29.81	11.54	21.15
>50 y.o.	7.2	31.67	21.67	20.00	26.67
<1000 lei **	22.9	27.03	29.19	20.54	23.24
1001–2000 lei	13.8	27.93	36.94	13.51	21.62
2001–3000 lei	16.7	33.09	40.44	14.71	11.76
3001–4000 lei	15	36.51	28.57	15.08	19.84
4001–5000 lei	11.2	35.16	24.18	16.48	24.18
5001–6000 lei	6.5	30.19	32.08	7.55	30.19
> 6000 lei	13.9	39.66	20.69	10.34	29.31

* y.o. = years old; ** lei = monthly income. Source: authors own interpretation of data.

Considering the distribution of the respondents by consumption frequency, we may see that, from the total number of respondents which consume milk and dairy (818), the majority has a frequent consumption. The percentages were obtained by reporting the number of respondents in a gender, age, or income frequency group to the total number of respondents in that category. The results are in line with previous studies [10,46]; Romania is a Balkan country, and therefore has a long-standing tradition of consuming milk and dairy.

There are some differences that may be observed between gender groups, with females having a higher percentage for daily consumption than men, who register the highest percentage in the 2–3 times/week category.

Considering the differences between age groups, the 30–39 years old category registers the highest percentage of respondents in the occasional frequency, followed by the above 50 years old category. Additionally, the 30–39 years old category has the highest percentage of respondents in the daily frequency group. An interesting observation emerges from the age groups distribution; the categories above 30 years old have the highest percentages in the daily and occasional frequency groups. This may be due to a better knowledge of the personal body and its tolerances and needs that come along with age.

Regarding the income groups, the above 6000 lei per month group registers the highest percentage of respondents in the daily frequency group, followed by the 3001–4000 lei/month income group. In addition, the higher income groups, above 3001 lei/month, register increasing percentages for the occasional frequency group.

Considering the preferred type of milk by animal species, the results of the study are presented in Figure 2.

The preferred source for milk and dairy is cows' milk, with more than 70% of the consumers participating in this study declaring they like it very much. The results are in line with previous studies [43,65]. The high preference for cows' milk is also supported by the higher availability in stores compared to milk and dairy from other species, as well as lower prices [65,70,87]. Additionally, the low national production of milk from other species [88] raises questions regarding the provenance of the products found in stores.

The goats' milk is the second most popular in the respondents' preference list, with more than 13% liking it much and very much [10]. The sheep and buffalos register less than 5% of the respondents who prefer it much or very much, while donkey milk registered insufficient answers to be taken into consideration in the analysis, supporting the idea that the lack of availability in stores is a prime barrier in consumption [65,70].

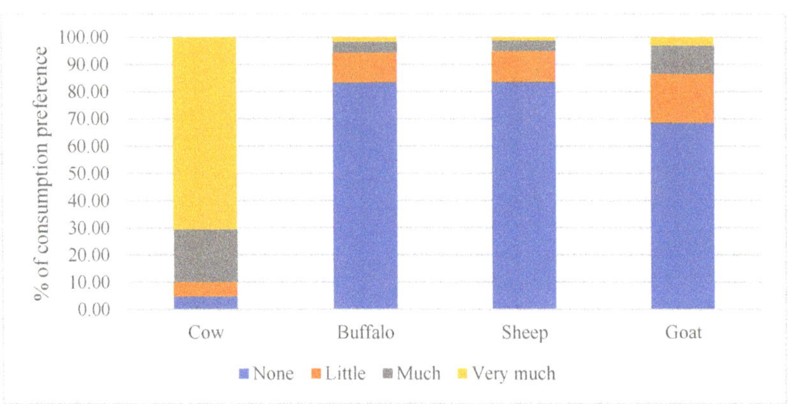

Figure 2. Preferred source for milk and dairy by animal species (%). Source: authors own interpretation of data.

The possible differences between men and women considering diverse selection criteria for milk and dairy products are presented in Figure 3.

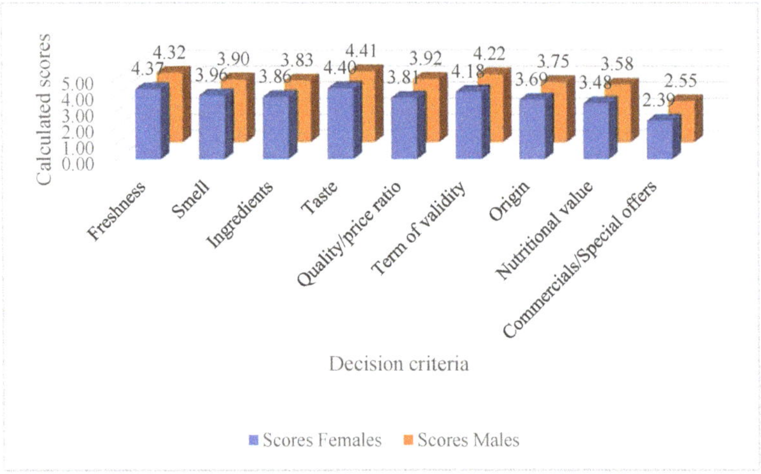

Figure 3. Selection criteria for milk and dairy. Source: authors own interpretation of data.

By using the five-point semantical differential scale [85], the general preference scores were calculated for the female and male respondents of this study.

By calculating the chi-squared test for the correlation of gender and the different choice criteria, with four degrees of freedom and a significance level of 9,49, we point out that there are significant differences between the calculated chi (spread from 0.647 to 9.52 for the different criteria) and the theoretical chi (9,49). Therefore, there is no significant correlation between gender and choice criteria, which refutes the first hypothesis of the study (H1) and presents different results than previous studies [51–53,67].

Considering the descriptive statistics, the most important selection criteria are taste, followed by freshness and term of validity, while the least important are commercials and store offers. The lowest score offers an important insight in the changes of consumer behavior, which is now less influenced by the price offers and pays a higher attention to other selection criteria, pointing to a more educated consumer.

While the freshness, smell, and ingredients have a slightly higher importance for women, the taste, quality/price ratio, term of validity, origin, nutritional value, and price offers are more important for men. The high importance given to the sensorial properties (taste, smell, fat percentage) are important pointers for the fact that consumers are accustomed with this type of product and are able to determine their quality through the sensorial properties, the results supporting previous studies [2].

The willingness to pay for milk and dairy products which present specific environmental or social benefits differentiated by income categories is presented in Table 2.

Table 2. The willingness to pay for better milk and dairy (%).

	Have an Ecologic Certification	Have a Superior Quality	Are Traditional Products	Mixed Answer (Two Options)	All Three Options	Not Available to Pay
<1000 lei **	5.67	34.54	10.82	22.16	21.13	5.67
1000–2000 lei	7.69	28.21	12.82	26.50	19.66	5.13
2001–3000 lei	13.38	21.83	18.31	18.31	21.13	7.04
3001–4000 lei	7.87	22.83	14.96	29.13	21.26	3.94
4001–5000 lei	8.42	36.84	8.42	21.05	20.00	5.26
5001–6000 lei	5.45	25.45	12.73	23.64	25.45	7.27
>6000 lei	6.84	23.08	5.98	25.64	36.75	1.71

** monthly income. Source: authors own interpretation of data.

The willingness to pay higher prices for products that respond to new social or environmental criteria, therefore proving the respondents' involvement in supporting the community it lives in through traditional products, for example, or the care for the environment through ecologic certificated products, or just wanting a higher quality of the products for its own health, are becoming important aspects studied through consumer behavior changes [1,7].

It is important to notice that the non-willingness to pay has the smallest percentage of the respondents' categories of monthly income. Yet, for the 2001–3000 lei/month and 5001–6000 lei/month, these percentages are above 7%. From the three single options, the perceived superior quality is of the highest appeal to the respondents, meaning that the personal gain is more priced than the social or the environmental one, for all income groups. However, the cumulated answers and for two or all three options register more options than the single ones. More importantly, all three options register higher percentages with the higher income groups, which implies a higher income allows a person to consider the social and environmental implications of its purchasing options, the results being in line with previous studies [16,33,75].

There is a significant link between respondents' income and their willingness to buy milk and dairy products at higher prices, with a probability of 95%. The calculated chi-square has a value of 47.68, being higher than theoretical chi of 43.77 for a significance threshold of 0.05. Therefore, the first part of the second hypothesis is confirmed.

The average amount declared to be spent by the respondents for milk and dairy, by groups of prices and incomes, may be seen in Table 3.

The average amount declared to be spent weekly on milk and dairy by the respondents of this study are medium, between 26 and 75 lei/week, being followed by the lesser amount, less than 25 lei/week. Only few respondents spend amounts higher than 75 lei/week for this type of product. Through the chi-square testing, it was found that the calculated chi of 15.56 is less than the theoretical chi of 21.03 for a significance threshold of 0.05, so there is no significant influence of the respondents' income on the amount allocated for the purchase of such products. Therefore, hypothesis H2b is rejected; there is no correlation between the level of income and the weekly expenditure for milk and dairy.

Table 3. Average amount declared to be spent by the respondents for milk and dairy (% of respondents).

	<25 lei	25–75 lei	>75 lei
<1000 lei **	37	52	11
1000–2000 lei	28	57	15
2001–3000 lei	37	54	9
3001–4000 lei	26	58	16
4001–5000 lei	32	55	14
5001–6000 lei	38	44	18
>6000 lei	32	62	7

** monthly income. Source: authors own interpretation of data.

The importance of the country of origin for milk and dairy by income categories, calculated through the semantic differential [85], is presented in Figure 4.

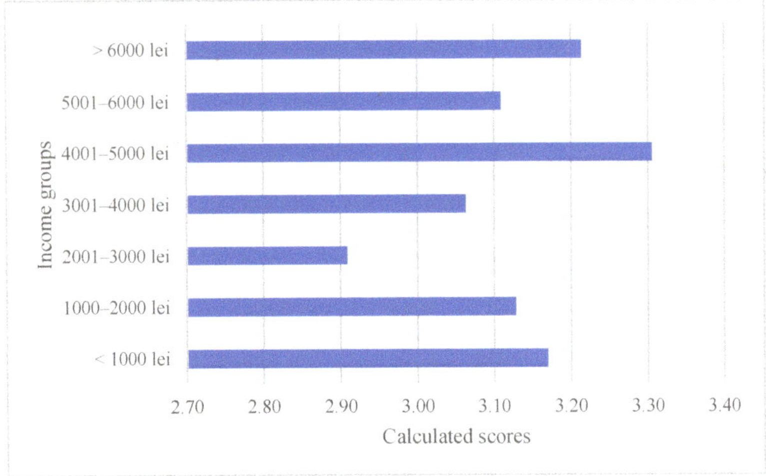

Figure 4. Consideration of country of origin in milk and dairy purchasing by income categories. Source: authors own interpretation of data.

For all income groups the score is higher than 2.9, meaning that the majority take this criterion into consideration. Surprisingly, the lowest and highest income categories have the highest scores, and therefore the highest consideration for the country of origin for the purchased products—the national provenance being preferred by the majority of the respondents.

Considering the chi-squared testing, the calculated chi value of 36.57 is exceeding the theoretical chi value of 36.42 for the significance threshold of 0.05, calculated for 24 degrees of freedom. Income has a significant correlation with the importance that respondents attach to the country of origin of the products they purchase.

Therefore, the third hypothesis is confirmed by the results of the study.

It is observed that the income influences the decision to buy these products depending on the country of origin, with a probability of 95%. Additionally, the origin of the products (industrial farming, traditional farming, ecological farming, own production) presents a high importance for the respondents, being in line with previous studies [33,40,67]. However, it is more important for income categories higher than 4001 lei/month.

The importance of national production is also confirmed by the top ten brands mentioned by the respondents to this study as being their preferred ones. The results are presented in Figure 5. The results oppose that of Yang et al. [69], who presents a higher preference for imported milk and dairy than for the local production for Chinese consumers.

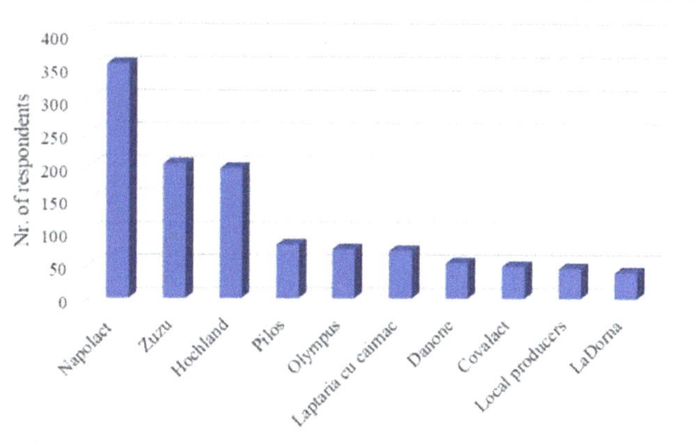

Figure 5. Top ten milk and dairy brands preferred by the respondents. Source: authors own interpretation of data.

All mentioned brands, except "local producers" which refers to small producers who sell their products without any brand, but are particularly present in local food markets, have factories spread on the Romanian territory, and therefore have a national understanding for the respondents. However, most of them are part of multinational chains which adapt their production to suite their different local markets and more, and the local origin of the fresh milk is not guaranteed by the factory location. The potential of local production may be extracted from the multiple mentions from the respondents of small local producers as the preferred sources for milk and dairy, being in line with previous studies [38,76].

The comparison of the importance between milk and dairy product availability in large stores and health recommendations (if any nutritional or health-related benefits of milk and dairy consumption coming from clinical physicians are taken into consideration in the choice of products) for them is presented in Tables 4 and 5.

Table 4. The importance of products availability in large stores (%).

	Not Important	Low Importance	Neutral	Important	High Importance
<1000 lei **	12.37	3.09	9.28	27.84	47.42
1000–2000 lei	5.98	5.98	10.26	25.64	52.14
2001–3000 lei	4.23	4.93	11.27	19.01	60.56
3001–4000 lei	11.81	7.09	7.09	22.05	51.97
4001–5000 lei	6.32	6.32	8.42	33.68	45.26
5001–6000 lei	3.64	7.27	18.18	23.64	47.27
>6000 lei	11.11	4.27	10.26	23.08	51.28

** monthly income. Source: authors own interpretation of data.

Considering the descriptive statistics, the highest percentages of respondents in each income category considers that the availability of milk and dairy in large stores, such as supermarkets and hypermarkets is very important. Therefore, the unavailability in large stores of a specific product does not mean the customer would not buy anything at all, but it would adapt to the store offer, being in line with previous research [70].

By comparison, the importance of the health-related recommendations in choosing milk and dairy products is much lower. Around a quarter of the respondents consider this criterion to be of some importance in their choice, no matter the income category. While we would expect that the importance of these recommendations would grow along with the increase of income, this appears not to be the case. What is noticeable is the high percentage of neutral respondents, meaning those who have not given this criterion any thought before participating in this study.

Table 5. The importance of health recommendations in milk and dairy consumption (%).

	Not Important	Low Importance	Neutral	Important	High Importance
<1000 lei **	18.04	15.46	23.20	25.26	18.04
1000–2000 lei	17.09	12.82	25.64	20.51	23.93
2001–3000 lei	23.24	14.79	21.83	24.65	15.49
3001–4000 lei	20.47	9.45	29.92	18.11	22.05
4001–5000 lei	21.05	10.53	29.47	23.16	15.79
5001–6000 lei	21.82	18.18	20.00	25.45	14.55
>6000 lei	17.09	13.68	23.93	25.64	19.66

** monthly income. Source: authors own interpretation of data.

When assessing the two sides of the fourth hypothesis using the chi-squared test, we observed that for the income influence on large stores availability, the calculated chi value of 30.73 is less than the theoretical chi value of 36.42 for the significance threshold of 0.05, calculated for 24 degrees of freedom. We found that there is no significant correlation between the income of the respondents and the supermarket or hypermarket availability of milk and dairy products.

When considering the second part of the fourth hypothesis, the correlation between income and taking into consideration the health-related recommendations for milk and dairy consumption, the calculated chi value of 16.48 is lower than the theoretical chi value of 36.42 for the significance threshold of 0.05, calculated for 24 degrees of freedom. It is found that there is no association between the respondents' income and the importance given to the health recommendations in the decision to buy milk and dairy products.

Therefore, there is no significant correlation between the respondents' income and the store availability or between income and health-related recommendations, and the fourth hypothesis is infirmed. By seeing the descriptive statistics, the Romanian consumers who participated in this study seem to consider store availability as more important than health-related recommendations when choosing a milk or dairy product. However, more research in this area needs to be done before providing a clear correlation.

Regarding the recommendation expressed by the respondents through an open answered question, the main results were grouped by type of suggestion in Figure 6.

Adding to the quantitative results, the recommendation map shows some similar ideas from the respondents. First, in green, the idea relates to the possibility of increasing the market power of local producers, so they may have contracts with restaurants, hotels, or school cafeterias, through diversification of the product line and an increase of promotion activity. These are pointers to the fact that the local production has considerable development potential if it can keep up with the new tastes and needs of the respondents to this study, being in line with previous studies [76,79]. The second recommendation, in yellow, is a general one, referring to the quality of the raw material used in production. The respondents request that the producers keep the quality of the natural milk and not diminish it through industrial practices or enhance it with artificial additives. In blue, the recommendation goes to small farmers. The respondents suggest that these farmers should cooperate so to have a higher market influence and to sell their products directly to the consumers, not through collectors or industrial dairy factories, being in line with other studies [8,9,79]. The suggestion in red is an environment-related one; the respondents ask for increased attention to recyclable or reusable packaging like glass, as new criteria for sustainable product choice [1,6,7,72,78]. Additionally, one of the preferred local brands stands out especially through their glass packaging. The recommendation in purple is an economic-related one; the respondents suggest fair prices, related to the quality of the products [68], and more care in advertising rather than the aggressive marketing methods that are sometimes used. Another general recommendation is related to the care for the consumers' health that is expected from the producers, the general feeling of the respondents being that this natural care is missing, with the products put on the market being sometimes perceived as low quality or unhealthy.

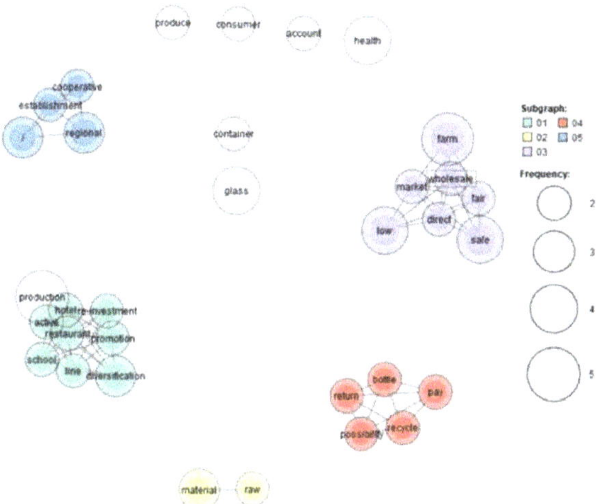

Figure 6. Respondents' recommendations for milk and dairy producers. Source: authors own interpretation with KH Coder [83].

5. Conclusions, Limitations, and Implications

The results of the current study point out the market potential of the local Romanian milk and dairy products, with particular attention on the traditional and ecological products, which turn out to be especially important for the Romanian respondents, even for the categories with lower monthly incomes.

The importance given to the sensorial properties of the milk and dairy products is proven by the results and the respondents' recommendations regarding the preservation of the natural qualities of raw milk, pointing out that they have the capacity to recognize products with additives, and therefore such practices might lead to a loss in the market share. Even more, some recommendations refer to products with respect to the consumers' health, raising some signals regarding threats to food safety, which should be looked into carefully by responsible authorities and also producers. Moreover, the food policies aimed at alleviating the income inequalities [51] should include serious considerations on ensuring quality basic food products for the people.

The modifications in consumer behavior are present in the results, with the level of income significantly influencing the willingness to pay for products with higher perceived quality, traditional characteristics, or ecologic certified products. In addition, reusable or recyclable packaging raises particular attention, since the consumers are tending to become more aware of their personal impact over the environment, and therefore tending to engage in making sustainable food choices.

Nevertheless, income has a low correlation with milk and dairy store availability, health-related recommendations for consuming milk and dairy, or the price of the products. Furthermore, gender has no correlation with the proposed selection criteria for milk and dairy. Therefore, income has a lesser influence than we assumed for these basic products, and gender does not differentiate the respondents' selection criteria.

Some limitations for this study come from the choice of a single studied country, as well as the study of a particular moment in time instead of longitudinal research. However, these are future paths for developing the research which we also invite fellow scholars to pursue.

The significance of this study resides in the considerable number of respondents, their answers serving well in forming research hypotheses for a larger, statistically significant study, both in number of respondents and in structure of the respondents. Nevertheless, the results should be of particular interest to marketers and producers in the milk and dairy industry, as knowing the needs and expectations and the purchasing power of the consumers is important in their activity. Therefore, offering a variety of qualitative milk and dairy products of national origin at fair prices and available in large stores should be a priority for the market players in this field. Other actors who should be interested in the results of this study are the public authorities, especially in the food sector. Knowing the fears or uncertainties regarding the food safety of the people is particularly important, as are the practical solutions of providing support for low-income people to have access to quality food products.

Author Contributions: Conceptualization, G.-R.L., M.S. and M.C.D.; methodology, M.C.D., G.-R.L. and D.M.I.; software, D.M.I. and M.C.D.; validation, D.M.I. and M.C.D.; formal analysis, G.-R.L. and D.M.I.; investigation, G.-R.L., D.M.I. and M.C.D.; resources, G.-R.L. and D.M.I.; writing—original draft preparation, M.C.D.; writing—review and editing, M.C.D., G.-R.L., M.S. and D.M.I. All authors have read and agreed to the published version of the manuscript.

Funding: This research was funded by ADER 24.1.1 project: "Research on development of studies and analyzes, for the substantiation of public decisions and policies, for the marketing of agri-food products in order to ensure the food security and safety for the population".

Data Availability Statement: The data presented in this study are available on request from the corresponding author. The data are not publicly available due to privacy.

Conflicts of Interest: The authors declare no conflict of interest. The funders had no role in the design of the study; in the collection, analyses, or interpretation of data; in the writing of the manuscript, or in the decision to publish the results.

References

1. White, K.; Habib, R.; Hardisty, D.J. How to SHIFT Consumer Behaviors to be More Sustainable: A Literature Review and Guiding Framework. *J. Mark.* **2019**, *83*, 22–49. [CrossRef]
2. Bekele, A.D.; Beuving, J.; Ruben, R. Food choices in Ethiopia: Does nutritional information matter? *Int. J. Consum. Stud.* **2016**, *40*, 625–634. [CrossRef]
3. Givens, D.I. MILK Symposium review: The importance of milk and dairy foods in the diets of infants, adolescents, pregnant women, adults, and the elderly. *J. Dairy Sci.* **2020**, *103*, 9681–9699. [CrossRef] [PubMed]
4. Werner, L.B.; Flysjö, A.; Tholstrup, T. Greenhouse gas emissions of realistic dietary choices in Denmark: The carbon footprint and nutritional value of dairy products. *Food Nutr. Res.* **2014**, *58*, 20687. [CrossRef] [PubMed]
5. Bahna, S.L. Cow's milk allergy versus cow milk intolerance. *Ann. Allergy Asthma Immunol.* **2002**, *89*, 56–60. [CrossRef]
6. Trudel, R. Sustainable consumer behavior. *Consum. Psychol. Rev.* **2018**, *2*, 85–96. [CrossRef]
7. Hoek, A.; Pearson, D.; James, S.; Lawrence, M.; Friel, S. Healthy and environmentally sustainable food choices: Consumer responses to point-of-purchase actions. *Food Qual. Prefer.* **2017**, *58*, 94–106. [CrossRef]
8. Bórawski, P.; Pawlewicz, A.; Parzonko, A.; Jayson, K.H.; Holden, L. Factors Shaping Cow's Milk Production in the EU. *Sustainability* **2020**, *12*, 420. [CrossRef]
9. Brodam, C. Milk Producers Reacted Differently at Quota Expiration. Arhaus University. 2021. Available online: https://dca.au.dk/en/current-news/news/show/artikel/milk-producers-reacted-differently-at-quota-expiration/ (accessed on 11 August 2021).
10. Teneva-Angelova, T.; Balabanova, T.; Boyanova, P.; Beshkova, D. Traditional Balkan fermented milk products. *Eng. Life Sci.* **2018**, *18*, 807–819. [CrossRef] [PubMed]
11. International Trade Center. Trade Balance for Milk and Dairy Products in Romania. 2021. Available online: https://www.trademap.org/Product_SelCountry_TS.aspx?nvpm=1%7c642%7c%7c%7c%7c04%7c%7c%7c4%7c1%7c1%7c3%7c2%7c1%7c1%7c1%7c1%7c1 (accessed on 11 August 2021).
12. Eurostat. Real GDP per Capita. 2021. Available online: https://ec.europa.eu/eurostat/databrowser/view/sdg_08_10/default/table?lang=en (accessed on 26 October 2021).
13. Eurostat. HICP-Annual Data (Average Index and Rate of Change). 2021. Available online: https://appsso.eurostat.ec.europa.eu/nui/show.do?dataset=prc_hicp_aind&lang=en (accessed on 26 October 2021).

14. Eurostat. Mean and Median Economic Resources of Households by Income, Consumption and Wealth Quantiles-Experimental Statistics. 2021. Available online: https://ec.europa.eu/eurostat/databrowser/view/icw_res_02/default/table?lang=en (accessed on 26 October 2021).
15. Blythe, J. *Consumer Behaviour: SAGE Publications*; SAGE: Newcastle, UK, 2013.
16. Nam, K.; Lim, H.; Ahn, B.-I. Analysis of Consumer Preference for Milk Produced through Sustainable Farming: The Case of Mountainous Dairy Farming. *Sustainability* **2020**, *12*, 3039. [CrossRef]
17. Brundtland, G.H. Report of the World Commission on Environment and Development: Our Common Future. 20 March 1987. Available online: https://sustainabledevelopment.un.org/content/documents/5987our-common-future.pdf (accessed on 18 July 2020).
18. United Nations. THE 17 GOALS | Sustainable Development. 2021. Available online: https://sdgs.un.org/goals (accessed on 28 July 2021).
19. Priyadarshini, P.; Abhilash, P.C. Policy recommendations for enabling transition towards sustainable agriculture in India. *Land Use Policy* **2020**, *96*, 104718. [CrossRef]
20. Harwood, R.R. A history of sustainable agriculture. In *Sustainable Agricultural Systems*; CRC Press: Boca Raton, FL, USA, 2020; pp. 3–19. ISBN 9781003070474.
21. Kylili, A.; Fokaides, P.A.; Vaiciunas, J.; Seduikyte, L. Integration of Building Information Modelling (BIM) and Life Cycle Assessment (LCA) for sustainable constructions. *J. Sustain. Arch. Civ. Eng.* **2016**, *13*, 28–38. [CrossRef]
22. Ideland, M.; Malmberg, C. 'Our common world' belongs to 'Us': Constructions of otherness in education for sustainable development. *Crit. Stud. Educ.* **2014**, *55*, 369–386. [CrossRef]
23. Kamble, S.S.; Gunasekaran, A.; Gawankar, S.A. Sustainable Industry 4.0 framework: A systematic literature review identifying the current trends and future perspectives. *Process Saf. Environ. Prot.* **2018**, *117*, 408–425. [CrossRef]
24. Beber, C.L.; Langer, G.; Meyer, J. Strategic Actions for a Sustainable Internationalization of Agri-Food Supply Chains: The Case of the Dairy Industries from Brazil and Germany. *Sustainability* **2021**, *13*, 10873. [CrossRef]
25. Feil, A.A.; da Silva Cyrne, C.C.; Sindelar FC, W.; Barden, J.E.; Dalmoro, M. Profiles of sustainable food con-sumption: Consumer behavior toward organic food in southern region of Brazil. *J. Clean. Prod.* **2020**, *258*, 120690. [CrossRef]
26. Park, S.; Lee, Y. Scale Development of Sustainable Consumption of Clothing Products. *Sustainability* **2020**, *13*, 115. [CrossRef]
27. Garrison, S.; Wallace, C. Media Tourism and Its Role in Sustaining Scotland's Tourism Industry. *Sustainability* **2021**, *13*, 6305. [CrossRef]
28. Henchion, M.; Moloney, A.; Hyland, J.; Zimmermann, J.; McCarthy, S. Review: Trends for meat, milk and egg consumption for the next decades and the role played by livestock systems in the global production of proteins. *Animal* **2021**, 100287. [CrossRef] [PubMed]
29. Scutariu, A.L. The Evolution of Economic Territorial Disparities in Romania. *USV Ann. Econ. Public Adm.* **2017**, *17*, 21–27. Available online: http://annals.seap.usv.ro/index.php/annals/article/view/1034/899 (accessed on 15 September 2021).
30. Berton, M.; Bittante, G.; Zendri, F.; Ramanzin, M.; Schiavon, S.; Sturaro, E. Environmental impact and efficiency of use of resources of different mountain dairy farming systems. *Agric. Syst.* **2020**, *181*, 102806. [CrossRef]
31. Glover, J. The dark side of sustainable dairy supply chains. *Int. J. Oper. Prod. Manag.* **2020**, *40*, 1801–1827. [CrossRef]
32. Becskei, Z.; Savić, M.; Ćirković, D.; Rašeta, M.; Puvača, N.; Pajić, M.; Đorđević, S.; Paskaš, S. Assessment of Water Buffalo Milk and Traditional Milk Products in a Sustainable Production System. *Sustainability* **2020**, *12*, 6616. [CrossRef]
33. Gao, Z.; Li, C.; Bai, J.; Fu, J. Chinese consumer quality perception and preference of sustainable milk. *China Econ. Rev.* **2020**, *59*, 100939. [CrossRef]
34. Food and Agriculture Organization (FAO). Sustainable Food and Agriculture. 2021. Available online: https://www.fao.org/sustainability/background/en/ (accessed on 20 October 2021).
35. Aschemann-Witzel, J.; Gantriis, R.F.; Fraga, P.; Perez-Cueto, F.J.A. Plant-based food and protein trend from a business perspective: Markets, consumers, and the challenges and opportunities in the future. *Crit. Rev. Food Sci. Nutr.* **2021**, *61*, 3119–3128. [CrossRef] [PubMed]
36. Megido, R.C.; Gierts, C.; Blecker, C.; Brostaux, Y.; Éric, H.; Alabi, T.; Francis, F. Consumer acceptance of insect-based alternative meat products in Western countries. *Food Qual. Prefer.* **2016**, *52*, 237–243. [CrossRef]
37. Röös, E.; Carlsson, G.; Ferawati, F.; Hefni, M.; Stephan, A.; Tidåker, P.; Witthöft, C. Less meat, more legumes: Prospects and challenges in the transition toward sustainable diets in Sweden. *Renew. Agric. Food Syst.* **2020**, *35*, 192–205. [CrossRef]
38. Vargas, A.M.; de Moura, A.P.; Deliza, R.; Cunha, L.M. The Role of Local Seasonal Foods in Enhancing Sustainable Food Consumption: A Systematic Literature Review. *Foods* **2021**, *10*, 2206. [CrossRef]
39. Verain, M.C.; Snoek, H.M.; Onwezen, M.C.; Reinders, M.J.; Bouwman, E.P. Sustainable food choice motives: The development and cross-country validation of the Sustainable Food Choice Questionnaire (SUS-FCQ). *Food Qual. Prefer.* **2021**, *93*, 104267. [CrossRef]
40. Tandon, A.; Dhir, A.; Kaur, P.; Kushwah, S.; Salo, J. Why do people buy organic food? The moderating role of environmental concerns and trust. *J. Retail. Consum. Serv.* **2020**, *57*, 102247. [CrossRef]
41. Aschemann-Witzel, J.; Zielke, S. Can't Buy Me Green? A Review of Consumer Perceptions of and Behavior Toward the Price of Organic Food. *J. Consum. Aff.* **2015**, *51*, 211–251. [CrossRef]
42. Zaharia, A.; Diaconeasa, M.-C.; Maehle, N.; Szolnoki, G.; Capitello, R. Developing Sustainable Food Systems in Europe: National Policies and Stakeholder Perspectives in a Four-Country Analysis. *Int. J. Environ. Res. Public Heal.* **2021**, *18*, 7701. [CrossRef]

43. Park, Y.W. *Overview of Bioactive Components in Milk and Dairy Products*; John Wiley & Sons: Hoboken, NJ, USA, 2009; pp. 1–12. ISBN 978-0-8138-1982-2.
44. Thorning, T.K.; Raben, A.; Tholstrup, T.; Soedamah-Muthu, S.S.; Givens, I.; Astrup, A. Milk and dairy products: Good or bad for human health? An assessment of the totality of scientific evidence. *Food Nutr. Res.* **2016**, *60*, 32527. [CrossRef]
45. Ratajczak, A.; Zawada, A.; Rychter, A.; Dobrowolska, A.; Krela-Kaźmierczak, I. Milk and Dairy Products: Good or Bad for Human Bone? Practical Dietary Recommendations for the Prevention and Management of Osteoporosis. *Nutrients* **2021**, *13*, 1329. [CrossRef] [PubMed]
46. Weaver, C.M. How sound is the science behind the dietary recommendations for dairy? *Am. J. Clin. Nutr.* **2014**, *99*, 1217S–1222S. [CrossRef] [PubMed]
47. Górska-Warsewicz, H.; Rejman, K.; Laskowski, W.; Czeczotko, M. Milk and Dairy Products and Their Nutritional Contribution to the Average Polish Diet. *Nutrients* **2019**, *11*, 1771. [CrossRef] [PubMed]
48. Filho, M.C.C.; Falcao, R.P.; Motta, P.C.D.M. Brand loyalty among low-income consumers? *Qual. Mark. Res. Int. J.* **2021**, *24*, 260–280. [CrossRef]
49. Tricarico, J.M.; Kebreab, E.; Wattiaux, M.A. MILK Symposium review: Sustainability of dairy production and consumption in low-income countries with emphasis on productivity and environmental impact. *J. Dairy Sci.* **2020**, *103*, 9791–9802. [CrossRef] [PubMed]
50. Wu, B.; Shang, X.; Chen, Y. Household dairy demand by income groups in an urban Chinese province: A multi-stage budgeting approach. *Agribusiness* **2021**, *37*, 629–649. [CrossRef]
51. Wardle, J.; Haase, A.M.; Steptoe, A.; Nillapun, M.; Jonwutiwes, K.; Bellisie, F. Gender differences in food choice: The contribution of health beliefs and dieting. *Ann. Behav. Med.* **2004**, *27*, 107–116. [CrossRef] [PubMed]
52. Davies, A.; Titterington, A.J.; Cochrane, C. Who buys organic food? *Br. Food J.* **1995**, *97*, 17–23. [CrossRef]
53. Levi, A.; Chan, K.K.; Pence, D. Real Men Do Not Read Labels: The Effects of Masculinity and Involvement on College Students' Food Decisions. *J. Am. Coll. Health* **2006**, *55*, 91–98. [CrossRef]
54. Ionut, J.; Andrada, G.I.; Elena, I.S.; Hrebenciuc, A. Income Inequalities and Their Social Determinants: An Analysis over Developed vs. Developing EU Member States. *Econ. Comput. Econ. Cybern. Stud. Res.* **2021**, *55*, 125–142. [CrossRef]
55. Eysteinsdottir, T.; Halldorsson, T.I.; Thorsdottir, I.; Sigurdsson, G.; Sigurðsson, S.; Harris, T.; Launer, L.J.; Gudnason, V.; Gunnarsdottir, I.; Steingrimsdottir, L. Milk consumption throughout life and bone mineral content and density in elderly men and women. *Osteoporos. Int.* **2013**, *25*, 663–672. [CrossRef]
56. EU Commission. Mozzarella di Bufala Campana DOP. 2021. Available online: https://ec.europa.eu/info/food-farming-fisheries/food-safety-and-quality/certification/quality-labels/eu-quality-food-and-drink/mozzarella-di-bufala-campana_ro (accessed on 11 August 2021).
57. Zicarelli, L. Buffalo Milk: Its Properties, Dairy Yield and Mozzarella Production. *Vet. Res. Commun.* **2004**, *28*, 127–135. [CrossRef] [PubMed]
58. Boyazoglu, J.; Morand-Fehr, P. Mediterranean dairy sheep and goat products and their quality: A critical review. *Small Rumin. Res.* **2001**, *40*, 1–11. [CrossRef]
59. Polidori, P.; Beghelli, D.; Mariani, P.; Vincenzetti, S. Donkey milk production: State of the art. *Ital. J. Anim. Sci.* **2009**, *8*, 677–683. [CrossRef]
60. Aspri, M.; Economou, N.; Papademas, P. Donkey milk: An overview on functionality, technology, and future prospects. *Food Rev. Int.* **2017**, *33*, 316–333. [CrossRef]
61. Madhusudan, N.C.; Ramachandra, C.D.; Udaykumar, N.D.; Sharnagouda, H.D.; Nagraj, N.D.; Jagjivan, R.D. *Composition, Characteristics, Nutritional Value and Health Benefits of Donkey Milk—A Review*; Dairy Science & Technology: Les Ulis, France, 2017.
62. Gjøstein, H.; Øystein, H.; Weladji, R.B. Milk production and composition in reindeer (Rangifer tarandus): Effect of lactational stage. *Comp. Biochem. Physiol. Part A Mol. Integr. Physiol.* **2004**, *137*, 649–656. [CrossRef] [PubMed]
63. Dobrotă, C.E.; Marcu, N.; Siminică, M.; Nețoiu, L.M. Disparities, gaps and evolution trends of innovation, as a vector. *Rom. J. Econ. Forecast.* **2019**, *22*, 174.
64. Sethi, S.; Tyagi, S.K.; Anurag, R.K. Plant-based milk alternatives an emerging segment of functional beverages: A review. *J. Food Sci. Technol.* **2016**, *53*, 3408–3423. [CrossRef]
65. Kempen, E.; Kasambala, J.; Christie, L.; Symington, E.; Jooste, L.; Van Eeden, T. Expectancy-value theory contributes to understanding consumer attitudes towards cow's milk alternatives and variants. *Int. J. Consum. Stud.* **2017**, *41*, 245–252. [CrossRef]
66. Kotler, P.; Keller, K.L. *A Framework for Marketing Management*; Pearson: Boston, MA, USA, 2016.
67. Hysen, B.; Mensur, V.; Muje, G.; Hajrip, M.; Halim, G.; Iliriana, M.; Njazi, B. Analysis of consumer behavior in regard to dairy products in Kosovo. *J. Agric. Res.* **2008**, *46*, 281–290.
68. Kapsdorferová, Z.; Nagyová, Ľ. Consumer behavior at the Slovak dairy market. *Agric. Econ.* **2012**, *51*, 362–368. [CrossRef]
69. Yang, R.; Ramsaran, R.; Wibowo, S. An investigation into the perceptions of Chinese consumers towards the country-of-origin of dairy products. *Int. J. Consum. Stud.* **2018**, *42*, 205–216. [CrossRef]
70. Lanfranchi, M.; Zirilli, A.; Passantino, A.; Alibrandi, A.; Giannetto, C. Assessment of milk consumer preferences. *Br. Food J.* **2017**, *119*, 2753–2764. [CrossRef]

71. Aizaki, H.; Nanseki, T.; Zhou, H. Japanese consumer preferences for milk certified as good agricultural practice. *Anim. Sci. J.* **2013**, *84*, 82–89. [CrossRef]
72. Kumar, A.; Babu, S. Factors influencing consumer buying behavior with special reference to dairy products in Pon-dicherry state. *Int. Mon. Refereed J. Res. Manag. Technol.* **2014**, *3*, 65–73, ISSN:2320-0073.
73. Vermeir, I.; Weijters, B.; De Houwer, J.; Geuens, M.; Slabbinck, H.; Spruyt, A.; Van Kerckhove, A.; Van Lippevelde, W.; De Steur, H.; Verbeke, W. Environmentally Sustainable Food Consumption: A Review and Research Agenda From a Goal-Directed Perspective. *Front. Psychol.* **2020**, *11*, 1603. [CrossRef] [PubMed]
74. Tait, P.; Saunders, C.; Guenther, M.; Rutherford, P. Emerging versus developed economy consumer willingness to pay for environmentally sustainable food production: A choice experiment approach comparing Indian, Chinese and United Kingdom lamb consumers. *J. Clean. Prod.* **2016**, *124*, 65–72. [CrossRef]
75. Román, S.; Sanchez-Siles, L.M.; Siegrist, M. The importance of food naturalness for consumers: Results of a systematic review. *Trends Food Sci. Technol.* **2017**, *67*, 44–57. [CrossRef]
76. De-Magistris, T.; Gracia, A. Consumers' willingness-to-pay for sustainable food products: The case of organically and locally grown almonds in Spain. *J. Clean. Prod.* **2016**, *118*, 97–104. [CrossRef]
77. Zander, K.; Feucht, Y. Consumers' Willingness to Pay for Sustainable Seafood Made in Europe. *J. Int. Food Agribus. Mark.* **2017**, *30*, 251–275. [CrossRef]
78. Merlino, V.M.; Brun, F.; Versino, A.; Blanc, S. Milk packaging innovation: Consumer perception and willingness to pay. *AIMS Agric. Food* **2020**, *5*, 307–326. [CrossRef]
79. Schott, L.; Bernard, J. Comparing consumer's willingness to pay for conventional, non-certified organic and organic milk from small and large farms. *J. Food Distrib. Res.* **2015**, *46*, 186–205. [CrossRef]
80. Lewin, C. Elementary quantitative methods. In *Research Methods in the Social Sciences*; SAGE: Newcastle, UK, 2005; pp. 215–225. ISBN 0-7619-4401-X.
81. Etikan, I.; Musa, S.A.; Alkassim, R.S. Comparison of Convenience Sampling and Purposive Sampling. *Am. J. Theor. Appl. Stat.* **2016**, *5*, 1–4. [CrossRef]
82. Adam, A.M. Sample Size Determination in Survey Research. *J. Sci. Res. Rep.* **2020**, 90–97. [CrossRef]
83. National Institute of Statistics. Resident Population. 2021. Available online: http://statistici.insse.ro:8077/tempo-online/#/pages/tables/insse-table (accessed on 3 August 2021).
84. Bryman, A.; Cramer, D. *Quantitative Data Analysis with SPSS 14, 15 & 16: A Guide for Social Scientists*; Routledge/Taylor & Francis Group: London, UK, 2009. [CrossRef]
85. Stoklasa, J.; Talasek, T.; Stoklasová, J. Semantic differential for the twenty-first century: Scale relevance and uncertainty entering the semantic space. *Qual. Quant.* **2018**, *53*, 435–448. [CrossRef]
86. Higuchi, K. *KH Coder 3 Reference Manual*; Ritsumeikan University: Kioto Japan, 2016; Available online: http://khcoder.net/en/manual_en_v3.pdf (accessed on 25 July 2021).
87. National Institute of Statistics. Agricultural Products Average Prices. 2021. Available online: http://statistici.insse.ro:8077/tempo-online/#/pages/tables/insse-table (accessed on 2 September 2021).
88. National Institute of Statistics. Milk Production by Animal Species. 2021. Available online: http://statistici.insse.ro:8077/tempo-online/#/pages/tables/insse-table (accessed on 2 September 2021).

Review

Consumer Perception and Understanding of European Union Quality Schemes: A Systematic Literature Review

Alexandra-Ioana Glogoveţan [1], Dan-Cristian Dabija [2], Mariantonietta Fiore [3] and Cristina Bianca Pocol [1,*]

1 Department of Animal Production and Food Safety, University of Agricultural Sciences and Veterinary Medicine of Cluj-Napoca, 400372 Cluj-Napoca, Romania; alexandra.glogovetan@gmail.com
2 Department of Marketing, Babeş-Bolyai University Cluj-Napoca, 400591 Cluj-Napoca, Romania; dan.dabija@ubbcluj.ro
3 Department of Economics, University of Foggia, 71121 Foggia, Italy; mariantonietta.fiore@unifg.it
* Correspondence: cristina.pocol@usamvcluj.ro

Abstract: Food, agriculture, and labeling, affecting the environment are well connected concepts, the balance between them being determined not only by pedological and climatic factors or the development level of agricultural techniques, but also by national governments and international organizations' food processing, trade policies and regulations. In this context, the European Union (EU) encourages the use of different food quality schemes: "Protected Designation of Origin" (PDO), "Protected Geographical Indication" (PGI), and "Traditional Specialty Guaranteed" (TSG) to protect producers of special-quality foods and assist consumers in their purchasing decisions. This review examines existing studies on the impact of these labels on customers behavior. A total of 32 studies were found and systematized. The papers were selected if they featured unique empirical research on consumer perceptions of any of PDO, PGI and TSG labels. Using the search strategy, a literature analysis was performed based on papers extracted from Web of Science, Springer Link, Emerald Insights, and Science Direct. Although these papers highlight quite diversified findings, the internationally used labels play an increasing role in contemporary society and pandemic conditions caused by COVID-19, thus making the quality schemes relevant in consumer decision-making processes.

Keywords: "Protected Designation of Origin" (PDO); "Protected Geographical Indication" (PGI); "Traditional Specialty Guaranteed" (TSG); "Geographical Indication" (GI); EU quality labels; consumer behavior

Citation: Glogoveţan, A.-I.; Dabija, D.-C.; Fiore, M.; Pocol, C.B. Consumer Perception and Understanding of European Union Quality Schemes: A Systematic Literature Review. *Sustainability* **2022**, *14*, 1667. https://doi.org/10.3390/su14031667

Academic Editors: Mariarosaria Lombardi, Vera Amicarelli and Erica Varese

Received: 21 December 2021
Accepted: 27 January 2022
Published: 31 January 2022

1. Introduction

Agriculture is of vital importance to the society, environment, and economy of the European Union [1]. Proper environmental conditions support agricultural activities, allowing farmers to use natural resources, create products and earn their living. In addition, agrarian income sustains farmers and families in rural communities, while agri-food strengthens society [1,2].

The next decade, starting with 2021, represents the transition to "smart" food that is more efficient, healthier, and greener, as it is obtained from the "smart" agriculture system [3]. The agricultural policies of the EU are based on specific measures regarding the development of entire food chains, from production and distribution to consumption, aiming at reducing food waste [4]. Public policies will have a pivotal role in protecting the availability, accessibility, and quality of agri-food products [2]. Therefore, agri-food products that are certified with quality schemes represent an ideal food product because they are manufactured from raw materials, being developed according to specific production methods, and technologies in a well-defined geographical area. These products are characterized by natural factors of production, traditions and/or specific historical procedures developed over centuries that cannot be replaced [5].

Thus, the supply chain of environmentally friendly products becomes a preferential reference point for both producers and consumers and allows a redefinition of financial support instruments to increase the efficiency of production and distribution processes, especially those affecting the environment [6]. Therefore, small and medium-sized companies located in different areas of the EU represent the ideal framework for quality food production (such as Geographical Indication or organic), which could move towards an economically and socially sustainable solution [6].

In the European Union, product names are protected by registering them in so called "quality schemes", which means that they entail unique characteristics such as a certain geographical origin, traditional manufacturing technologies and/or long-lasting practices [7]. Quality schemes have the following features: (1) most production stages must be implemented in a delimited geographical area, (2) the recipe after they are manufactured is authentic (mixture of ingredients); the raw materials are original, the production process is traditional and/or contains specific features for that region, (3) are available on the market for at least 50 years and (4) share a part of the gastronomic heritage of a society/community [7,8].

The EU's geographical indication system thus provides protection for products names from various regions around the world, which have some unique features or enjoys a stable reputation, depending on the territory where they are produced. Geographical indications include "PDO—Protected Designation of Origin" (food and wine), "PGI—Protected Geographical Indication" (food and wine), "GI—Geographical Indication" (for alcoholic beverages) [7]. Other quality certification systems highlight the traditional production process ("Traditional Specialty Guaranteed")—TSG—or some products that are made in more challenging areas, such as mountains (mountain products). When considering the characteristics of PDO and PGI, the main differences relate to the proportion of raw materials (at least 85 percent) that are usually common for the area where they come from, but also on the production stages, that must be implemented in the considered geographical region. GI is typical for spirits and aromatic wines [7].

The PDO quality label represents a proper reference for the manufacturing place of agri-food products. Thus, all transformation stages from the raw materials to the final product must take place in a particular region. As for wines, the essential condition is that the raw material (grapes) comes exclusively from the site where the wine will be produced [7]. The PGI label pinpoints the connection between a certain geographical region and a certain product brand. In this situation, at least one of the production steps must be implemented at the place of origin. Concerning PGI-certified wines, 85% of the raw materials (grapes) must have their origin only in the geographical area where the wine will be produced [7].

The "Traditional Specialty Guaranteed" (TSG) emphasizes many traditional aspects, such as the composition and ingredients, a specific recipe, without being necessary connected to any specific geographical area. The name of a registered TSG product protects it from being falsified or misused [7]. TSG certified agri-food products could be manufactured by any producer who respects this production method. Their 'specific' character refers to the characteristics that differentiate them from other foodstuffs belonging to the same category. Even if agri-food products certified with the TSG quality scheme often come from a particular country or region, their international reputation might result in the interest of producers from other countries in them [7].

By allowing producer groups to mark and label the origin of their products, quality schemes provide a means to protect traditional products' integrity and prevent and avoid abuse and counterfeiting [9]. Each of these certifications is represented graphically through logos, after which the certified products can be recognized (Table 1).

Through these logos (Table 1), agricultural producers can communicate the product's characteristics and quality attributes to consumers, thus ensuring fair competition, intellectual property rights, and an integrated internal market [10,11]. Consequently, the main benefits for consumers are identified as follows: producers of agri-food products certified with quality schemes are required to provide reliable information on the origin of their products. They must guarantee that the products are authentic goods, not fakes or imitations (confirmed to the final consumer by the logo attached to the product packaging and charging a higher price than other foods in the same category). Thus, by purchasing certified agri-food products, the consumer can recognize products from their region or other regions [12,13]. In Europe, there are numerous agricultural products and alcoholic beverages certified with European quality schemes. The table below (Table 2) provides an official statistic containing the number of products registered and protected with quality schemes from each country. The first position is occupied by Italy. Figure 1 shows the situation of PDO/PGI/GI/TSG products by country in descending order (status—registered, all application type).

As Figure 2 shows, the interest in consumer-focused studies is concentrated across European countries. This fact is because most of the agri-food products and the alcoholic beverages certified with European quality schemes are from the territory of the European Union.

In the light of the above-mentioned arguments, the purpose of this paper is to provide an outline of what is acknowledged about the perception, willingness to pay, and buying behavior of food products certified with PDO, PGI, and TSG schemes. At the same time, there is a lack of studies linking the origin of PDO/PGI/TSG to healthy eating in the context of COVID 19-pandemic today. This review can serve as a starting point for discussions about the utility and advantage of these quality schemes as a marketing tool for the stakeholders involved (from producers to final consumers) to promote market transparency and food quality in pandemic times.

The following section discusses the materials and methods employed. The third section describes the results, divided between the jurisdiction and methodologies used by the reviewed studies. They are sorted according to the declared perception of consumers, preferences, recognition, and willingness to pay for certified agri-food products, purchasing and consumption behaviors towards certified agri-food products, and online purchasing of certified agri-food products. The fourth section presents critical discussions, while the final section pinpoints the conclusions for theory, the implications for market participants and public institutions, along with the limitations and further research directions.

Table 1. The different quality schemes of the EU.

EU Quality Schemes	Label
"Protected Designation of Origin" (PDO)	PROTECTED DESIGNATION OF ORIGIN
"Protected Geographical Indication" (PGI)	PROTECTED GEOGRAPHICAL INDICATION
"Traditional Specialty Guaranteed" (TSG)	TRADITIONAL SPECIALITY GUARANTEED

Source: [7].

Table 2. Agricultural Products, Foodstuffs and Alcoholic Beverages—Status: Registered.

Country	Number of Agricultural Products and Foodstuffs Registered *		Number of Alcoholic Beverages Registered *		
	PDO/PGI	TSG	PDO/PGI	GI	TSG
			Wine	Spirit Drinks	Beers
Austria	16	3	27	9	0
Belgium	16	0	10	10	5
Bulgaria	3	6	54	12	0
Croatia	33	0	18	6	0
Republic of Cyprus	9	0	11	2	0
Czech Republic	30	1	13	0	0
Denmark	8	0	5	0	0
Estonia	1	0	0	1	0
Finland	7	2	0	2	1
France	258	2	437	53	0
Germany	93	0	45	35	0
Greece	113	1	147	15	0
Hungary	28	2	38	12	0
Ireland	8	0	0	3	0
Italy	313	6	526	34	0
Latvia	3	4	0	0	0
Lithuania	7	2	0	7	0
Luxembourg	4	0	1	0	0
Malta	0	0	3	0	0
Norway	2	0	0	2	0
Netherlands	11	5	18	5	0
Poland	34	11	0	2	0
Portugal	140	2	40	11	0
Romania	9	1	53	9	0
Slovakia	13	3	9	1	0
Slovenia	23	4	17	4	0
Spain	200	4	140	19	0
Sweden	8	2	0	3	0
Turkey	7	0	0	0	0
United Kingdom	69	6	5	5	0
TOTAL	1466	67	1617	262	6

Note: * Agricultural Products, Foodstuffs and Alcoholic Beverages—Status: Registered until 29 January 2022. Andorra and Iceland: 1 food PDO/PGI quality scheme; the Russian Federation: 1 Spirit Drinks quality scheme; Serbia and Switzerland: 1 Wine quality schemes; Belarus: 2 food PDO/PGI quality scheme. Source: [14].

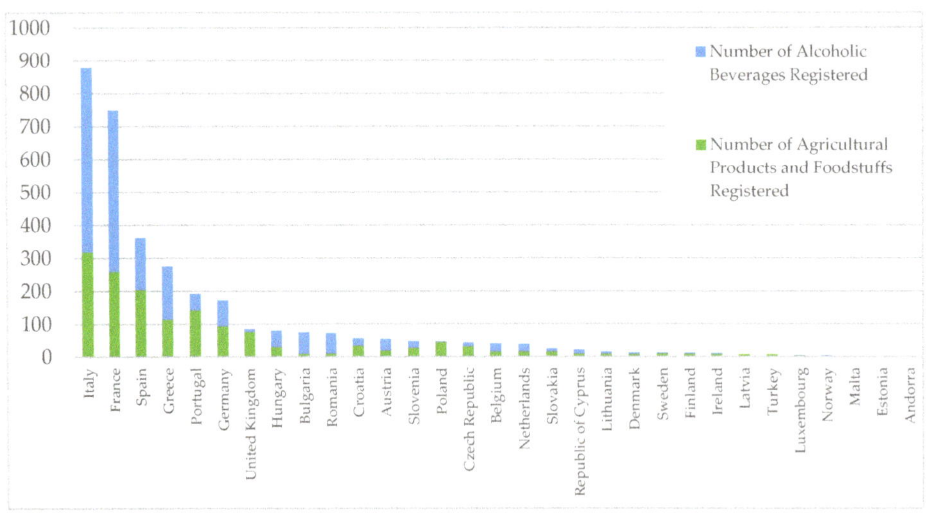

Figure 1. Statistic of PDO/PGI/GI/TSG products sorted by country. Source: Own development. Note: Agricultural products, foodstuffs and alcoholic beverages registered until 16 January 2022.

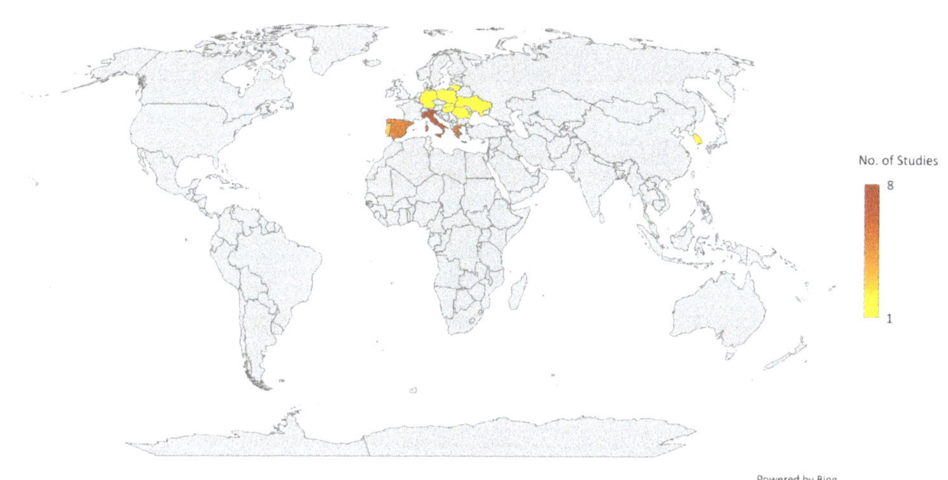

Figure 2. Geographic heatmap based on revised literature jurisdictions. Source: Own development. Note: The range of 1 by 8 is the maximum number of revised studies from a country.

2. Materials and Methods

Using the search strategy reported by Campos et al. [15] and Grunert and Aaachman [16], a literature analysis was performed through a combination of the following keywords (Figure 3) in Web of Science, and others relevant international databases according to their international visibility and authors library access (Springer Link, Emerald Insight, Science Direct). The main research directions identified are composed of the consumers' perceptions about certified agri-food products, preference, recognition, willingness to pay, and purchasing and consumption behavior of certified agri-food products. More recent studies discussing the online purchasing of certified agri-food products are also reviewed.

Figure 3. Keywords cloud. Source: Own development.

The initial search generated 79 papers, of which 37 titles fit the considered criteria (see Table 3), therefore being further analyzed. The publications were evaluated to pinpoint if they deal with one or more of the registered EU quality schemes. The papers were selected only if they featured unique empirical research on consumer perceptions of any of PDO, PGI and TSG labels. These publications were retained for further analysis only if they fulfilled simultaneously the eight methodological criteria proposed by Campos et al. [15]

and Olbrich et al. [16]. The final set of papers included in the present systematic literature review consist of 32 publications.

Table 3. The methodological criteria.

Criterion	Possible Outcome
1. "Is the research question well stated?"	Y/N
2. "Is the sample/population identified and appropriate?"	Y/N
3. "Are the inclusion/exclusion criteria described and appropriate?"	Y/N or N/A
4. "If applicable, is the participation rate reported and appropriate?"	Y/N or N/A
5. "Is the same data collection method used for all respondents?"	Y/N
6. "Are important the variables, well measured, valid, and reliable?"	Y/N or N/A
7. "Is the outcome defined and measurable?"	Y/N
8. "Is the statistical analysis appropriate?"	Y/N or N/A

Note: Y states for Yes; N states for No; Y/N states for Not applicable. Source: [15,16].

3. Results

In implementing the research scope, the main results of the conceptual framework are reviewed. The results are divided between the jurisdiction and methodologies used by the reviewed studies; and they are sorted according to the following: the declared perception of consumers from the identified papers about certified agri-food products, preferences, recognition, and willingness to pay for certified agri-food products, purchasing and consumption behaviors towards certified agri-food products, and online purchasing of certified agri-food products.

3.1. Jurisdiction and Methodologies

The 32 revised articles (Table 4) originate from the following jurisdictions: Italy, Poland, Lithuania, Slovakia, Romania, Ukraine, Hungary, Spain, Portugal, Greece, Germany, and South Korea. The online questionnaire represents the most used study tool. The papers also used PAPI and CAWI surveys, Eurobarometer surveys, online consumer databases, and household journals conceived by consumers participating in the study. Regarding the analyses applied, cross-sectional analysis, Partial Least Square path modelling, multi-group analysis, Structural Equations Modelling, Web content analysis, ANOVA, and eMICA analysis were mainly used. The samples on which the studies were conducted are various and range from 150 respondents to 35,000 respondents. The certified agri-food products with quality schemes on which the studies were carried out are mainly olive oil, wine, meat, and cheese. Most of the studies reviewed focused on all three labels: PDO, PGI, and TSG. The rest of the studies performed analyses based on agri-food products certified either with PDO or TSG.

Table 4. Journals and citations of the reviewed literature.

Title	Authors	Journal of Publication	Publication Year	Total Citations *
How Much Do Consumers Value Protected Designation of Origin Certifications? Estimates of Willingness to Pay for PDO Dry-Cured Ham in Italy	Garavaglia, C.; Mariani, P.	*Agribusiness (New York)*	2017	29
PDO Labels and Food Preferences: Results from a Sensory Analysis	Savelli, E.; Bravi, L.; Francioni, B.; Murmura, F.; Pencarelli, T.	*Br. Food J.*	2021	3
Premium Private Labels Products: Drivers of Consumers' Intention to Buy	Martinelli, E.; De Canio, F.	*Int. J. Bus. Manag.*	2019	1
Consumers' Trust in Greek Traditional Foods in the Post COVID-19 Era	Skalkos, D.; Kosma, I. S.; Vasiliou, A.; Guine, R. P. F.	*Sustainability*	2021	1
Perceived Risk Factors Affecting Consumers' Online Shopping Behaviour	Tham, K. W.; Dastane, O.; Johari, Z.; Ismail, N.B.	*J. Asian Finance Econ. Bus.*	2019	60
Consumer Reactions to the Use of EU Quality Labels on Food Products: A Review of the Literature	Grunert, K.G.; Aachmann, K.	*Food Control*	2016	183

Table 4. *Cont.*

Title	Authors	Journal of Publication	Publication Year	Total Citations *
Gastronomy as a tourism resource in the province of Alicante	Martínez, A.A.; Fernández-Poyatos, M.D.	*Int. J. of Sci. Mgmt. and Tourism*	2017	N/A
Promotion of Regional and Traditional Products	Oleksiuk, I.; Werenowska, A.	*Środ. Stud. Polit.*	2019	1
Premium Private Labels and PDO/PGI Products: Effects on Customer Loyalty	Martinelli, E.; De Canio, F.; Marchi, G.; Nardin, G.	*Advances in National Brand and Private Label Marketing.*	2017	4
Organic and Online Attributes for Buying and Selling Agricultural Products in the E-Marketplace in Spain	Robina-Ramírez, R.; Chamorro-Mera, A.; Moreno-Luna, L.	*Electron. Commer. Res. Appl.*	2020	13
The Importance of Websites for Organic Agri-Food Producers	Fernández-Uclés, D.; Bernal-Jurado, E.; Mozas-Moral, A.; Medina-Viruel, M.J.	*Econ. Res.-Ekon. Istraž.*	2020	14
Understanding the Role of Purchasing Predictors in the Consumer's Preferences for PDO Labelled Honey	Di Vita, G.; Pippinato, L.; Blanc, S.; Zanchini, R.; Mosso, A.; Brun, F.	*J. Food Prod. Mark.*	2021	1
Generation X versus Millennials Communication Behaviour on Social Media When Purchasing Food versus Tourist Services	Dabija, D.-C.; Bejan, B.M.; Tipi, N.	*E+M Ekon. Manag.*	2018	99
EU Quality Label vs Organic Food Products: A Multigroup Structural Equation Modeling to Assess Consumers' Intention to Buy in Light of Sustainable Motives	De Canio, F.; Martinelli, E.	*Food Res. Int.*	2021	16
Food tourism and regional development: A systematic literature review	Rachão, S.; Breda, Z.; Fernandes, C.; Joukes, V.	*Eur. J. of Tourism Research*	2019	68
Social media and consumer buying behavior decision: what entrepreneurs should know?	Palalic, R.; Ramadani, V.; Mariam Gilani, S.; Gërguri-Rashiti, S.; Dana, L.	*Mgmt. Decision*	2021	24
Online Shopping: Factors That Affect Consumer Purchasing Behaviour	Bucko, J.; Kakalejčík, L.; Ferencová, M.	*Cogent bus. manag.*	2018	56
Expanding the PGI Certification Scheme as a Marketing Tool in the Olive Oil Industry: A Perspective on Consumer Behavior	Di Vita, G.; Cavallo, C.; Del Giudice, T.; Pergamo, R.; Cicia, G.; D'Amico, M.	*Br. Food J.*	2021	3
Rural Cooperatives in the Digital Age: An Analysis of the Internet Presence and Degree of Maturity of Agri-Food Cooperatives' e-Commerce	Cristobal-Fransi, E.; Montegut-Salla, Y.; Ferrer-Rosell, B.; Daries, N.	*J. Rural Stud.*	2020	46
A Study on Agrifood Purchase Decision-making and Online Channel Selection according to Consumer Characteristics, Perceived Risks, and Eating Lifestyles	Lee, M.K.; Park, S.H.; Kim, Y.J.	*Asia-Pacific j. of Bus. Venturing and Entrepreneurship.*	2021	4
Protected Designation of Origin (PDO), Protected Geographical Indication (PGI) and Traditional Speciality Guaranteed (TSG): A Bibiliometric Analysis	Dias, C.; Mendes, L.	*Food Res. Int.*	2018	68
Understanding the Real-World Impact of Geographical Indications: A Critical Review of the Empirical Economic Literature	Török, Á.; Jantyik, L.; Maró, Z. M.; Moir, H. V. J.	*Sustainability*	2020	14
The importance of "origin" for online agrifood products	Scuderi, A.; Sturiale, L.; Timpanaro, G.	*Quality—Access to Success*	2015	16
Geographical Indications, Public Goods, and Sustainable Development: The Roles of Actors' Strategies and Public Policies	Belletti, G.; Marescotti, A.; Touzard, J.-M.	*World Dev.*	2017	211
Product versus Region of Origin: Which Wins in Consumer Persuasion?	Luceri, B.; Latusi, S.; Zerbini, C.	*Br. Food J.*	2016	30
Importance of Regional and Traditional EU Quality Schemes in Young Consumer Food Purchasing Decisions	Angowski, M.; Jarosz-Angowska, A.	*Eur. Res. Stud.*	2020	1

Table 4. *Cont.*

Title	Authors	Journal of Publication	Publication Year	Total Citations *
A Study on Consumer Characteristics According to Social Media Use Clusters When Purchasing Agri-food Online	Lee, M.K.; Park, S.H.; Kim, Y.J.	*Asia-Pacific j. of Bus. Venturing and Entrepreneurship.*	2021	N/A
Estimating the Market Share and Price Premium of GI Foods—the Case of the Hungarian Food Discounters	Jantyik, L.; Török, Á.	*Sustainability*	2020	10
Consumers' Awareness of the EU's Protected Designations of Origin Logo	Goudis, A.; Skuras, D.	*Br. Food J.*	2021	4
The Role of Intrinsic and Extrinsic Characteristics of Honey for Italian Millennial Consumers	Blanc, S.; Zanchini, R.; Di Vita, G.; Brun, F.	*Br. Food J.*	2021	3
Consumer Preferences Regarding National and EU Quality Labels for Cheese, Ham and Honey: The Case of Slovenia	Kos Skubic, M.; Erjavec, K.; Klopčič, M.	*Br. Food J.*	2018	22
Social Marketing: A New Marketing Tool for the Food Sector	Elghannam, A.; Mesías, F. J.	*Advances in Business Strategy and Competitive Advantage*	2017	1

* Total citations in Google Scholar on 20 January 2022. N/A if no citation was reported.

3.2. *Perception about Certified Agri-Food Products*

To obtain market success, products must benefit from a positive overall image among target segments, exhibiting a proper added value and/or providing certain qualities that meet or exceed consumers' expectations [5,17]. Consumers are regarded as a subject more interested in the symbolic or cultural value of certified agri-food products than in their intrinsic functions and utilities [18]. Consumers are considered active players in the market, where they exercise their freedom to move in search of products, but also gain experiences through which they can express their identity [5]. Looking for options to fulfill their expectations and desires, consumers are looking for food quality in terms of product origin, uniqueness, respect for the environment, animal welfare, traditional manufacturing process, taste, providing growth opportunities for small businesses operating in the niche market, the so-called "restricted food", a term that refers to local, certified foods [19]. Other papers [20–22] reflect that agri-food products certified with quality schemes are perceived positively by consumers, as they contribute to improving their health condition, their quality of life, strengthening them and ensuring that with increasing age consumers are still fit. As regards the geographical delimitations, consumers from southern European regions tend to associate more often the term "traditional food" with their culture or history [23]. Agri-food products are consumed on some typical occasions, like on certain holidays and/or seasons, knowledge about that being generally transmitted from one generation to another. Such products are usually manufactured precisely after some certain procedures, being part of the gastronomic heritage of a region or an ethnic group [19], with little or no processing/handling of the original receipt and known for its sensory properties. Furthermore, these products are often associated with a clear delimited geographical area [19]. On the other hand, consumers from central and northern Europe tend to focus mainly on practical issues, such as convenience, health, or the ease and speed of purchasing food [24]. Some consumers consider PDO/PGI labels to be organic, while every second consumers are unsure whether PDO/PGI certified foods are produced without fertilizers and other chemicals [25,26].

3.3. *Preference, Recognition, and Willingness to Pay for Certified Agri-Food Products*

The recent literature indicates a renewed consumer interest in certified agri-food products [19,27]. A concern about consumers' perception of certified products is the willingness to pay higher prices than for the non-certified alternatives [28]. These consumers realize that "origin" cannot be always considered a determining factor in consumer choices compared to cost, safety, and nutrition [29]. The concepts that help explain the correlation between quality labels and willingness to pay depend on factors such as the geographical area investigated, the consumer's residence concerning the production area, consumer demographics, GI label awareness, and product type [29]. At the same time, consumers who know the region to which the certified products refer or feel a certain attachment towards them, tend to be more optimistic about the products labeled PDO/PGI/TSG, thus also exerting a higher willingness of paying even a price premium [30–32]. Because consumers identify certified products with customs and heritage passed down from generation to generation, traditions cannot be exported. These certified products outside their "area of influence" may not have the emotional attachment of experience [31,33]. Studies measuring the awareness and recognition of quality labels among European consumers conclude that consumers from Sweden, Denmark, Finland, and the Netherlands, but also France, Italy, Spain, Greece, Portugal have a higher likelihood of recognizing quality labels and their logos [34,35]. This is due to the collaborative activities between European producers, which target consumers in these countries.

3.4. Purchasing and Consumption Behavior of Certified Agri-Food Products

The reviewed studies set out the different concepts that explain consumers' motivations to buy traditionally produced agri-food products, certified with quality schemes. Regarding the decision to purchase certified agri-food products by young consumers, while recent studies reveal a relatively small significance of EU quality schemes in food purchasing decisions taken by young consumers [36,37]. Young consumer behavior is strongly influenced by globalization, social media, online behavior, and current trends, as they frequently do not differentiate between quality schemes such as PDO, PGI, and TSG [37]. Young consumers' most important determinants of food choice are product prices, freshness, and shelf life of products, but also convenience [33,37]. Consumers attribute a higher value to a PDO label than to a PGI. The preference for buying the PDO label over the PGI one might be explained by the fact that consumers tend to perceive PDO as a certification that firmly guarantees the production, processing, and preparation of agri-food products in a well-established geographical area [38,39]. Perceived quality associated with extrinsic attributes (such as quality, brands, labels, design, information on use and benefits, authenticity, commitment to the environment, cultural ties) significantly influences the purchasing of certified agri-food products [22,27]. Older consumers with higher education and above-average incomes show an increased preference for certified products with quality schemes [21].

3.5. Online Purchasing of Certified Agri-Food Products

Information and Communication Technology (ICT) has developed as the new frontier for organizations in various industries, as well as agriculture, thus being considered a strategy that will bridge the gap between producers and consumers [40,41]. Therefore, the agri-food sector needs to increase its competitiveness, and be able to respond quickly to the ever-changing consumer needs and desires, thus satisfying him/her and properly communicating the extrinsic and intrinsic added values of certified agri-food products through online stores developed for modern customers [42,43]. Social networks represent a channel from which consumers take relevant information for their next purchase decisions; consumers are often more influenced and trust strangers and online influencers than official representative of companies [44,45]. The shopping decision is strongly influenced by online reviews and recommendations from blogs, forums and/or social networks [46]. Agricultural cooperatives take information about consumers, which they integrate into their communication strategies, and inform customers about certified agri-food products [47–49]. Such organizations are usually aware of the importance that the territory of origin (physical, sensory, and cultural) of certified agri-food products and production techniques plays for consumers, thus representing strong values that might trigger consumers preferences [42]. Furthermore, agricultural cooperatives must go beyond their traditional presence, thus encouraging online interaction and collaboration, connectivity, and giving consumers the possibility to find and share information and gain knowledge about certified agri-food products [50]. For instance, cooperatives that produce and sell olive oil, fruit, and wine are more recently aware of the importance that their online communication plays for consumers purchase intention [50–52]. Online shopping causes consumers to behave differently concerning the intangibility of the product [53,54].

While in on-site shopping, the information comes from the sensory examination of the product, online shopping is determined by other factors: the customer's intention to buy, the influence of friends and family, consumer personality, but also knowledge and curiosity [53,55]. The attitude of buying online food products is also improved by extrinsic factors, such as the quality of the website (design, content, and navigation), product availability, ease of use, which positively affect the purchase intention [54,56–58]. Consumers who purchase certified agri-food products online would like to have access to information on the environmental impact and sustainability of products, in addition to the unique properties and characteristics of agri-food products [53,59–61].

The aspects and findings presented in the previous sub-sections of the Results are summarized in Table 5.

Table 5. Overview of representative papers according to different assessment criteria.

Publication Year	Authors	Type of Paper			Type of Study		Quality Schemes			Consumer Concepts			
		Case Study	Original Research	Literature Review	Qualitative Study	Quantitative Study	PDO	PGI	TSG	Perception	Preference and Willingness to Pay	Purchasing and Consumption	Online Purchasing
2015	Scuderi, A.; Sturiale, L.; Timpanaro, G.	-	✓	-	✓	-	✓	✓	-	-	-	✓	-
2016	Grunert, K.G.; Aachmann, K.	-	-	✓	✓	-	✓	✓	✓	✓	✓	-	-
2016	Luceri, B.; Latusi, S.; Zerbini, C.	-	✓	-	-	✓	✓	✓	-	✓	✓	-	-
2017	Garavaglia, C.; Mariani, P.	-	✓	-	-	✓	✓	-	-	-	✓	-	-
2017	Martínez, A.A.; Fernández-Poyatos, M.D.	-	✓	-	✓	-	✓	✓	-	-	-	✓	-
2017	Martinelli, E.; De Canio, F.; Marchi, G.; Nardin, G.	-	✓	-	-	✓	✓	✓	-	✓	-	-	-
2017	Belletti, G.; Marescotti, A.; Touzard, J.-M.	✓	-	-	✓	-	✓	✓	✓	✓	✓	✓	-
2017	Elghannam, A.; Mesías, F. J.	-	✓	-	✓	-	✓	✓	✓	-	-	-	-
2018	Dabija, D.-C.; Bejan, B.M.; Tipi, N.	-	✓	-	-	✓	-	-	-	-	-	-	-
2018	Bucko, J.; Kakalejčík, L.; Ferencová, M.	-	✓	-	-	✓	-	-	-	✓	✓	-	-
2018	Dias, C.; Mendes, L.	-	-	✓	✓	-	✓	✓	✓	✓	✓	✓	-
2018	Kos Skubic, M.; Erjavec, K.; Klopčič, M.	-	✓	-	-	✓	✓	✓	-	✓	✓	-	-
2019	Martinelli, E.; De Canio, F.	✓	-	-	-	✓	✓	✓	-	-	✓	✓	-
2019	Tham, K. W.; Dastane, O.; Johari, Z.; Ismail, N.B.	-	✓	-	-	✓	-	-	-	-	-	-	✓

Table 5. *Cont.*

Publication Year	Authors	Type of Paper			Type of Study		Quality Schemes			Consumer Concepts			
		Case Study	Original Research	Literature Review	Qualitative Study	Quantitative Study	PDO	PGI	TSG	Perception	Preference and Willingness to Pay	Purchasing and Consumption	Online Purchasing
2019	Oleksiuk, I.; Werenowska, A.	-	✓	-	-	✓	✓	✓	✓	✓	✓	✓	-
2019	Rachão, S.; Breda, Z.; Fernandes, C.; Joukes, V.	-	-	✓	-	✓	✓	✓	-	✓	✓	✓	-
2020	Robina-Ramírez, R.; Chamorro-Mera, A.; Moreno-Luna, L.	✓	-	-	-	✓	-	-	-	-	✓	-	✓
2020	Fernández-Uclés, D.; Bernal-Jurado, E.; Mozas-Moral, A.; Medina-Viruel, M.J.	-	✓	-	✓	-	-	-	-	-	-	-	✓
2020	Cristobal-Fransi, E.; Montegut-Salla, Y.; Ferrer-Rosell, B.; Daries, N.	-	✓	-	✓	-	✓	✓	✓	-	-	-	✓
2020	Török, Á.; Jantyik, L.; Maró, Z. M.; Moir, H. V. J.	-	-	✓	✓	-	✓	✓	✓	✓	✓	✓	✓
2020	Angowski, M.; Jarosz-Angowska, A.	-	✓	-	-	✓	✓	✓	✓	✓	✓	✓	-
2020	Jantyik, L.; Török, Á.	-	✓	-	-	✓	✓	✓	✓	-	✓	-	-
2021	Savelli, E.; Bravi, L.; Francioni, B.; Murmura, F.; Pencarelli, T.	-	✓	-	-	✓	✓	-	-	✓	-	-	-
2021	Skalkos, D.; Kosma, I. S.; Vasiliou, A.; Guine, R. P. F.	-	✓	-	-	✓	✓	✓	✓	✓	✓	✓	-
2021	Di Vita, G.; Pippinato, L.; Blanc, S.; Zanchini, R.; Mosso, A.; Brun, F.	-	✓	-	-	✓	✓	-	-	-	✓	-	-
2021	De Canio, F.; Martinelli, E.	-	✓	-	-	✓	✓	✓	-	-	✓	-	-
2021	Palalic, R.; Ramadani, V.; Mariam Gilani, S.; Gërguri-Rashiti, S.; Dana, L.	-	✓	-	-	✓	-	-	-	-	-	-	✓
2021	Di Vita, G.; Cavallo, C.; Del Giudice, T.; Pergamo, R.; Cicia, G.; D'Amico, M.	-	✓	-	-	✓	-	✓	-	-	-	✓	-
2021	Lee, M.K.; Park, S.H.; Kim, Y.J.	-	✓	-	-	✓	✓	✓	✓	-	-	✓	✓
2021	Lee, M.K.; Park, S.H.; Kim, Y.J.	-	✓	-	-	✓	✓	✓	✓	✓	✓	-	-
2021	Goudis, A.; Skuras, D.	-	✓	-	-	✓	✓	✓	-	✓	-	-	-
2021	Blanc, S.; Zanchini, R.; Di Vita, G.; Brun, F.	-	✓	-	-	✓	-	-	-	✓	-	-	-

Note: ✓ is marking the presence of the criteria; - is marking the absence of the criteria.

4. Discussion

Even though the studies covered use various methodologies and provide contradictory results, this systematic review reveals several common features that stand in line with previous research [16,62–65], indicating that the understanding about certified agri-food products are mixed. In line with previous studies, were identified consumers that consider that the food quality is not verified [25,26]. Thus, there are consumers that trust the meaning of these certifications and choose to buy a more traditional healthy food product [20,21,23,24,35,66].

The "area of influence" is one of the most crucial factors for selling certified products; the emotional attachment of experience that each product comes with could help consumers refine the natural taste. This represents a major objective for certified product, to keep its taste, smell, and/or nutritional qualities. Several studies [19,23,24,67] showed that the culture of the geographical delimitations influences the perceptions about certified agri-food products. The certified agri-food products have an advantage for consumers who know the product's region, so the certification proves that the product is created strictly in that region it kept its originality. In southern Europe tend to associate them with the terms "traditional food" and "brand-name"; this is seen more often in combination with the concepts, culture, or even history, heritage, and customs passed down from one generation to another. From the past, we can learn about the types of food that our ancestors were eating without any chemicals for growing. The central and northern Europe regions tend to focus more on the practical benefits of product convenience, health, or purchase access in another area of Europe.

The "origin" of the product is not always the determining factor in consumer choices. Many consumers consider that the nutritional aspects, cost, and safety sometimes come first when choosing the right product for their needs. Education, income, and globalization are factors that influence the consumption behavior of certified agri-food products. Consumers with above-average income and higher education show more interest in the certified product with quality schemes. On the other hand, we have the "young generation" the consumers strongly influenced by globalization and the current trends. They do not differentiate between certified agri-food products. The most critical factors that determine the young consumer to purchase are nutritional factors, freshness, and price. Young consumers caring about their health choose the most suitable product to pay as economical as possible and get the best outcome for their budget [21,27,36,37,68].

In both the on-site and online environment, we can find different factors that help the consumers choose the right product for their needs. In the on-site situation, we see distinct influences from extrinsic and intrinsic influential factors. Most of the time, the extrinsic factors that influence the purchasing decision of the certified agri-food product are the purchasing environment around the products, such as the shelf arrangement and even the type of store. Regarding intrinsic factors, we have the smell, package, nutritional information about the product, the price, the colors. On the other hand, in the Online, we have a different set of influential factors that are much more of a technical nature, such as the User Experience (UX) of the website, the speed, the colors, and most important aspects like the delivery duration, information about the product (description of the quality schemes and logos, area of production, etc.), the online support of the website [47–49,51–54]. These are some of the factors that help in choosing the right product online. One of the essential elements that online shopping offers to customers is package delivery. In 2022, the world is starting to change towards a new era of packaging where cheap and efficient is not enough anymore. A package should be ergonomic, safe, recyclable, and, most important, a storyteller for the brand and its products.

During the COVID-19 pandemic, consumer preferences have leant toward certified foods whose origin is known. Thus, the PDO label begins to become a choice for consumers concerned about their health and a diet that supports their immune system [66,69–71]. Moreover, the traditional shopping system has been altered, so consumers tend to buy healthy food online [72,73]. Although the price of certified agri-food was higher, there is a preference for certified food products with quality schemes among the consumers [71,74].

Also, by consuming products of this type—of controlled origin, certified by the EU, the health can be maintained, and the body's immunity can be increased. In COVID-19 and pandemic restrictions, the consumers' food must be safe when human movement is restricted due to regulations. Thus, they must have an appropriate quality, respectively, to have a controlled origin [66,69,70].

5. Conclusions

In 2021, perception of success in the food market is about exceeding the consumer expectations, providing them with much better quality than they have asked, providing package, information, and a premium feeling about the product. These details help to reach a positive opinion about the certified agri-food products. EU quality labels were introduced as a consumer decision-making tool. Still, they are also a way of controlling food, as the logo's appearance ensures that the product can be traced back to a specific manufacturing area and to a specified know-how process. EU quality schemes can thereby potentially reduce confusion about food purchases, assuring the customer of the certified agri-food products' uniqueness and nutritional qualities. The on-site and online environment is trying to draw attention to more specific aspects that can bring quality to food products, such as certifications, animal welfare standards, and respect for the environment. In both climates, one can highlight different types of influence trying to make the final customers self-generate the mindset that "eating healthy" might be understood as "living healthy". The influential factors are all about sharing as much quality information as possible with the customer: nutritional information, region of production/origin, price, package, colors. The "young generation" is powerfully influenced by globalization, social media, Internet, green behavior, and current trends, through which they can be educated about the importance of consuming quality products, what effect it has over their body in the long term, and what conduct they should adopt to have a healthier life in a healthfuller community.

5.1. Implications for Market Participants and Public Institutions

Nationally sustained by different post-COVID-19 strategies, the PDO, PGI, and TSG certifications would have, as a result, the increased level of health of the population. One of the solutions would be to encourage local producers to apply for this certification. The food products with the certification PDO, PGI, TSG have a better impact on consumers' health because of their pure ingredients and the lack of artificial chemicals. Consuming a healthy, non-altered, and natural product is one of the leading health benefits of these products. Moreover, these review results are helpful to different government agencies and companies to improve their promotion strategies towards these types of certifications that verify quality and tradition.

5.2. Limitations and Further Directions of Research

There are certain limitations to our research. The search strategy may have omitted pertinent material that brings the possibility that removed articles include information that could affect our conclusions. Given the prevalence of the PGI and PDO certification schemes, more research into the TSG quality certification scheme is required. More research is necessary on consumer behavior regarding PGI, PDO, and TSG food products, considering the variances between nations or areas. Since there is a focus on examining certified products susceptible to some form of agro-industrial production, such as meat, cheese, wine, and olive oil, perception and consumption behavior of certified fruit or vegetable varieties could provide a viable path for further directions of research. In addition, more research is needed to link certified food products with the European quality schemes to the health benefits they can provide in pandemic times, relying on educating consumers about the value and benefits of these certified products with quality schemes.

Author Contributions: Conceptualization, A.-I.G. and C.B.P.; methodology, A.-I.G. and C.B.P.; validation, C.B.P., D.-C.D. and M.F.; formal analysis, A.-I.G.; investigation, A.-I.G.; resources, A.-I.G. and C.B.P.; writing—original draft preparation, A.-I.G.; C.B.P.; D.-C.D.; writing—review and editing,

A.-I.G.; C.B.P., D.-C.D. and M.F.; visualization, C.B.P.; D.-C.D. and M.F.; supervision, C.B.P. and D.-C.D. All authors have read and agreed to the published version of the manuscript.

Funding: This research received no external funding.

Institutional Review Board Statement: Not applicable.

Informed Consent Statement: Not applicable.

Data Availability Statement: Not applicable.

Acknowledgments: The authors would like to express their gratitude to all reviewers who provided helpful suggestions for improving our work.

Conflicts of Interest: The authors declare no conflict of interest.

References

1. Di Giacomo, M.G.G. The relationship between Food-Agriculture-Environment compared with the new Common Agricultural Policy. *AGEI Geotema* **2020**, *52*, 8–17.
2. Vilke, R.; Vidickiene, D.; Gedminaite-Raudone, Z.; Simonaityte, V.; Ribasauskiene, E. *Rural Economic Developments and Social Movements: A New Paradigm*; Springer Nature: Cham, Switzerland, 2021; ISBN 978-303-071-983-8.
3. Bu, F.; Wang, X. A Smart Agriculture IoT System Based on Deep Reinforcement Learning. *Future Gener. Comput. Syst.* **2019**, *99*, 500–507. [CrossRef]
4. Sikora, A. European Green Deal–legal and financial challenges of the climate change. *ERA Forum* **2021**, *21*, 681–697. [CrossRef]
5. De Canio, F.; Martinelli, E. EU Quality Label vs Organic Food Products: A Multigroup Structural Equation Modeling to Assess Consumers' Intention to Buy in Light of Sustainable Motives. *Food Res. Int.* **2021**, *139*, 109846. [CrossRef] [PubMed]
6. European Green Deal, the Economic, Social and Environmental Sustainability, a Priority for the PDO PGI Supply Chains. Available online: http://www.lifettgg.eu/en/2020/12/16/the-economic-social-and-environmental-sustainability-a-priority-for-the-pdo-pgi-supply-chains/ (accessed on 23 September 2021).
7. European Commission. Quality Schemes Explained. Available online: https://ec.europa.eu/info/food-farming-fisheries/food-safety-and-quality/certification/quality-labels/quality-schemes-explained/ (accessed on 23 September 2021).
8. Gellynck, X.; Kühne, B. Innovation and Collaboration in Traditional Food Chain Networks. *J. Chain Netw. Sci.* **2008**, *8*, 121–129. [CrossRef]
9. Carbone, A.; Caswell, J.; Galli, F.; Sorrentino, A. The Performance of Protected Designations of Origin: An Ex Post Multi-Criteria Assessment of the Italian Cheese and Olive Oil Sectors. *J. Agric. Food Ind. Organ.* **2014**, *12*, 121–140. [CrossRef]
10. Regulamentul (UE) nr. 1151/2012 al Parlamentului European și al Consiliului din 21 Noiembrie 2012 Privind Sistemele din Domeniul Calității Produselor Agricole și Alimentare. Available online: https://eur-lex.europa.eu/legal-content/RO/TXT/?uri=CELEX%3A32012R1151 (accessed on 23 September 2021).
11. Cassago, A.L.L.; Artêncio, M.M.; de Moura Engracia Giraldi, J.; Da Costa, F.B. Metabolomics as a Marketing Tool for Geographical Indication Products: A Literature Review. *Eur. Food Res. Technol.* **2021**, *247*, 1–17. [CrossRef]
12. Marescotti, A. Typical products and rural development: Who benefits from PDO/PGI recognition. In Proceedings of the Food Quality Products in the Advent of the 21st Century: Production, Demand and Public Policy. 83rd EAAE Seminar, Chania, Greece, 9–10 September 2003; CIHEAM: Chania, Greece, 2003; Volume 4.
13. Mancini, M.; Menozzi, D.; Donati, M.; Biasini, B.; Veneziani, M.; Arfini, F. Producers' and Consumers' Perception of the Sustainability of Short Food Supply Chains: The Case of Parmigiano Reggiano PDO. *Sustainability* **2019**, *11*, 721. [CrossRef]
14. European Commission. The EU Geographical Indications Register. Available online: https://ec.europa.eu/info/food-farming-fisheries/food-safety-and-quality/certification/quality-labels/geographical-indications-register/ (accessed on 14 January 2022).
15. Campos, S.; Doxey, J.; Hammond, D. Nutrition Labels on Pre-Packaged Foods: A Systematic Review. *Public Health Nutr.* **2011**, *14*, 1496–1506. [CrossRef]
16. Olbrich, R.; Jansen, H.C.; Hundt, M. Effects of Pricing Strategies and Product Quality on Private Label and National Brand Performance. *J. Retail. Consum. Serv.* **2017**, *34*, 294–301. [CrossRef]
17. Arfini, F.; Bellassen, V. Correction to: Sustainability of European Food Quality Schemes. In *Sustainability of European Food Quality Schemes*; Springer International Publishing: Cham, Switzerland, 2020; p. C1, ISBN 978-303-027-507-5.
18. Verbeke, W.; Guerrero, L.; Almli, V.L.; Vanhonacker, F.; Hersleth, M. European Consumers' Definition and Perception of Traditional Foods. In *Traditional Foods*; Springer US: Boston, MA, USA, 2016; pp. 3–16. [CrossRef]
19. Likoudis, Z.; Sdrali, D.; Costarelli, V.; Apostolopoulos, C. Consumers' Intention to Buy Protected Designation of Origin and Protected Geographical Indication Foodstuffs: The Case of Greece: Greek Consumer Behaviour and PDO/PGI. *Int. J. Consum. Stud.* **2016**, *40*, 283–289. [CrossRef]
20. Prakash, V. Introduction: The Importance of Traditional and Ethnic Food in the Context of Food Safety, Harmonization, and Regulations. In *Regulating Safety of Traditional and Ethnic Foods*; Prakash, V., Martín-Belloso, O., Keener, L., Astley, S., Braun, S., McMahon, H., Lelieveld, H., Eds.; Academic Press: Waltham, MA, USA, 2016; pp. 1–6, ISBN 978-012-800-605-4.

21. Belletti, G.; Marescotti, A.; Touzard, J.-M. Geographical Indications, Public Goods, and Sustainable Development: The Roles of Actors' Strategies and Public Policies. *World Dev.* **2017**, *98*, 45–57. [CrossRef]
22. Barska, A.; Wojciechowska-Solis, J. Traditional and Regional Food as Seen by Consumers–Research Results: The Case of Poland. *Br. Food J.* **2018**, *120*, 1994–2004. [CrossRef]
23. Martínez, A.A.; Fernández-Poyatos, M.D. Gastronomy as a tourism resource in the province of Alicante. *Int. J. Sci. Manag. Tour.* **2017**, *3*, 25–45.
24. Fandos-Herrera, C. Exploring the Mediating Role of Trust in Food Products with Protected Designation of Origin. The Case of 'Jamón de Teruel'. *Span. J. Agric. Res.* **2016**, *14*, 2. [CrossRef]
25. Magagnoli, S. *The Construction of Planetary Taste*; Routledge: London, UK, 2021; pp. 173–185, ISBN 978-100-305-441-2.
26. Martinelli, E.; De Canio, F. Premium Private Labels Products: Drivers of Consumers' Intention to Buy. *Int. J. Bus. Manag.* **2019**, *14*, 36. [CrossRef]
27. Jantyik, L.; Török, Á. Estimating the Market Share and Price Premium of GI Foods—the Case of the Hungarian Food Discounters. *Sustainability* **2020**, *12*, 1094. [CrossRef]
28. Garavaglia, C.; Mariani, P. How Much Do Consumers Value Protected Designation of Origin Certifications? Estimates of Willingness to Pay for PDO Dry-Cured Ham in Italy: How Much Do Consumers Value Protected Designation of Origin Certifications? *Agribusiness* **2017**, *33*, 403–423. [CrossRef]
29. Luceri, B.; Latusi, S.; Zerbini, C. Product versus Region of Origin: Which Wins in Consumer Persuasion? *Br. Food J.* **2016**, *118*, 2157–2170. [CrossRef]
30. Balogh, P.; Békési, D.; Gorton, M.; Popp, J.; Lengyel, P. Consumer Willingness to Pay for Traditional Food Products. *Food Policy* **2016**, *61*, 176–184. [CrossRef]
31. Muça, E.; Pomianek, I.; Peneva, M. The Role of GI Products or Local Products in the Environment—Consumer Awareness and Preferences in Albania, Bulgaria and Poland. *Sustainability* **2021**, *14*, 4. [CrossRef]
32. Russo, V.; Zito, M.; Bilucaglia, M.; Circi, R.; Bellati, M.; Marin, L.E.M.; Catania, E.; Licitra, G. Dairy Products with Certification Marks: The Role of Territoriality and Safety Perception on Intention to Buy. *Foods* **2021**, *10*, 2352. [CrossRef]
33. Oleksiuk, I.; Werenowska, A. Promotion of Regional and Traditional Products. *Środ. Stud. Polit.* **2019**, *2*, 135–149. [CrossRef]
34. Goudis, A.; Skuras, D. Consumers' Awareness of the EU's Protected Designations of Origin Logo. *Br. Food J.* **2021**, *123*, 1–18. [CrossRef]
35. Blanc, S.; Zanchini, R.; Di Vita, G.; Brun, F. The Role of Intrinsic and Extrinsic Characteristics of Honey for Italian Millennial Consumers. *Br. Food J.* **2021**, *123*, 2183–2198. [CrossRef]
36. Angowski, M.; Jarosz-Angowska, A. Importance of Regional and Traditional EU Quality Schemes in Young Consumer Food Purchasing Decisions. *Eur. Res. Stud.* **2020**, *23*, 916–927. [CrossRef]
37. Martinelli, E.; De Canio, F.; Marchi, G.; Nardin, G. Premium Private Labels and PDO/PGI Products: Effects on Customer Loyalty. In *Advances in National Brand and Private Label Marketing*; Springer International Publishing: Cham, Switzerland, 2017; pp. 65–72. [CrossRef]
38. Di Vita, G.; Cavallo, C.; Del Giudice, T.; Pergamo, R.; Cicia, G.; D'Amico, M. Expanding the PGI Certification Scheme as a Marketing Tool in the Olive Oil Industry: A Perspective on Consumer Behavior. *Br. Food J.* **2021**, *123*, 3841–3856. [CrossRef]
39. Scuderi, A.; Sturiale, L.; Timpanaro, G. The importance of "origin" for online agrifood products. *Qual.–Access Success* **2015**, *16*, 260–266.
40. Pegan, G.; Vianelli, D.; de Luca, P. Online Channels and the Country of Origin. In *International Marketing Strategy*; Springer International Publishing: Cham, Switzerland, 2020; pp. 149–180, ISBN 978-303-033-588-5.
41. Watts, D.C.; Ilbery, B.; Maye, D. Making reconnections in agro-food geography: Alternative systems of food provision. In *The Rural*; Routledge: London, UK, 2017; pp. 165–184, ISBN 978-131-523-721-3.
42. Lee, M.K.; Park, S.H.; Kim, Y.J. A Study on Agrifood Purchase Decision-making and Online Channel Selection according to Consumer Characteristics, Perceived Risks, and Eating Lifestyles. *Asia-Pac. J. Bus. Ventur. Entrep.* **2021**, *16*, 147–159.
43. Lee, M.K.; Park, S.H.; Kim, Y.J. A Study on Consumer Characteristics According to Social Media Use Clusters When Purchasing Agri-food Online. *Asia-Pac. J. Bus. Ventur. Entrep.* **2021**, *16*, 195–209.
44. Palalic, R.; Ramadani, V.; Mariam Gilani, S.; Gërguri-Rashiti, S.; Dana, L. Social media and consumer buying behavior decision: What entrepreneurs should know? *Manag. Decis.* **2021**, *59*, 1249–1270. [CrossRef]
45. Dabija, D.-C.; Bejan, B.M.; Tipi, N. Generation X versus Millennials Communication Behaviour on Social Media When Purchasing Food versus Tourist Services. *E+M Ekon. Manag.* **2018**, *21*, 191–205. [CrossRef]
46. Hofacker, C.F.; Belanche, D. Eight Social Media Challenges for Marketing Managers. *Span. J. Mark.-ESIC* **2016**, *20*, 73–80. [CrossRef]
47. Elghannam, A.; Mesías, F.J. Social Marketing: A New Marketing Tool for the Food Sector. In *Advances in Business Strategy and Competitive Advantage*; IGI Global: Hershey, PA, USA, 2017; pp. 91–106. [CrossRef]
48. Brečić, R.; Tomić Maksan, M.; Đugum, J. The Case of the PDO and PGI Label in the Croation Market. In Proceedings of the 7th International M-Sphere Conference for Multidisciplinarity in Business and Science, Zagreb, Croatia, 24–26 October 2019.
49. Cristobal-Fransi, E.; Montegut-Salla, Y.; Ferrer-Rosell, B.; Daries, N. Rural Cooperatives in the Digital Age: An Analysis of the Internet Presence and Degree of Maturity of Agri-Food Cooperatives' e-Commerce. *J. Rural Stud.* **2020**, *74*, 55–66. [CrossRef]

50. Bernal-Jurado, E.; Mozas-Moral, A.; Fernández-Uclés, D.; Medina-Viruel, M.J. Online Popularity as a Development Factor for Cooperatives in the Winegrowing Sector. *J. Bus. Res.* **2021**, *123*, 79–85. [CrossRef] [PubMed]
51. Karanikolas, P.; Martinez-Gomez, V.; Galli, F.; Prosperi, P.; Hernández, P.A.; Arnalte-Mur, L.; Rivera, M.; Goussios, G.; Fastelli, L.; Oikonomopoulou, E.; et al. Food System Integration of Olive-Oil-Producing Small Farms in Southern Europe. *Glob. Food Sec.* **2021**, *28*. [CrossRef]
52. Robina-Ramírez, R.; Chamorro-Mera, A.; Moreno-Luna, L. Organic and Online Attributes for Buying and Selling Agricultural Products in the E-Marketplace in Spain. *Electron. Commer. Res. Appl.* **2020**, *42*, 1–12. [CrossRef]
53. Xu, H.; Zhang, K.Z.K.; Zhao, S.J. A dual systems model of online impulse buying. *Ind. Manag. Data Syst.* **2020**, *120*, 845–861. [CrossRef]
54. Bucko, J.; Kakalejčík, L.; Ferencová, M. Online Shopping: Factors That Affect Consumer Purchasing Behaviour. *Cogent Bus. Manag.* **2018**, *5*, 1535751. [CrossRef]
55. Ariff, M.S.M.; Yan, N.S.; Zakuan, N.; Bahari, A.Z.; Jusoh, A. Web-Based Factors Affecting Online Purchasing Behaviour. *IOP Conf. Ser. Mater. Sci. Eng.* **2013**, *46*, 10. [CrossRef]
56. Tham, K.W.; Dastane, O.; Johari, Z.; Ismail, N.B. Perceived Risk Factors Affecting Consumers' Online Shopping Behaviour. *J. Asian Financ. Econ. Bus.* **2019**, *6*, 246–260. [CrossRef]
57. Fernández-Uclés, D.; Bernal-Jurado, E.; Mozas-Moral, A.; Medina-Viruel, M.J. The Importance of Websites for Organic Agri-Food Producers. *Econ. Res.-Ekon. Istraž.* **2020**, *33*, 2867–2880. [CrossRef]
58. Di Vita, G.; Pippinato, L.; Blanc, S.; Zanchini, R.; Mosso, A.; Brun, F. Understanding the Role of Purchasing Predictors in the Consumer's Preferences for PDO Labelled Honey. *J. Food Prod. Mark.* **2021**, *27*, 42–56. [CrossRef]
59. Grunert, K.G.; Aachmann, K. Consumer Reactions to the Use of EU Quality Labels on Food Products: A Review of the Literature. *Food Control* **2016**, *59*, 178–187. [CrossRef]
60. Teuber, R. Consumers' and Producers' Expectations towards Geographical Indications: Empirical Evidence for a German Case Study. *Br. Food J.* **2011**, *113*, 900–918. [CrossRef]
61. González-Azcárate, M.; Cruz Maceín, J.L.; Bardají, I. Why Buying Directly from Producers Is a Valuable Choice? Expanding the Scope of Short Food Supply Chains in Spain. *Sustain. Prod. Consum.* **2021**, *26*, 911–920. [CrossRef]
62. Dias, C.; Mendes, L. Protected Designation of Origin (PDO), Protected Geographical Indication (PGI) and Traditional Speciality Guaranteed (TSG): A Bibiliometric Analysis. *Food Res. Int.* **2018**, *103*, 492–508. [CrossRef]
63. Rachão, S.; Breda, Z.; Fernandes, C.; Joukes, V. Food tourism and regional development: A systematic literature review. *Eur. J. Tour. Res.* **2019**, *21*, 33–49. [CrossRef]
64. Török, Á.; Jantyik, L.; Maró, Z.M.; Moir, H.V.J. Understanding the Real-World Impact of Geographical Indications: A Critical Review of the Empirical Economic Literature. *Sustainability* **2020**, *12*, 9434. [CrossRef]
65. Dimitrakopoulou, M.-E.; Vantarakis, A. Does Traceability Lead to Food Authentication? A Systematic Review from A European Perspective. *Food Rev. Int.* **2021**, 1–23. [CrossRef]
66. Skalkos, D.; Kosma, I.S.; Vasiliou, A.; Guine, R.P.F. Consumers' Trust in Greek Traditional Foods in the Post COVID-19 Era. *Sustainability* **2021**, *13*, 9975. [CrossRef]
67. Savelli, E.; Bravi, L.; Francioni, B.; Murmura, F.; Pencarelli, T. PDO Labels and Food Preferences: Results from a Sensory Analysis. *Br. Food J.* **2021**, *123*, 1170–1189. [CrossRef]
68. Kos Skubic, M.; Erjavec, K.; Klopčič, M. Consumer Preferences Regarding National and EU Quality Labels for Cheese, Ham and Honey: The Case of Slovenia. *Br. Food J.* **2018**, *120*, 650–664. [CrossRef]
69. Skalkos, D.; Kosma, I.S.; Chasioti, E.; Bintsis, T.; Karantonis, H.C. Consumers' Perception on Traceability of Greek Traditional Foods in the Post-COVID-19 Era. *Sustainability* **2021**, *13*, 12687. [CrossRef]
70. Petrescu-Mag, R.M.; Vermeir, I.; Petrescu, D.C.; Crista, F.L.; Banatean-Dunea, I. Traditional Foods at the Click of a Button: The Preference for the Online Purchase of Romanian Traditional Foods during the COVID-19 Pandemic. *Sustainability* **2020**, *12*, 9956. [CrossRef]
71. Brugarolas, M.; Martínez-Carrasco, L.; Rabadán, A.; Bernabéu, R. Innovation Strategies of the Spanish Agri-Food Sector in Response to the Black Swan COVID-19 Pandemic. *Foods* **2020**, *9*, 1821. [CrossRef] [PubMed]
72. Alaimo, L.S.; Fiore, M.; Galati, A. Measuring Consumers' Level of Satisfaction for Online Food Shopping during COVID-19 in Italy Using POSETs. *Socioe-Econ. Plan. Sci.* **2021**, 101064. [CrossRef]
73. Alaimo, L.S.; Fiore, M.; Galati, A. How the Covid-19 Pandemic Is Changing Online Food Shopping Human Behaviour in Italy. *Sustainability* **2020**, *12*, 9594. [CrossRef]
74. Vidaurreta, I.; Orengo, J.; Fe, C.; González, J.M.; Gómez-Martín, Á.; Benito, B. Price Fluctuation, Protected Geographical Indications and Employment in the Spanish Small Ruminant Sector during the COVID-19 Crisis. *Animals* **2020**, *10*, 2221. [CrossRef]

Article

Assessment of the Relations for Determining the Profitability of Dairy Farms, A Premise of Their Economic Sustainability

Rodica Chetroiu [1,*], Ana Elena Cişmileanu [2], Elena Cofas [3], Ionut Laurentiu Petre [1,4,*], Steliana Rodino [1], Vili Dragomir [1], Ancuța Marin [1] and Petruța Antoneta Turek-Rahoveanu [1]

[1] Research Institute for Agriculture Economy and Rural Development, 011464 Bucharest, Romania; steliana.rodino@yahoo.com (S.R.); dragomir.vili@iceadr.ro (V.D.); marin.ancuta@iceadr.ro (A.M.); turek.petruta@iceadr.ro (P.A.T.-R.)

[2] National Research and Development Institute for Animal Biology and Nutrition, 077015 Balotesti, Romania; ana_cismileanu@yahoo.com

[3] Faculty of Management and Rural Development, University of Agronomic Sciences and Veterinary Medicine, 011464 Bucharest, Romania; cofasela@yahoo.com

[4] Department of Agrifood and Environmental Economics, Bucharest University of Economic Studies, 010374 Bucharest, Romania

[*] Correspondence: rodica.chetroiu@iceadr.ro (R.C.); laurentiu.petre@eam.ase.ro (I.L.P.)

Abstract: The profitability of dairy farms is a broadly addressed issue in research, for different farming systems and even more so now, when it comes to the issue of sustainability in different agricultural fields. The present study presents an evaluation of the relations used for the determination of profitability of various categories of dairy farms, in terms of size, geographical area, and total milk production. In order to analyze the associated influence exerted on the level of profitability by the selected technical and economic indicators, regression functions were applied. The TableCurve program was used to determine the ideal equation that describes the data entered in a two- or three-dimensional representation. The research results showed that the size of farms and the level and value of milk production are directly correlated with profitability, and the unit cost is inversely correlated with it.

Keywords: farm profitability; milk production; regression functions

Citation: Chetroiu, R.; Cişmileanu, A.E.; Cofas, E.; Petre, I.L.; Rodino, S.; Dragomir, V.; Marin, A.; Turek-Rahoveanu, P.A. Assessment of the Relations for Determining the Profitability of Dairy Farms, A Premise of Their Economic Sustainability. *Sustainability* **2022**, *14*, 7466. https://doi.org/10.3390/su14127466

Academic Editor: Giuseppe Todde

Received: 18 May 2022
Accepted: 13 June 2022
Published: 18 June 2022

1. Introduction

Economic efficiency is one of the key prerequisites for ensuring the competitiveness of any business regardless of the economic sector of production or position in the value chain [1]. Kingwell R. (2011) [2] showed that profitable farming systems are often large, complex, highly technologized, and involve time-consuming activities even for high-skilled managers. The farm productivity derived from production technology properly adapted to given conditions determines the financial results, and these influence strategic decisions regarding further development or, in some cases, to cease operations [3].

Previous studies [4] have demonstrated that higher intensification of agricultural activities significantly increases production efficiency. Profitability of the farm can be achieved by improving the input–output ratio and also by increasing income based on expanding production capacity, thus aiming to achieve competitive agricultural systems [5].

The modern farmer must be a skilled manager, selecting different investment opportunities so as to obtain as high a profit as possible, while fully developing human capital and observing environmental protection rules, all at the same time [6]. The available resources and the existing capacities of a given farm determine its development plans [7]. In order to be competitive, farmers need to be constantly aware of changing circumstances and have the ability to adapt to changes in the economic environment [8]. Proper management strategies can only be implemented based on detailed analysis of farm indicators.

The economic sustainability of milk production, as in other economic activities, is measured using the net profit indicator [9,10]. A good understanding of the influence of each cow's contribution to farm profitability can lead to improved dairy farm management [11]. There are a variety of interconnected factors that affect the efficiency of dairy farms, including management decisions, genetic factors, feed self-sufficiency, and animal welfare [12].

The profitability of dairy farms also depends on the efficiency of feeding, associated with the milk production obtained [13]. Farm profitability is influenced by fluctuations in prices for various inputs, especially feedstock, which have the highest share on expenditure, as well as the volatility of finished-product prices [14]. Low-performing farms have low milk production, unbalanced feed ratios, and low forage area. Large farms have higher turnovers and are more productive because they use better technology; at the same time, they are more specialized in production activity [15].

Economies of scale are one of the factors influencing the economic efficiency of milk production and economic sustainability [16]. Small farmers have limited bargaining power, so in order to become more competitive in the market a change of scale and the development of innovative capacity are needed [17].

Economic sustainability can also be achieved by limiting the number of dairy cows to those that can be fed mainly with forages from the farmer's own farm [18]. Another important factor influencing the economic performance is the labor force and its productivity [19].

The aim of this study was to evaluate the relations for determining the profitability of dairy farms of various sizes, with different levels of milk production, with different allocations of expenditure categories, and located in different areas.

2. Materials and Methods

Data from 54 farms from 20 counties located in all 8 development regions of Romania were used. Most of the farms (23) were located in the South-Muntenia Region of Romania.

The total sample of dairy cows from the 54 farms was 3966 heads, calculated as the average number of milking cows for the end of the years 2018, 2019, and 2020, without taking into account other age and production categories of cattle, which were not the subject of the study. A share of 51.41% were located in the South-Muntenia Region, included in the largest plain area in Romania. The rest of the livestock composition was as follows: South-West Oltenia Region—4.56% West Region—1.52%, North-West Region—4.83%, Central Region—7.72%, North-East Region—18.01 %, South-East Region—1.24%, and Bucharest–Ilfov Region—10.70%. The average farm size calculated for the period 2018–2020 was 73.44 heads, with a minimum of 5.0 cows and a maximum of 568.3 cows.

Total milk production from the 54 farms (calculated as an average of 2018, 2019, and 2020) was 264,465 hectoliters, distributed by development regions as follows: in the North-East Region 59,080.9 hL (22.34%), in the North-West Region 7261.6 hL (2.75%), in the West Region 3753.8 hL (1.42%), in the South-West Oltenia Region 6248.7 hL (2.36%), in the South-West Region Muntenia 143,477.8 hL (54.26%), in the Central Region 17,416.1 hL (6.59%), in the South-East Region 1925.7 hL (0.73%), and in the Bucharest–Ilfov Region 25,280.4 hL (9.56%) (Figure 1). The average milk production on the farm in the period 2018–2020 was 4554.94 L/cow, with a minimum of 2600 L/cow and a maximum of 9633.3 L/cow.

Data collection from farms encountered some difficulties, primarily due to the fact that it took place during the COVID-19 pandemic with restrictions on mobility and social distancing, so that the originally planned interviews could not be conducted directly on farms, but were conducted mostly by phone. Another challenge was related to the availability of farmers to provide information on different categories of expenditure or delivery prices of products, even though their identity was anonymized. The questions in the questionnaire referred to the landform of the area where the farm was located, the livestock number for the 3 years, milk production, maintenance system, farm equipment, feed rations, different categories of expenses, sale of production, etc.

Figure 1. Milk production from case studies, by counties. Source: authors' illustration, using map chart on geographical regions in Excel.

For each of the 54 farms, the annual estimates of expenditures and the annual budgets of revenues and expenditures of the farms were calculated. The average estimate and the average budget for the 3-year period were calculated.

The structure of the estimated variable expenses included the following elements: feedstock, biological material (heifer value), energy and fuel, medicines and medical supplies, other material expenses, supply quota, and insurance. In addition, the fixed costs included labor costs, general costs, interest on loans, and depreciation.

Based on the elements of the estimate and the data provided by the farms on the capitalization of the main production (milk) and secondary production (calf, manure, and animal slaughtering), the revenue and expenditure budget was prepared.

The technical–economic indicators calculated were: value of production, value of main production, total costs, costs for main production, variable costs, fixed costs, unit cost, profit or loss per unit of product, taxable income rate, threshold in value units, threshold in physical units, and exploitation risk rate, using the following relationships:

Value of production $VQ = VQm + VQs$, in which: *VQm*—value of main production, *VQs*—value of secondary production.

Total costs $TC = VC + FC$, in which: *VC*—variable costs, *FC*—fixed costs.

Costs for main production $MC = TC - VQs$

Variable costs $VC = FoC + EnC + MedC + OC + SupC$, in which: *FoC*—forages costs, *EnC*—energy and fuel costs, *MedC*—medicines costs, *OC*—other material costs, *SupC*—supply costs.

Fixed costs $FC = LabC + GC$, in which: *LabC*—labor costs, *GC*—general costs.

$UC = MC/MP$, in which: *MP*—main production

Total profit $TPr = VQ - TC$

Profit or loss per unit of product $Pr/l = TPr/MP$, in which: *TPr*—total profit.

Net profitability rate $NPrR = (TPr/MC) \times 100$

Margin on variable costs $MgVC = VQ - VC$

Margin on variable costs $MgVC\% = MgVC/VQ \times 100$

Profitability threshold in value units $PrThv = (FC/MgVC\%) \times 100$

Profitability threshold in physical units $PrTrph = PrThv/UP$, in which: *UP*—unit price.

Exploitation risk rate $ERR = PrThv/VQm$

In order to analyze the associated influence of different technical and economic indicators of dairy farms on the results regarding profitability, the TableCurve program was used, which can determine the ideal equation, and, respectively, the representative regression, which describes the data entered. Thus, the relationships between two calculated indicators were illustrated by the resulting curves, and the relationship that includes three indicators was integrated into a spatial model.

3. Findings and Discussions

3.1. Distribution of Farms in Case Studies

In order to study the dairy farm size distribution, to compare them with the normal distribution (Gaussian curve) and to highlight the strength of dairy farm size, a graphical representation of the sample was performed, as well as statistical analysis of data.

As can be seen from Figure 2, the physical size of dairy farms in the sample analyzed in the case studies showed a different distribution than normal, with most farms measuring herds between 5 and 100 heads.

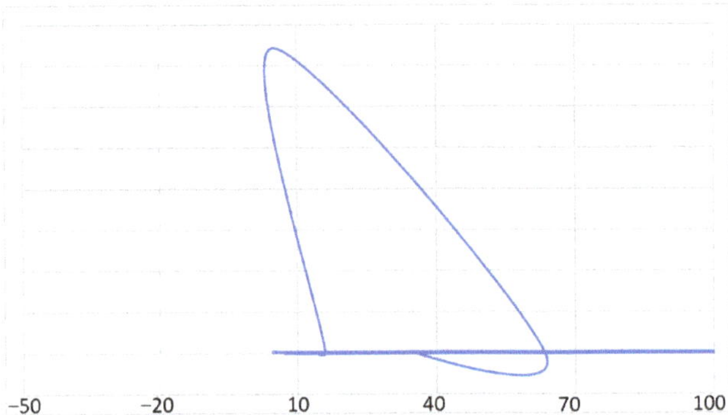

Figure 2. Distribution of the physical size of dairy farms. Source: authors' own elaboration.

Figure 3 shows the clustering of farm size in the sample.

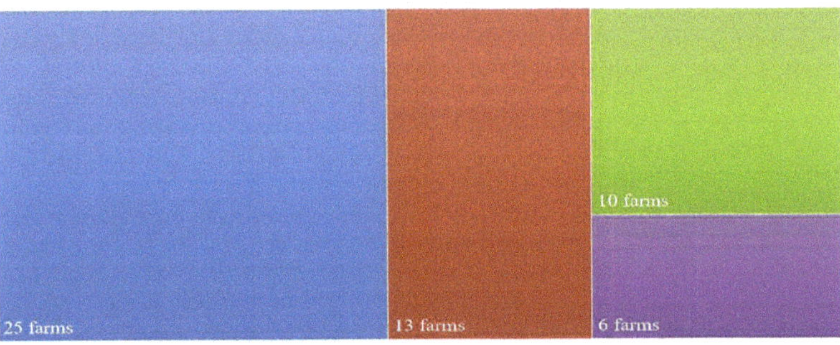

Figure 3. Clustering the size of farms in the sample.

By farm size segments, the average milk production was as follows: in the category below 20 heads, in which 25 farms were included, the average milk production was 3910.67 L/cow, at 21–50 heads (13 farms) was 4471.79 L/cow, at 51–100 heads (10 farms) it was 4328.33 L/cow, and in the category over 100 heads (6 farms), it was 7797.22 L/cow. The smallest size segments, below 100 heads, with yields below 4000 L, generally had the lowest values of profitability indicators, high operating risk rates, and negative safety indices. They also had among the highest unit costs and the lowest labor productivity.

Data related to the size of the farm were analyzed and interpreted with descriptive statistical indicators. Thus, were determined in Table 1 these indicators related to the data string, and, respectively, physical size of the farms.

Table 1. Determination of descriptive statistical indicators for farm size.

	Farm Size
Mean	73.44444
Standard Error	17.71771
Median	24.5
Mode	18.66667
Standard Deviation	130.1981
Sample Variance	16,951.53
Kurtosis	6.98522
Skewness	2.81767
Range	563.3333
Minimum	5
Maximum	568.3333
Sum	3966
Count	54

Source: authors' own elaboration.

Regarding the average of the farm segment taken into analysis, it was of 73.4 heads per farm, with a standard error of 17.7. However, the median was 24.5 heads. Regarding the homogeneity of the data, they were not homogeneous, with a standard deviation of ± 130 heads, which caused very large variation. However, the study aimed to cover as many classes of farm size as possible.

The indicators that study the data distribution, the vaulting (Kurtosis) and the asymmetry (Skewness), were aligned, and at the same time confirmed the graphical distribution in Figure 4. The vaulting coefficient showed a positive value, well above the zero value of 6.98, which describes a leptokurtic distribution. Similarly, the symmetry coefficient confirmed the graphical representation, reaching a value of 2.81, which causes asymmetry to the left.

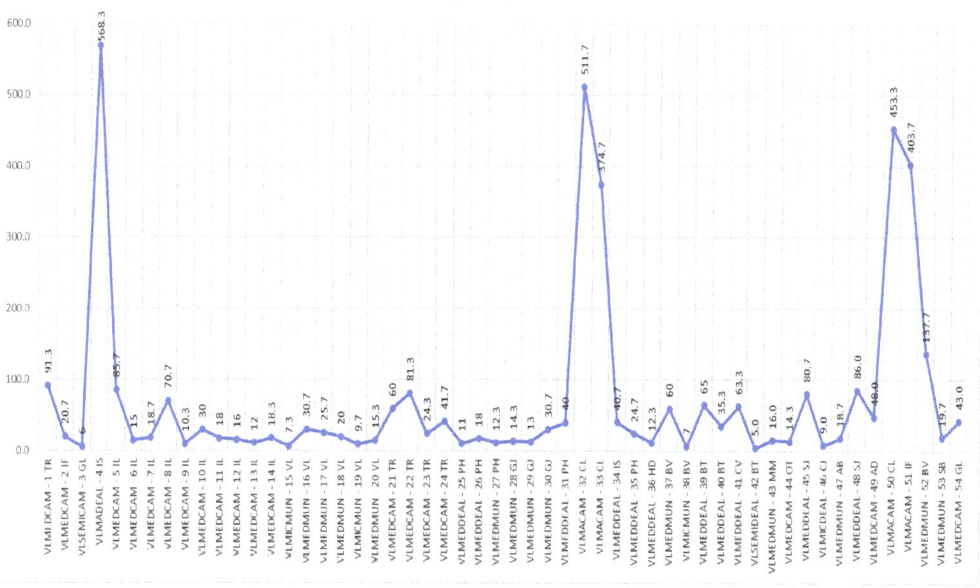

Figure 4. Farm size in the case studies. Source: authors' own elaboration.

3.2. Centralized Data Analysis

Following the analysis of the 54 dairy farms, it was possible to centralize the technical and economic indicators with the help of the simple arithmetic mean, as well as the standard deviation (Table 2).

Table 2. Determining the averages of technical–economic indicators.

Specification	Unit	Avrg	Standard Deviation
Farm size	**cows**	73.44	130.2
Average production	**L/cow**	4554.94	1809.3
Value of main production	USD/L	0.38	0.12
Costs for the main production	USD/L	0.37	0.10
Variable costs	USD/L	0.32	0.05
Material costs	USD/L	0.30	0.05
Fixed costs	USD/L	0.10	0.05
Labor costs	USD/L	0.08	0.05
Labor productivity in physical expression	Man-hours/L	0.06	0.0
Labor productivity in value expression	USD/man-hours	10.52	10.96
Labor costs at 1000 RON total production	USD	48.08	21.35
Material costs at 1000 RON total production	USD	178.12	23.37
Expenses per 1000 RON main production	USD	243.28	24.20
Profit or loss per unit of product	USD	0.00	0.05
Taxable income rate	%	0.2	10.0
Net income rate	%	−0.1	9.4
Profitability threshold in value units	USD	1937.84	761.98
Profitability threshold in physical units	L	5506	3048.7
Exploitation risk rate	%	146.6	132.7
Security index		−0.5	1.3

Source: authors' own elaboration. Note: AVRG—average, L—liter.

The size of the farms in the analyzed segment varied between 5.0 heads per farm and 568.3 heads per farm, registering an average of 73.44 heads per farm, with a variation of 130.2 heads (Figure 4).

In terms of per capita yield, there was an average milk production of a minimum 2600 L of cow's milk per head and 9633.3 L of cow's milk per head, with an average of all the farms in the study of 4554.94 L/cow, and a standard deviation from this average of 1809.3 L.

Differences in the prices obtained from the sale of milk relate both to milk sold to the dairy processing industry [20] and to milk marketed directly on the market, as drinking milk, as cheese, or through milk dispensers. The value of milk production, determined per unit of product, ranged between 0.27 USD (1.10 RON)/L and 0.88 USD (3.67 RON)/L, with an average value of 0.37 USD (1.56 RON)/L, and a standard deviation of 0.12 USD (0.5 RON)/L.

Analyzing the expenses, there was a variation between 0.27 USD (1.13 RON)/L and 0.71 USD (2.94 RON)/L. On average, the level of expenses was 0.37 USD (1.55 RON)/L, with a deviation of 0.09 USD (0.4 RON). Thus, it was possible to identify an increase in the lower limit of expenditures compared to the value of production, exceeding the latter. Farms with the lowest production values run the risk of not being economically sustainable. Comparing the standard deviation for the value of production (indicator related to price) and the standard deviation for the expenses related to a liter of milk (indicator related to cost), it was found that there were no significant differences, with the deviation for the value of production being ±0.12 USD/L, and in the case of expenses being ±0.09 USD/L. Thus, even if the price varied quite a bit (±32%), unfortunately the costs also varied similarly by ±25.8%, which indicated that the production technologies were influenced fairly high by both external factors and by the cost elements, and the cost was also influenced by the level of production, being in an almost linear relationship with it [21]. Nutrition strategies and good breeding practices can also contribute to increasing the efficiency of animal production [22].

The structure by elements of expenditures, depending on the farm size—small-, large-, and medium-sized farms—is illustrated in Figure 5.

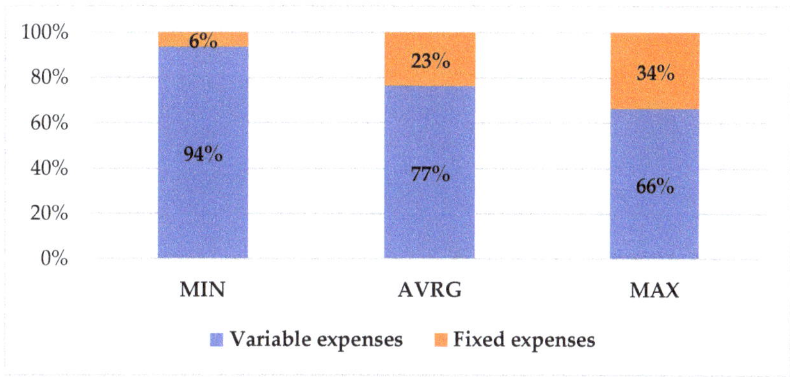

Figure 5. Expenditure structure according to the minimum, maximum, and average size of the farm. Source: authors' own elaboration.

When analyzing the structure of costs, it could be observed that, for the smallest farm in the sample (five dairy cows), the share of variable costs represented 94% of total costs. On the other hand, for the largest farm in the sample (568 dairy cows), the share of variable expenditures was 66% of total expenditures. Management costs for large farms were much higher than for small farms. On average, which was 73 dairy cows, the share of variable expenditures per farm and per unit of product was around 77% of total expenditures and the share of fixed expenditure was 23%.

As viability and economic sustainability indicate the ability of the farm to operate longer and to grow, labor productivity indicators in relation to output are also important [23]. Directing funds to investments that improve labor productivity encourages sustainable practices on dairy farms [24]. Labor productivity in dairy farms is determined by a number of factors, including, for example, the volume of manual labor and the degree of mechanization. Large-scale dairy farms have higher labor productivity than other farms [25]. The indicator can be expressed in physical units of product, or in value units. The productivity of work in physical expression ranged between 0.01 man-hours per liter and 0.17 man-hours per liter, with an average working time to obtain a liter of milk of 0.06 man-hours. The productivity of labor in value terms ranged between 8.27 RON/man-hours and 208.37 RON/man-hours, but, on average, in one hour of work a worker produced milk in value of 43.56 RON. The size of the farms in the analyzed segment varied between 5.0 heads per holding and 568.3 heads per holding, registering an average of 73.44 heads per farm, with a variation from the average of 130.2 heads (Figure 4).

In order to ensure economic sustainability in conditions of market competition, a proper decision making plays a key role [26]. Economic sustainability can also be determined on the basis of the costs related to the value of the main production. In this situation, there are three indicators, shown in Figure 6.

Expenses per 1000 RON main production characterizes more strongly the degree of economic sustainability. This indicator shows the share of expenditure in the value of production, the rest representing the share of profit. Labor costs ranged from 3.07% to 48.66%, with an average of 19.9%. The high shares of this indicator were affected by the extreme data from certain case studies in the sample, in which the average production was only 2600–2700 L/cow, with farm sizes below 12 heads.

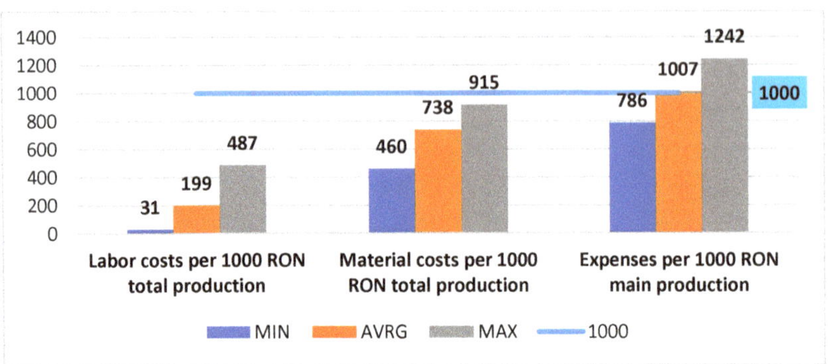

Figure 6. Determining the economic sustainability of farms based on costs and the value of production. Source: authors' own elaboration.

Cheng, S., Zheng, Z., and Henneberry, S. (2019) [27] showed that, compared to large farms, smaller farms consume more labor force, and for higher yields, more labor efforts, inputs, and precision technology are necessary. Productivity changes are more important for smaller farms and require further modernization of technology, with a certain balance between own and borrowed capital [28].

Analyzing the expenses with materials, they oscillated, with weights between 46% and 91.5%. A key indicator associated with maximizing farm-level profitability is the proportion of forages purchased [29], as the forages accounts for the largest share of material costs. An increase in feed prices increases the cost of milk [30], and thus profitability will be negatively affected.

Finally, analyzing the total expenses related to 1000 RON main production, it was observed that the most efficient farm registered a level of expenses of 786 RON to obtain a value of milk production of 1000 RON, which can be concluded as having an added value of 21.4%. On the other hand, the most economically inefficient farm was the one that had to make a financial effort of 1242 RON to produce milk worth 1000 RON, which obviously led to a loss for that farm. In general, on average, it was observed that the level of expenses incurred to obtain a milk production of 1000 RON was higher than this threshold by 7 RON, which suggested that, on average, the farms studied do not make a profit per unit of product, being at a slight loss, mainly due to low levels of milk production.

3.3. Correlation of Farm Size with Production, by Landforms

In order to determine the influence that dairy farm size may have on total production, a regression equation can be applied between these two variables, with the farm size being the independent variable and total production as the dependent variable. Thus, following the graphical representation of data and the point cloud, the regression line and the corresponding equation can be identified. This correlation was made for each geographical area included in the case study farms (plain, hill, mountain).

Regarding the influence that the farm size can have on the milk production for the 24 farms located in the plain area, it was observed that the Pearson correlation coefficient between variables was very high, being 0.97, and the coefficient of determination was 0.949 as can be seen from Figure 7. This suggested that the dependent variable (milk production) is explained in a proportion of 94.8% by the independent variable (farm size in the plain area).

Analyzing the regression equation, it can be observed that the value of the independent variable coefficient is 8228.5 units. Thus, it was estimated that at an increase of one unit in the independent variable, the dependent variable will increase by 8228.5 units. In other words, for farms located in the plain area, an increase in the size of the farm by one cow results in an increase in total production by 8228.5 L of milk.

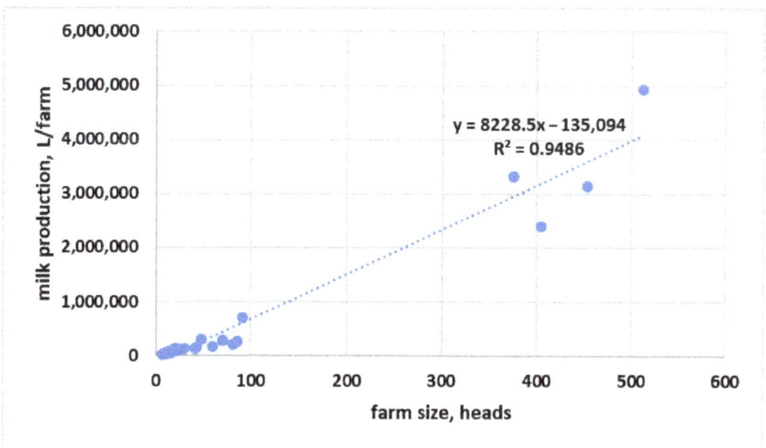

Figure 7. The correlation between farm size and total production for the 24 farms located in the plain area Source: authors' own elaboration.

Regarding the influence that the size of the farm can have on the milk production for the 14 farms located in the hill area, it was observed that Pearson correlation coefficient between variables was very high, being 0.99, and the coefficient of determination was 0.986, as can be seen in Figure 8. This suggested that the dependent variable is explained in a proportion of 98.5% by the independent variable.

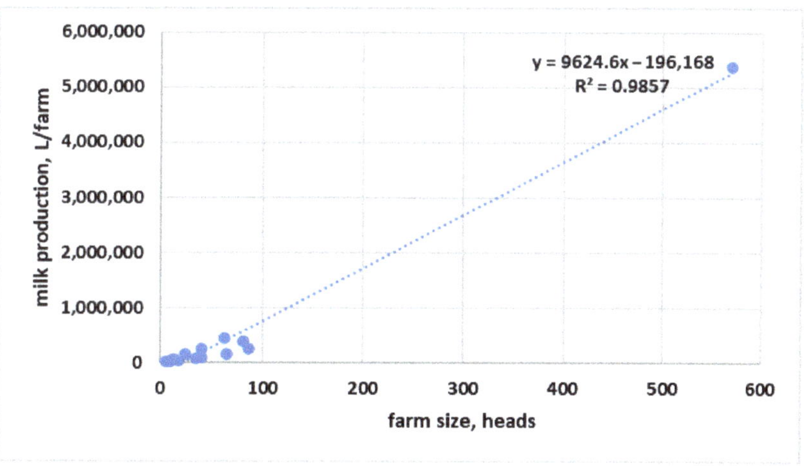

Figure 8. The correlation between farm size and total production for the 14 farms located in the hill area. Source: authors' own elaboration.

In the regression equation, the value of the independent variable coefficient was 9624.6 units. It can be estimated that at an increase of one unit in the independent variable, the dependent variable will increase by 9624.6 units. In other words, for the farms in the hilly areas, an increase in the farm size by one cow results in an increase in total production by 9624.6 L of milk.

Regarding the influence that the farm size can have on the milk production for the 16 farms located in the mountain area, it was observed that, between the variables, the Pearson correlation coefficient was very high, 0.99, and the coefficient of determination was 0.987, as can be seen from Figure 9, suggesting that the dependent variable is 98.7% explained by the independent variable.

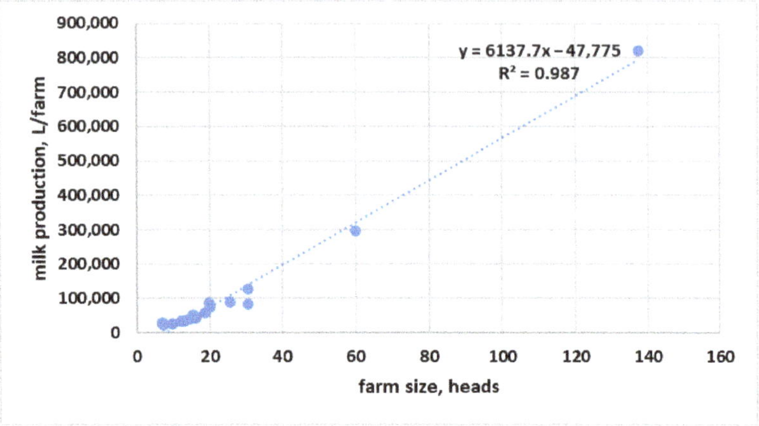

Figure 9. The correlation between farm size and total production for 16 farms located in the mountain area. Source: authors' own elaboration.

The regression equation in this situation presented the value of the coefficient of the independent variable of 6137.7 units. This means that at an increase of one unit in the independent variable, the dependent variable will increase by 6137.7 units for this model.

The influence of the main production value on the farm profit level was illustrated using the applications in the TableCurve program, in which a nonlinear regression was used (Figure 10), described by the ideal equation:

$$y = \frac{a + bx + cx^2 ln_x + dx}{ln_x + ex^{0.5}} \tag{1}$$

with 95% confidence limits. The value of the coefficient of determination (r^2) was very high, given the objective of the program, namely, to identify the function that passes through most points, so this coefficient was 0.94, and r^2 adjusted of 0.93 assumes, in this case, that the dependent variable (profit) is explained by the independent variable (the value of the main production) in a proportion of at least 93%. Such a high coefficient of determination determines a very strong correlation coefficient (r) of 0.969, indicating a strong relation between variables (Figure 10). The value of the statistical parameter Fstat is approximately 194.9, being much higher than the value of the parameter Fcritical, in this case $F_{0.05;\,1;\,53}$ being 4.023. Therefore, the null hypothesis of equal means between variables is rejected, the quadratic mean inter-group being higher than the quadratic mean intra-group, and it can be concluded that there is a statistically significant difference between the means of the sample.

The resulting curve illustrated that as the value of the main production increases, so does the size of the farm's profit. In any agricultural activity, farmers pursue the efficient use of factors of production in order to maximize profits [31,32]. Furthermore, the welfare conditions of cows, associated with a higher level of milk production, are reflected in higher economic margins for the farm [33]. However, technical conditions are not the most important determinant of the level of profitability and price fluctuations also influence farm profits [34]. Prices are the main contributor to income risk, along with the level of milk production [35].

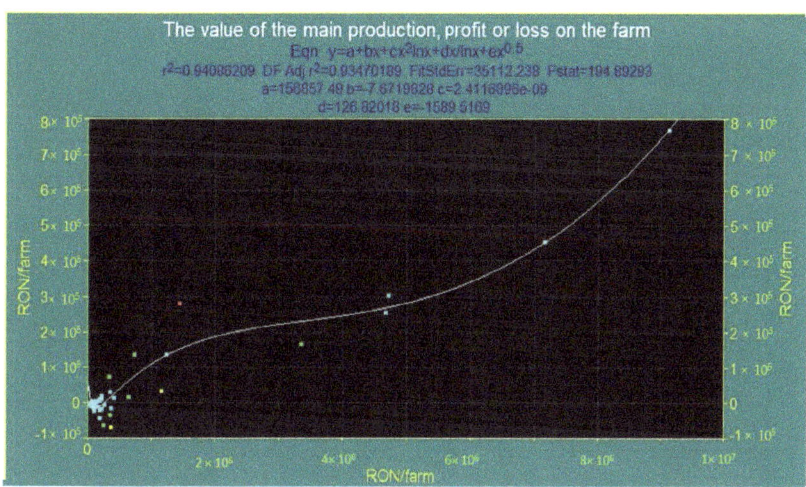

Figure 10. The equation of the value of main production influence on the level of farm profit. Source: authors' own elaboration.

As the net income of the farm is also influenced by its size, the comparison of farms of different sizes can be problematic if this aspect is not taken into account [36]. The influence of farm size on the level of financial results, namely, profit or loss, was described by the ideal equation:

$$y = a + bx^{0.5} + cx + dx^{1.5} + ex^2 + fx^{2.5} + gx^{2.5} + hx^{3.5} \tag{2}$$

with 95% confidence limits. The value of the coefficient of determination (r^2) was very high, given the objective of the program to identify the function that passes through most points, so this coefficient was 0.867 and r^2 adjusted of 0.84, which means, in this case, that the dependent variable (profit) is explained by the independent variable (farm size) in a proportion of at least 84%. Such a high coefficient of determination results in a very strong correlation coefficient (r) of 0.931, which indicates a strong link between the variables.

Yan, J., Chen, C., and Hu, B. (2019) [37] found that the relation between farm size and profit efficiency in agricultural production is illustrated by a U-shaped curve. In the present study, the curve of this equation indicates that the profit of the farm is in a directly proportional relationship to the size of the farm (Figure 11). In fact, large dairy farms have higher economic sustainability. Therefore, they are more likely to operate for medium and long periods of time [38]. However, in the case studies, there were also smaller cow farms which obtained comparable profits to larger farms [39], which indicates that the farm size is not the sole factor in determining the level of profitability.

Ferrazza, R.A., Lopes, M.A., Prado, D.G.O., Lima, R.R., and Bruhn, F.R.P. (2020) [40] concluded that the intensification of activities is the main determinant of economic results, milk production per cow being the most positive indicator correlated with profitability. In addition, the above-mentioned authors pointed out that the profitability of milk production depends in particular on the price of milk, so that it is particularly important to allocate inputs efficiently, thus contributing to the economic sustainability of dairy farms.

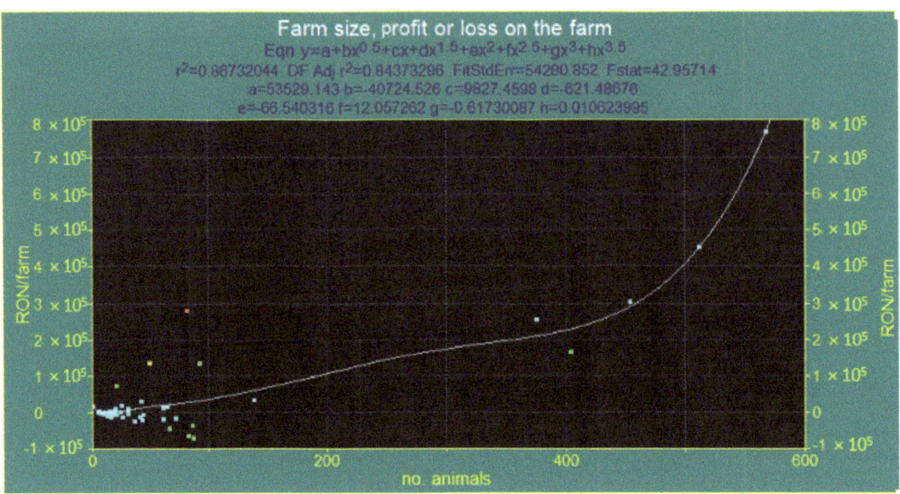

Figure 11. The equation of the influence of farm size on the level of profit. Source: authors' own elaboration.

Illustrating the correlation between the total milk production of the farm and its profit, the curve of the regression equation alternates two convex segments with two concave segments, but on an ascending path, according to the relation:

$$y = a + bx^{0.5} + cx + dx^{1.5} + ex^2 + fx^{2.5} + gx + hx^{3.5} + ix^4 \quad (3)$$

with a probability of 95% (Figure 12). The value of the coefficient of determination (r^2) was very high, given the objective of the program, namely, to identify the function that passes through most points, so that this coefficient was 0.907 with an r^2 adjusted of 0.88, which means, in this case, that the dependent variable is explained by the independent variable in a proportion of at least 88%. Such a high coefficient of determination results in a very close correlation coefficient (r) of 0.952, which indicates a strong link between the variables.

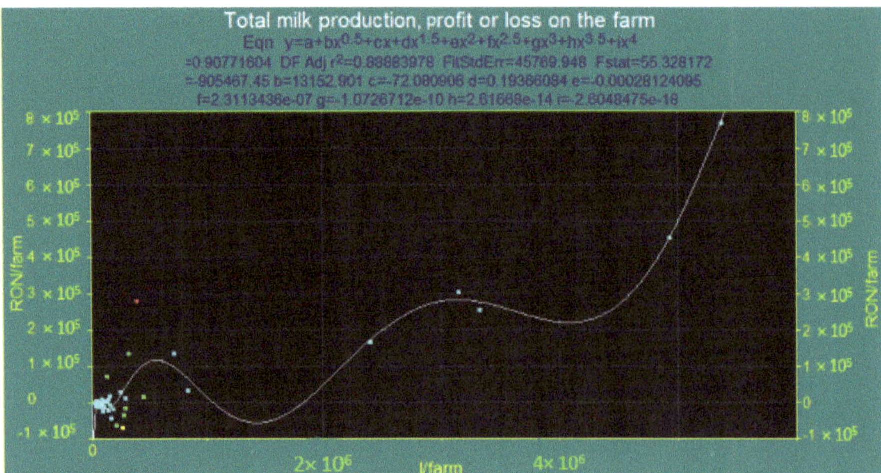

Figure 12. The equation of the influence of total milk production on the level of profit. Source: authors' own elaboration.

Hadrich, J.C. and Olson, F. (2011) [41] demonstrated that a single indicator may not capture the aspects of farm size and performance and that several indicators should be used. Therefore, studying the concomitant influence of two variables, namely, farm size and total milk production, on the farm profit level, a three-dimensional illustration of the regression equation is obtained as:

$$z = a + bx + cln_x + dx^2 + e(ln_y)^2 + fxln_y + gx^3 + h(ln_y)^3 + ix(ln_y)^2 + jx^2ln_y \quad (4)$$

with r^2 calculated of 0.92, r^2 adjusted of 0.90, and 95% probability, indicating that farm profit increases in direct proportion to farm size and total milk production (Figure 13). The value of the statistical parameter Fstat is approximately 57.86, being much higher than the value of the parameter Fcritical, in this case $F_{0.05;\,2;\,52}$ being 3.18. Therefore, the null hypothesis of equal means is rejected and it can be concluded that there is a statistically significant difference between the means of the sample.

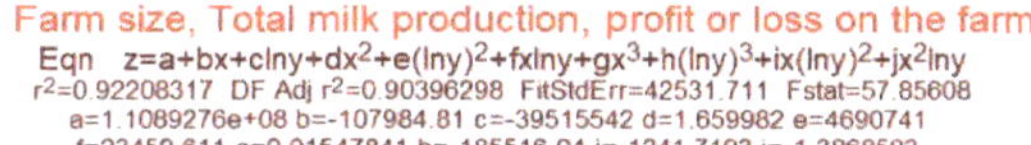

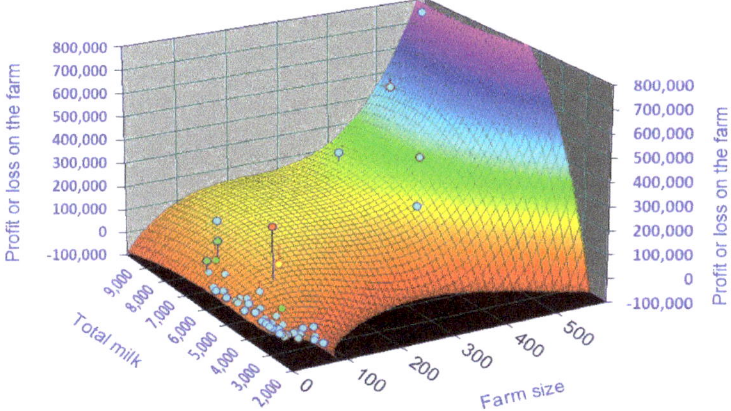

Figure 13. The equation of the influence of farm size and total milk production on the level of profit. Source: authors' own elaboration.

The judicious use of production management factors, such as farm size and milk production, has a positive impact on farm profitability [42].

The application of the TableCurve program to highlight the correlation between farm size, unit cost, and profit level produces a three-dimensional illustration of the regression equation:

$$z = a + bc + cy + dx^2 + ey^2 + fxy + gx^3 + hy^3 + ixy^2 + jx^2y \quad (5)$$

with r^2 calculated of 0.94, r^2 adjusted of 0.93, and 95% probability, indicating that farm profit increases in direct proportion to farm size and is inversely related to unit cost (suggested by the concavity of the graphical representation) (Figure 14). The value of the statistical parameter Fstat is about 80, being much higher than the value of the parameter Fcritical, in this case $F_{0.05;\,2;\,52}$ being 3.18. Therefore, the null hypothesis of equal means between variables is rejected, the quadratic mean inter-group being higher than the quadratic mean intra-group. Thus, we conclude that there is a statistically significant difference between the means of the sample.

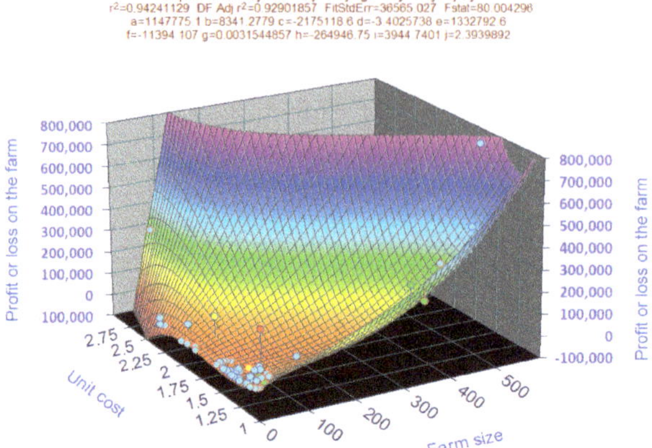

Figure 14. The equation of the influence of farm size and unit cost on the level of profit. Source: authors' own elaboration.

Lukas Kiefer, Friederike Menzel, and Enno Bahrs (2014) [43] have shown that efficiently managed milk production creates the potential to optimize farm income. The calculation of efficiency in milk production should account for unit costs [44] and their minimizing. Dairy farms need to find ways to ensure that their production cost is lower than the market price of milk, and that the strategy to increase the farm size allows reduction in production costs [45]. It is necessary for farmers to periodically analyze milk production, production costs, and profit in order to identify those favorable factors that may contribute to increasing the profitability of their activities [46]. The exact knowledge of the cost of production by the farmer is a management tool [47]. In terms of unit cost of production, large farms have much lower costs, on average, than smaller farms [48].

The difference in production technology and inputs could be an explanation for the difference in productivity between large and small farms, given the same prices relative to inputs [49]. Studies by Yu Sheng, Alistair Davidson, Keith Fuglie, and Dandan Zhang (2016) [50] show that farmers who respond to changing technologies and prices by replacing different inputs thus gain "income effects". In order to ensure economic sustainability, managerial effort and technological investment is needed to increase the daily average of milk production without increasing the average variable cost [51]

4. Conclusions

Analyzing from the perspective of profitability, there are rates of return between about −20% and +10%, and in the sample analyzed, thus, it can be concluded that several dairy farms were not profitable in the analyzed period.

The increase in the physical size of the farm, no matter the geographical area, positively influenced the milk production. However, in the mountain area the increase in production was slower than for plain and hill areas.

A graphical representation of the profitability of dairy farms was elaborated. The farms with a low value of main production had a small increase in profit, while when the value of main production increased, the profit growth became slower. Further, as the value of production increases, the curve indicates an exponential evolution of the profit.

In determining the farm's profit equation based on the farm size, it was found that in the case of small farms, the increase in livestock leads to a relatively small increase in farm profit, and subsequently, once the size of 400–450 cows is exceeded, the increase in numbers will lead to an exponential increase in farm profits.

The statistical analysis that describes the farm profit equation according to the total milk production led to an almost sinusoidal graph, actually formed of several connected Gaussian curves. Therefore, the profit of the farms increased with the increase in production, up to the moment when the increase in production involved a high level of costs to support it, so that the profit turned into a loss when the level of expenses exceeded that of income. Subsequently, the situation replicated, at a higher level of total production and profit, and so on.

The graphical representation of the multiple regression of farm profit indicated that the highest profit values were recorded when the farm size and milk production were as high as possible. This situation is usual for large and very large farms, but it must be pointed out that most farms in this study owned between 5 and 100 cows. Furthermore, most of the small farms had a fairly high unit cost, being in a situation of economic inefficiency, but the highest profit was recorded in terms of a low unit cost and a high physical size of the holding (ideal case, encountered in the case of large and very large holdings). At the same time, there are quite high profits in the case of medium-sized farms with the lowest possible unit costs.

Author Contributions: Conceptualization, R.C. and I.L.P.; methodology, R.C., I.L.P., A.E.C. and E.C.; software, I.L.P. and R.C.; validation, E.C., A.E.C., S.R. and V.D.; formal analysis, I.L.P., R.C., S.R., A.M. and P.A.T.-R.; investigation, S.R., A.M., V.D. and P.A.T.-R.; resources, R.C. and A.M.; data curation, E.C. and A.E.C.; writing—original draft preparation, R.C., I.L.P. and V.D.; writing—review and editing, R.C., I.L.P. and S.R.; visualization, R.C., I.L.P. and S.R.; project administration, R.C. All authors have read and agreed to the published version of the manuscript.

Funding: This research was funded by ADER 24.1.2. project: "Research on the economic efficiency of raising sheep, goats, dairy cows, cattle and buffalos".

Institutional Review Board Statement: This study did not require ethical approval.

Informed Consent Statement: Not applicable.

Data Availability Statement: The data presented in this study are available on request from the corresponding author. The data are not publicly available due to privacy.

Conflicts of Interest: The authors declare no conflict of interest.

References

1. Spicka, J.; Smutka, L. The Technical Efficiency of Specialised Milk Farms: A Regional View. *Sci. World J.* **2014**, *2014*, 985149. [CrossRef] [PubMed]
2. Kingwell, R. Managing complexity in modern farming. *Aust. J. Agric. Resour. Econ.* **2011**, *55*, 12–34. [CrossRef]
3. Bragg, L.; Dalton, T. Factors Affecting the Decision to Exit Dairy Farming: A Two-Stage Regression Analysis. *J. Dairy Sci.* **2004**, *87*, 3092–3098. [CrossRef]
4. Xie, H.; Huang, Y.; Chen, Q.; Zhang, Y.; Wu, Q. Prospects for Agricultural Sustainable Intensification: A Review of Research. *Land* **2019**, *8*, 157. [CrossRef]
5. Tey, Y.S.; Brindal, M. Factors Influencing Farm Profitability. In *Sustainable Agriculture Reviews*; Lichtfouse, E., Ed.; Springer: Cham, Switzerland, 2015; Volume 15. [CrossRef]
6. Bewley, J. Precision Dairy Farming: Advanced Analysis Solutions for Future Profitability. The First North American Conference on Precision Dairy Management. 2010. Available online: https://www.researchgate.net/profile/Jeffrey-Bewley/publication/267711814_Precision_Dairy_Farming_Advanced_Analysis_Solutions_for_Future_Profitability/links/582f650f08ae138f1c0356cd/Precision-Dairy-Farming-Advanced-Analysis-Solutions-for-Future-Profitability.pdf (accessed on 18 April 2022).
7. Hansson, H.; Ferguson, R. Factors influencing the strategic decision to further develop dairy production—A study of farmers in central Sweden. *Livest. Sci.* **2011**, *135*, 110–123. [CrossRef]
8. Ramsbottom, G.; Horan, B.; Berry, D.; Roche, J. Factors associated with the financial performance of spring-calving, pasture-based dairy farms. *J. Dairy Sci.* **2015**, *98*, 3526–3540. [CrossRef]
9. Bozoglu, M.; Saglam, O.; Topuz, B.K. Economic sustainability of family dairy farming within the scope of technical efficiency: A case study of Bafra District, Turkey. *Custos E Agronegocio* **2017**, *13*, 295–316.
10. Bijl, R.; Kooistra, S.R.; Hogeveen, H. The Profitability of Automatic Milking on Dutch Dairy Farms. *J. Dairy Sci.* **2007**, *90*, 239–248. [CrossRef]
11. Stott, A.; DeLorenzo, M. Factors Influencing Profitability of Jersey and Holstein Lactations1. *J. Dairy Sci.* **1988**, *71*, 2753–2766. [CrossRef]
12. Battaglini, L.; Bovolenta, S.; Gusmeroli, F.; Salvador, S.; Sturaro, E. Environmental Sustainability of Alpine Livestock Farms. *Ital. J. Anim. Sci.* **2014**, *13*, 3155. [CrossRef]

13. Macdonald, K.; Beca, D.; Penno, J.; Lancaster, J.; Roche, J. Short communication: Effect of stocking rate on the economics of pasture-based dairy farms. *J. Dairy Sci.* **2011**, *94*, 2581–2586. [CrossRef] [PubMed]
14. Gál, T. Efficiency analysis of dairy farms in the Northern Great Plain region using deterministic and stochastic DEA models. *Appl. Stud. Agribus. Commer.* **2012**, *6*, 113–122. [CrossRef]
15. Kryszak, Ł.; Guth, M.; Czyżewski, B. Determinants of farm profitability in the EU regions. Does farm size matter? *Agric. Econ.–Czech* **2021**, *67*, 90–100. [CrossRef]
16. Lovarelli, D.; Bava, L.; Zucali, M.; D'Imporzano, G.; Adani, F.; Tamburini, A.; Sandrucci, A. Improvements to dairy farms for environmental sustainability in Grana Padano and Parmigiano Reggiano production systems. *Ital. J. Anim. Sci.* **2019**, *18*, 1035–1048. [CrossRef]
17. Dervillé, M.; Allaire, G. Change of competition regime and regional innovative capacities: Evidence from dairy restructuring in France. *Food Policy* **2014**, *49*, 347–360. [CrossRef]
18. Fadul-Pacheco, L.; Wattiaux, M.A.; Espinoza-Ortega, A.; Sánchez-Vera, E.; Arriaga-Jordán, C.M. Arriaga-Jordán, Evaluation of Sustainability of Smallholder Dairy Production Systems in the Highlands of Mexico During the Rainy Season. *Agroecol. Sustain. Food Syst.* **2013**, *37*, 882–901. [CrossRef]
19. Atzori, A.; Tedeschi, L.; Cannas, A. A multivariate and stochastic approach to identify key variables to rank dairy farms on profitability. *J. Dairy Sci.* **2013**, *96*, 3378–3387. [CrossRef]
20. Gandini, G.; Maltecca, C.; Pizzi, F.; Bagnato, A.; Rizzi, R.M. Comparing Local and Commercial Breeds on Functional Traits and Profitability: The Case of Reggiana Dairy Cattle. *J. Dairy Sci.* **2007**, *90*, 2004–2011. [CrossRef]
21. Erba, E.; Aplin, R. Factors Affecting Labor Productivity and Cost per Gallon in Fluid Milk Plants. *J. Dairy Sci.* **1996**, *79*, 1304–1312. [CrossRef]
22. Salter, A. Improving the sustainability of global meat and milk production. *Proc. Nutr. Soc.* **2017**, *76*, 22–27. [CrossRef]
23. Spicka, J.; Hlavsa, T.; Soukupova, K.; Stolbova, M. Approaches to estimation the farm-level economic viability and sustainability in agriculture: A literature review. *Agric. Econ.* **2019**, *65*, 289–297. [CrossRef]
24. Griffith, K.; Zepeda, L. Farm level trade-offs of intensifying tropical milk production. *Ecol. Econ.* **1994**, *9*, 121–133. [CrossRef]
25. van der Meulen, H.A.B.; Dolman, M.A.; Jager, J.H.; Venema, G.S. The impact of farm size on sustainability of dutch dairy farms. *Int. J. Agric. Manag.* **2014**, *3*, 119–123. [CrossRef]
26. Phochanikorn, P.; Tan, C. A New Extension to a Multi-Criteria Decision-Making Model for Sustainable Supplier Selection under an Intuitionistic Fuzzy Environment. *Sustainability* **2019**, *11*, 5413. [CrossRef]
27. Cheng, S.; Zheng, Z.; Henneberry, S. Farm size and use of inputs: Explanations for the inverse productivity relationship. *China Agric. Econ. Rev.* **2019**, *11*, 336–354. [CrossRef]
28. Baležentis, T.; Galnaitytė, A.; Kriščiukaitienė, I.; Namiotko, V.; Novickytė, L.; Streimikiene, D.; Melnikiene, R. Decomposing Dynamics in the Farm Profitability: An Application of Index Decomposition Analysis to Lithuanian FADN Sample. *Sustainability* **2019**, *11*, 2861. [CrossRef]
29. Hanrahan, L.; McHugh, N.; Hennessy, T.; Moran, B.; Kearney, R.; Wallace, M.; Shalloo, L. Factors associated with profitability in pasture-based systems of milk production. *J. Dairy Sci.* **2018**, *101*, 5474–5485. [CrossRef] [PubMed]
30. Tapki, N. The comparison of dairy farms in different scales regarding milk production cost and profitability in Turkey: A case study from Hatay province. *Custos E@ Gronegócio Line* **2019**, *15*, 48–62. Available online: https://www.custoseagronegocioonline.com.br/numero2v15/OK%203%20cost.pdf (accessed on 19 April 2022).
31. Diakité, Z.; Corson, M.; Brunschwig, G.; Baumont, R.; Mosnier, C. Profit stability of mixed dairy and beef production systems of the mountain area of southern Auvergne (France) in the face of price variations: Bioeconomic simulation. *Agric. Syst.* **2019**, *171*, 126–134. [CrossRef]
32. Lawson, L.; Bruun, J.; Coelli, T.; Agger, J.; Lund, M. Relationships of Efficiency to Reproductive Disorders in Danish Milk Production: A Stochastic Frontier Analysis. *J. Dairy Sci.* **2004**, *87*, 212–224. [CrossRef]
33. Villettaz, R.M.; Rushen, J.; De Passillé, A.M.; Vasseur, E.; Orsel, K.; Pellerin, D. Associations between on-farm animal welfare indicators and productivity and profitability on Canadian dairies: I. On freestall farms. *J. Dairy Sci.* **2019**, *102*, 4341–4351. [CrossRef] [PubMed]
34. Zakova Kroupova, Z. Profitability development of Czech dairy farms. *Agric. Econ.* **2016**, *62*, 269–279. [CrossRef]
35. El Benni, N.; Finger, R. Gross revenue risk in Swiss dairy farming. *J. Dairy Sci.* **2013**, *96*, 936–948. [CrossRef] [PubMed]
36. Gloy, B.; Tauer, L.; Knoblauch, W. Profitability of Grazing Versus Mechanical Forage Harvesting on New York Dairy Farms. *J. Dairy Sci.* **2002**, *85*, 2215–2222. [CrossRef]
37. Yan, J.; Chen, C.; Hu, B. Farm size and production efficiency in Chinese agriculture: Output and profit. *China Agric. Econ. Rev.* **2019**, *11*, 20–38. [CrossRef]
38. Bánkuti, F.I.; Prizon, R.C.; Damasceno, J.C.; De Brito, M.M.; Pozza, M.S.S.; Lima, P.G.L. Farmers' actions toward sustainability: A typology of dairy farms according to sustainability indicators. *Animal* **2020**, *14*, s417–s423. [CrossRef]
39. Hanson, G.D.; Cunningham, L.C.; Morehart, M.J.; Parsons, R.L. Profitability of Moderate Intensive Grazing of Dairy Cows in the Northeast. *J. Dairy Sci.* **1998**, *81*, 821–829. [CrossRef]
40. Ferrazza, R.D.A.; Lopes, M.A.; Prado, D.G.D.O.; De Lima, R.R.; Bruhn, F.R.P. Association between technical and economic performance indexes and dairy farm profitability. *Rev. Bras. Zootec.* **2020**, *49*, e20180116. [CrossRef]

41. Hadrich, J.C.; Olson, F. Joint measurement of farm size and farm performance: A confirmatory factor analysis. *Agric. Financ. Rev.* **2011**, *71*, 295–309. [CrossRef]
42. Gloy, B.; Hyde, J.; LaDue, E. Dairy Farm Management and Long-Term Farm Financial Performance. *Agric. Resour. Econ. Rev.* **2002**, *31*, 233–247. [CrossRef]
43. Kiefer, L.; Menzel, F.; Bahrs, E. The effect of feed demand on greenhouse gas emissions and farm profitability for organic and conventional dairy farms. *J. Dairy Sci.* **2014**, *97*, 7564–7574. [CrossRef] [PubMed]
44. Michaličková, M.; Krupová, Z.; Polák, P.; Hetényi, L.; Krupa, E. Development of competitiveness and its determinants in Slovak dairy farms. *Agric. Econ.–Czech* **2014**, *60*, 82–88. [CrossRef]
45. Hemme, T.; Uddin, M.M.; Ndambi, O.A. Benchmarking cost of milk production in 46 countries. *J. Rev. Glob. Econ.* **2014**, *3*, 254–270.
46. Popescu, A. Research on profit variation depending on marketed milk and production cost in dairy farming. *Sci. Pap. Ser. Manag. Econ. Eng. Agric. Rural Dev.* **2014**, *14*, 223–230.
47. Camilo Neto, M.; Campos, J.M.D.S.; Oliveira, A.S.D.; Gomes, S.T. Identification and quantification of benchmarks of milk production systems in Minas Gerais. *Rev. Bras. Zootec.* **2012**, *41*, 2279–2288. [CrossRef]
48. MacDonald, J.M.; McBride, W.D.; O'Donoghue, E.; Nehring, R.F.; Sandretto, C.; Mosheim, R. Profits, Costs, and the Changing Structure of Dairy Farming. *USDA-ERS Econ. Res. Rep.* **2007**, *47*, 3. Available online: https://ssrn.com/abstract=1084458 (accessed on 20 April 2022). [CrossRef]
49. Sheng, Y.; Zhao, S.; Nossal, K.; Zhang, D. Productivity and farm size in Australian agriculture: Reinvestigating the returns to scale. *Aust. J. Agric. Resour. Econ.* **2014**, *59*, 16–38. [CrossRef]
50. Sheng, Y.; Davidson, A.; Fuglie, K.; Zhang, D. Input Substitution, Productivity Performanceand Farm Size. *Aust. J. Agric. Resour. Econ.* **2016**, *60*, 327–347. [CrossRef]
51. Pelegrini, D.F.; Lopes, M.A.; Demeu, F.A.; Rocha, Á.G.F.; Bruhn, F.R.P.; Casas, P.S. Effect of socioeconomic factors on the yields of family-operated milk production systems. *Semin. Ciências Agrárias* **2019**, *40*, 1199–1214. [CrossRef]

MDPI

Article

What about Responsible Consumption? A Survey Focused on Food Waste and Consumer Habits

Jurgita Paužuolienė [1], Ligita Šimanskienė [1] and Mariantonietta Fiore [2,*]

[1] Faculty of Social Sciences and Humanities, Klaipeda University, 92294 Klaipeda, Lithuania; jurgita.pauzuoliene@ku.lt (J.P.); ligita.simanskiene@ku.lt (L.Š.)
[2] Department of Economics, University of Foggia, 71121 Foggia, Italy
* Correspondence: mariantonietta.fiore@unifg.it

Abstract: The article analyses the problems of food waste and responsible consumption that include taking into account environmental-social-health and economic impacts of products and services. The study raises the research question related to whether people consume food responsibly. Analysis of research literature sources, systematization, synthesis, generalization, quantitative research and data processing methods were used in the article. The questionnaire was arranged on the pollimill.com website, and the link was shared with selected possible respondents. The survey was carried out in Lithuania and in European countries. The survey sample is equal to 1080 respondents (566 respondents from Lithuania and 514 from Italy, Poland, Latvia, Germany and France). A simple random sample was used in this research. The survey highlighted that the majority of respondents in the survey state that food is not often wasted. In addition, findings show that the population of Lithuania emits slightly less food than the population of the European countries participating in the survey. These findings could be crucial for the future green directions from the side of policymakers.

Keywords: sustainable production; responsible consumption; food waste; Europe

Citation: Paužuolienė, J.; Šimanskienė, L.; Fiore, M. What about Responsible Consumption? A Survey Focused on Food Waste and Consumer Habits. *Sustainability* **2022**, *14*, 8509. https://doi.org/10.3390/su14148509

Academic Editor: Gioacchino Pappalardo

Received: 7 June 2022
Accepted: 8 July 2022
Published: 12 July 2022

1. Introduction

The EU and the EU countries have to reach by 2030 their Sustainable Development Goal 12.3 target to halve per capita food waste at the retail and consumer level, and reduce food losses (according to the EU actions against food waste). Around 88 million tons of food waste are delivered yearly in the EU. According to preliminary calculations, every EU citizen throws away about 173 kg of food every year that could still be consumed [1]. For example, in Lithuania, the amount of food waste in the mixed municipal waste is about 15%, with an average of 41 kg of food waste per person per year; on the other hand. 75 million tons of bio-waste from municipal waste is created every year across Europe. It is crucial that recycling of bio-waste has to take place in order to meet the overall recycling target of 65% of municipal waste by 2035 [2].

The problem of food waste is relevant throughout the food supply chain, from the production of agricultural products to storage, processing, transport, trade and consumption [1]. Food waste poses environmental, ethical, and economical questions, and shows the need to change our food system.

Food waste prevention is included in the EU's plan for a circular economy, which the European Commission defines as where "the value of products, materials and resources is maintained in the economy for as long as possible, and the generation of waste [is] minimised". This strategy aims to improve competitiveness, promote sustainable growth, and create new jobs [3].

The revised 2018 Waste Framework Directive adopted on 30 May aims to reduce and monitor food waste and report back regarding progress made. Therefore, Member States have to:

- implement food waste prevention programs;
- promote food donation and other redistribution, prioritizing human use over animal feed and the reprocessing into non-food products as part of measures taken to prevent waste generation;
- deliver incentives for the application of the waste hierarchy, such as facilitation of food donation.

The crucial starting point of responsible consumption is awareness (behavior and attitude) of the impacts of consumption. Therefore, this paper aims to investigate whether people consume food responsibly.

The paper is presented in the following structure: the next section presents scientific analysis on the food waste issue. Then, research methods are described, and later main results and findings are presented. Finally, discussion and conclusions, with some suggestions for future research, close the paper.

2. Literature Review

Food waste is a global problem and it happens when food is left unused due to poor commercial appearance, or leftovers from uneaten food that are not composted. A common cause of food wastage is improper food storage—it spoils. Another is consumer shopping habits, and in some countries oversupply. Waste of food means any food lost due to spoilage or waste [4,5]. Thus, the term "wastage" encompasses both food loss and food waste [6]. Food waste is defined as food lost in any food supply chain. The food is then discarded and not used for any other productive use, e.g., animal feed or seeds. The FUSIONS framework defines food waste as "food and inedible parts of food removed from the food supply chain" that is to be disposed of (e.g., crops ploughed back into the soil, left unharvested or incinerated, food disposed of in sewers or landfill sites, or fish discarded at sea) or used for nutrient recovery or energy generation (e.g., through composting, or anaerobic digestion and other bioenergy pathways) [7]. Food is wasted in many ways, for example [8]:

- Fresh produce that deviates from what is considered optimal (e.g., size, shape or color) and is removed during sorting actions.
- Foods that are discarded by retailers or consumers when they are close to or beyond the best before date.
- Unused or leftover food that is thrown out from households or restaurants.

Food losses and waste also impact on other natural resources, many of which are scarce. Three key related resources are freshwater, cropland, and fertilizers [9]. The problem of food wastage is multifaceted, ranging from the misuse of arable land, the financial loss to restaurants and hotels and the discarding of prepared meals by households, while counting the working time of employees in cooking [10]. Another problem with food waste is that if food waste is not composted, it emits a lot of methane in landfills—more powerful greenhouse gases than even CO_2. Huge amounts of food emissions contribute to global warming and climate change. With agriculture accounting for 70 percent of global water consumption, food waste is also a huge waste of fresh and groundwater resources [11].

Comparing the extent of food waste according to the level of development of countries, more unused food is discarded in economically stronger countries [9]. However, there is also a significant amount of food wastage in developing countries, especially in the supply of food to retail chains. Researchers do not have exact data on how much unused food is lost to smallholder farms. Also, in developing countries, especially in Africa, storage losses on farms can be significant, although the exact nature of such losses is much debated [9]. But, some research shows that consumers are becoming more socially conscious and are including ethical considerations in their purchase decisions [12], as well as becoming increasingly interested in various forms of responsible consumption [13]. Consumers have more product choices and, therefore, have more opportunities to reveal their social preferences when making purchase decisions.

Analyzing food waste by different food groups, the authors found that vegetables (24%) and fruit (22%), followed by cereals (12%), meat (11%) and oil crops (10%) accounted for the largest share of food waste. The fish and eggs food groups, which make up the smallest parts of the food supply chain, also generate the lowest quantities of food waste in absolute terms, despite the fact that much of these food groups (50% and 31%, respectively) go to waste [14].

Based on the Lithuanian State Food and Veterinary Service [1], unreasonable food waste is promoted by:

- Improper planning of purchases and portions of food to be prepared;
- Promotional shopping;
- Lack of knowledge on how to use products marked with the terms 'best before' and 'use by';
- Standard portions of meals in restaurants and canteens, too little for the consumer (half a portion) or not a whole portion (omitting any ingredient);
- Challenges for restaurants, canteens, and other catering establishments in planning for the number of customers (surplus production);
- Exceptional quality requirements and marketing rules for the shape, color, consistency, etc., of foodstuffs, in particular fruit, vegetables and pastries in retail trade;
- Naturally occurring food surplus during the season;
- Damage to the products or their packaging that does not affect safety during production, packaging or transport;
- Improper storage/transportation of products;
- Underestimation of production volumes, poor management of raw materials, and production surplus, etc.

Nowadays, the concept of sustainable consumption is becoming more and more a major interest of the population and the latter make more conscious food purchasing decisions [15]. Indeed, increasing awareness towards environmental issues and climate change led society to the formation of sustainable consumption habits [16]. However, Ganglmair-Wooliscroft and Wooliscroft [17] argue that also external factors such as government regulations, business initiatives, and geographic characteristics determine consumers' behavior, including food consumption as well as recycling.

On the other hand, Block et al. [18] prove that consumers are often mistaken in estimating the consumption that is the basis of raising food waste. In the line with this, numerous initiatives have been launched in order to benefit from the remainder of these products. For instance, food waste valorization to hydrogen on the one hand reduces the harmful impact on nature, decreasing the quantity of spoiled food in the environment and, on the other hand, the alternative energy source is generated by transformation of biogas that can replace fossil fuel or produce electricity [19,20]. Additionally, it is economically feasible [21].

According to [22], the right tools for reducing food loss and waste have the potential to increase the sustainability of food supply chains. For this reason, authors suggest government to finance the relevant infrastructure for recycling disposed products and consumers' education for shifting towards responsible consumption including earlier food donation. Similarly, Sundin et al. [23] prove the environmental feasibility of food donation calculating a double of the benefit comparing to anaerobic digestion. Kumar and Dholakia [24] see the huge role of the firms to change consumers' behavior. Authors argue that firms have a power to promote innovative thinking, address consumers' environmental identity as well as brand assurance, and edit consumers' choices.

It is noteworthy that the COVID-19 pandemic has influenced consumers' food purchase decisions, their management and consumption that, in turn, has reduced the household waste [25,26]. Nowadays, 81% of consumers make a list before shopping and check the expiration date of the product that is almost double of the number before pandemic [27]. The increased awareness about food waste and its impact on the environment lead the reduction in the quantity of spoiled and thrown products even if purchases have increased during the COVID-19 pandemic [28–31].

In this context, as claimed by a recent research [32], a holistic 4Es Ethical, Equity, Ecological and Economic approach can be useful for better handling food loss issues along the agri-food chain from upstream to consumers by changing the entrepreneur and consumer approaches. Finally, the spread of the pandemic has been leading society to re-think the manner in which we produce and consume food by facing new future green global challenges [32,33].

3. Research Methodology

The quantitative research method was used in the research. The questionnaire was prepared on the pollimill.com website, and the link was sent to respondents. Regarding the criteria, only those respondents who had an internet connection could participate in the study. The research was guided by ethical principles: the principle of goodwill is ensured by the statements of the questionnaire, which are presented in a respectful style, without creating preconditions for respondents to lose privacy; applying the principle of respect to the individual, the purpose of the study was explained to the respondents; volunteering is the free will of study participants to participate or not to participate in a study; research participants were guaranteed anonymity and data confidentiality. The collected empirical data were processed using the SPSS 20.00 (Klaipeda university, Klaipėda, Lithuania) (Statistical Package for the Social Sciences). The data processing descriptive statistics were used, such as percentiles, mean, mode, and standard deviation. The data were also processed by independent samples *t-test* where significant differences are when $p \leq 0.05$. To assess the reliability, or internal consistency, of a set of scale, Cronbach's alpha coefficient was used.

The research population. The questionnaire items are based on the analysis of scientific literature and EU strategy on sustainable consumption [28,31,34,35]. The survey was done in February and March 2022. The respondents were reached during the third pandemic period by means of internal research mailing lists of the University of Klaipeda and Foggia [34,35]. The goal was to get as many responses as possible from different European countries, but in this study we were only able to collect data from these countries. The study selected this kind of online research to survey consumers in a fast manner, thus assuring safety and security under pandemic conditions [34,35]. The items of the questionnaire were corroborated by a virtual focus of experts in the agri-food-sustainable field. The survey sample is composed of 1080 respondents. In this survey, 566 respondents from Lithuania and 514 from other European countries (Italy, Poland, Latvia, Germany and France) participated. A simple random sample was used in the research. This kind of sample is a subset of a statistical population in which each member of the subset has an equal probability of being chosen. Preferably, using random sampling, the sample size should be larger than a few hundred in order to allow a simple random sampling to be applied correctly. This method has been selected in order to get as much information as possible on the analyzed topic.

Principles of compiling the questionnaire. The questionnaire consisted of five-point ranking scale questions [33]. First, we asked respondents where they usually buy food. The second and third questions in the questionnaire were designed to find out the respondents' buying habits, how often they pay attention to certain aspects when shopping and how often they throw away certain types of uneaten food. Respondents rated the questions on a five-point ranking scale from 1 to 5, with 1—very often and 5—never. The four questions explain whether respondents compost the waste, and the last one explains responsible food consumption habits of respondents. For this question, respondents rated the statements on a five-point ranking scale from 1 to 5, with 1—strongly agree and 5—strongly disagree [33].

The last four questions were designed to find out the demographics of the respondents, gender, age, monthly income and country of residence.

Demographic characteristics. The demographical data are provided in Table 1. From the table we can see that 52.4% of the respondents were from Lithuania, and 47.6% of the respondents (from Latvia 4.6%, from Poland 12.7%, from Germany 2.3%, from Italy

18.5%, from France 9.4%) were from other EU countries (see Table 1). As the number of respondents from different countries is quite different, the data will usually be analyzed together for other European countries, comparing the data with the Lithuanian data.

Table 1. Demographical data.

Home Country	Percent	Gender	Percent
Lithuania	52.4	Male	33.1
Other EU countries	47.6	Female	66.5
Monthly income	**Percent**	**Age**	**Percent**
Up to EUR 400	12.1	18–25	13.1
EUR 401–600	7.6	26–35	21.3
EUR 601–800	7.8	36–45	28.4
EUR 801–1000	15.5	46–55	19.4
EUR 1001–1400	19.7	56–60	8.9
EUR 1401–1600	7.3	More than 61 years	8.4
More than EUR 1601	30.0		

4. Results

The research evaluates the responsible food consumption habits in the daily life of Lithuanian and other European countries' respondents. Respondents of the research had to evaluate items about shopping habits with a range scale, where 1 means that respondents very often do that, and 5—they never do that. For the assessment of the question scale internal consistency, Cronbach's alpha coefficient was used, for a properly composed question scale should be greater than 0.7. In our case, Cronbach's alpha coefficient value ranged from 0.733 to 0.946 (see Table 2). A high Cronbach's alpha coefficient means that the items in the questionnaire are highly correlated.

Table 2. Cronbach's alpha coefficient for different questions groups.

Scale	Cronbach's Alpha Coefficient	Number of Statements
Respondents' shopping habits	0.809	19
Uneaten food waste	0.946	9
Food consumption habits	0.733	6

Analyzing respondents' shopping habits, we submitted mean and p value. p value shows the significant differences between countries (significance level is $p \leq 0.05$) (see Table 3). The standard deviation for the analyzed items ranges from 0.290 to 1.010.

Table 3. Respondents' shopping habits (mean).

Statements	Mean	Mean Lithuania	Mean European Countries	Significant Differences between Countries p-Value
I try to buy foods that are packaged in recyclable containers	2.72	2.70	2.53	**$p = 0.001 < 0.05$**
I prefer organic foods	2.47	2.26	2.48	**$p = 0.000 < 0.05$**
I am planning purchases, I am making a list	2.21	2.34	2.08	$p = 0.847 > 0.05$
I buy products with a promotion, even if it was not in my plan	2.98	3.05	2.81	**$p = 0.039 < 0.05$**
Buy products that are about to expire (because they are usually cheaper)	3.60	3.74	3.44	$p = 0.445 > 0.05$

Table 3. *Cont.*

Statements	Mean	Mean Lithuania	Mean European Countries	Significant Differences between Countries p-Value
When choosing foods, I pay attention to the composition of the product	2.20	2.22	2.18	$p = 0.086 > 0.05$
If only there is an opportunity to choose bulk foods	2.90	2.74	3.07	$p = 0.404 > 0.05$
When buying, it does not matter to me whether the appearance of the product is non-standard (e.g., cucumber curved, or tomato with a different shade but not spoilage)	2.62	2.46	2.80	$\mathbf{p = 0.001 < 0.05}$
I choose foods of local origin	2.19	2.14	2.24	$\mathbf{p = 0.000 < 0.05}$
I choose imported food	3.11	2.95	3.29	$\mathbf{p = 0.001 < 0.05}$
I choose more expensive foods	3.05	2.90	3.23	$p = 0.135 > 0.05$
I choose the foods advertised	3.44	3.45	3.42	$p = 0.805 > 0.05$
I buy fresh (unprocessed and not frozen) food	1.99	2.02	1.87	$\mathbf{p = 0.017 < 0.05}$
I buy long-life foods	2.88	2.93	2.82	$p = 0.642 > 0.05$
I buy fast food products (processed, semi-finished)	3.45	3.46	3.45	$p = 0.542 > 0.05$
I buy frozen food	3.19	3.08	3.31	$p = 0.296 > 0.05$
I shop as much as I need for the day	3.33	3.34	3.32	$p = 0.157 > 0.05$
I buy for a longer period (4–5 days)	2.03	2.03	2.13	$p = 0.705 > 0.05$
When buying food, I pay attention to special signs (e.g., Fair trade, green dot, etc.)	2.82	2.94	2.69	$p = 0.058 > 0.05$

Notes: Range scale where 1 means very often; 2 often; 3 rarely; 4 sometimes and 5 never. Significant differences are when $p \leq 0.05$.

The results show that the mean between Lithuania and other European countries are very similar, but shopping habits are quite different. Respondents were asked to evaluate their buying habits, how often they pay attention to certain aspects when shopping. Respondents rated the statements on a five-point ranking scale from 1 to 5, with 1—very often and 5—never. Research results reveal that most of the respondents only rarely buy products with a promotion, even if it was not in their plan (2.98); buy frozen food (mean 3.19) or shop as much as they need for the day (mean 3.33). Most of the respondents sometimes choose the foods advertised (mean 3.44); buy fast food products (processed, semi-finished) (mean 3.45) or buy products that are about to expire (because they are usually cheaper) (mean 3.60). Most of the respondents often plan purchases by making a list (mean 2.21); choosing foods and pay attention to the composition of the product (mean 2.20); choose foods of local origin (mean 2.19); buy fresh (unprocessed and not frozen) food (mean 1.99) and buy for a longer period (4–5 days) (mean 2.03) (see table).

The independent T sample test discloses the mean difference between country groups, significant data is in bold. The obtained data show that respondents from EU countries try to buy foods that are packaged in recyclable containers ($p = 0.001 < 0.05$); respondents from Lithuania prefer organic foods ($p = 0.000 < 0.05$). Respondents from EU countries are more likely to buy products with a promotion, even if it was not in their plan ($p = 0.039 < 0.05$). For a bigger part of respondents from Lithuania, the appearance of the product does not matter ($p = 0.001 < 0.05$). The Lithuanian respondents are more likely to choose foods of local origin ($p = 0.000 < 0.05$) and buy imported food ($p = 0.001 < 0.05$). The EU country respondents are more likely to buy fresh (unprocessed and not frozen) food compared with Lithuanian respondents ($p = 0.017 < 0.05$). For other items, country does not have a significant impact as the p-value is higher than 0.05, which indicates that there is no statistical difference.

If we look at the overall averages, the data ranges from 1.99 to 3.6 (mean). This shows that respondents are more likely to agree with the options available for purchase. The majority of respondents in both groups choose to buy fresh products and the minority buy products that will soon expire. It is interesting to note that in both groups, respondents said that they rarely choose the advertised products (average 3.44), which is somewhat surprising. This shows that respondents in the survey have an opinion about what they need when they go shopping, and it is difficult to change their opinion at the store. It is needed to mention that respondents from both groups go to the stores with a shopping list, plan to do so, and probably do not throw away unused food. Our respondents also look at the composition of the product and look if the packaging is recyclable.

We searched for whether there is a statistical relationship between the income received by the respondents and the place of shopping. However, no statistical dependencies have been identified. The responses of some higher-income respondents do not differ statistically from those of lower-income respondents. Respondents usually shop in supermarkets, it does not depend on the amount of income they receive. There is no statistical link between low-earning respondents growing their own vegetables or fruits.

We were interested in how often respondents buy food. The results reveal that most of the respondents buy food on average two to three times a week (Lithuania 26.9%, other EU countries 20.5%) (Figure 1). Also, a lot of respondents from Lithuania buy food once a week (16.9%). It is likely that the majority of respondents actually give priority to fresh produce when shopping several times per week, as mentioned in previous responses.

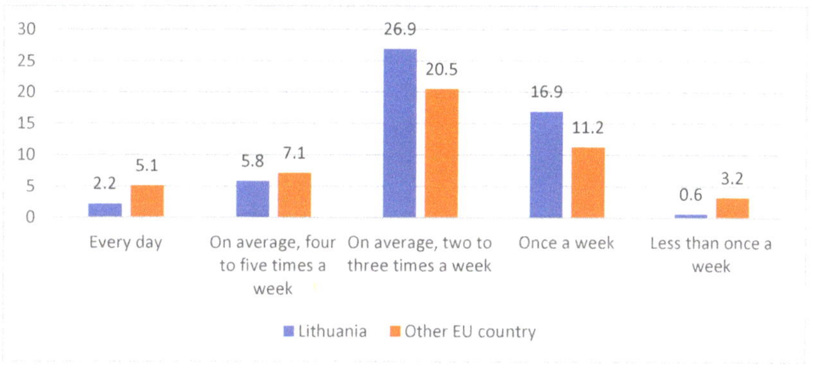

Figure 1. Respondents' frequency of food shopping; Significant differences are when $p \leq 0.05$. Chi-square test $p = 0.000 < 0.05$.

We searched for a statistical relationship between respondents' income and shopping frequency. However, no statistical dependencies have been identified. The responses of some higher-income respondents do not differ statistically from those of lower-income respondents. Respondents usually shop two or three times a week, which does not depend on the amount of income they receive.

The research data reveal that respondents sometimes or never throw away food. The mean ranges from 3.10 to 4.41. The obtained data show that there are statistically significant differences between respondents from EU countries and Lithuania ($p = 0.000 < 0.05$), analyzing the question of what products respondents throw away (see Table 4). Respondents from Lithuania more often throw away bread products, fruits and vegetables, while respondents from European countries more often throw away dairy products, meat, fish, pasta, eggs and sweets compared to respondents living in Lithuania. Both groups of respondents throw away eggs the least. These responses are also consistent with the responses where we asked respondents about their shopping habits, showing that respondents do not usually buy unplanned groceries.

Table 4. Data on food waste.

Statements	Mean	Mean Lithuania	Mean European Countries	Significant Differences between Countries *p*-Value
Milk and milk products	3.92	4.05	3.79	p **= 0.000 < 0.05**
Meat products	4.08	4.20	3.96	p **= 0.000 < 0.05**
Bread products	3.71	3.95	3.46	p **= 0.000 < 0.05**
Fruit	3.72	3.89	3.53	p **= 0.000 < 0.05**
Vegetables	3.70	3.86	3.53	p **= 0.000 < 0.05**
Fish	4.26	4.40	4.11	p **= 0.000 < 0.05**
Grains, pasta	4.31	4.51	4.08	p **= 0.000 < 0.05**
Eggs	4.41	4.60	4.20	p **= 0.000 < 0.05**
Sweets (cakes, biscuits, candies, etc.)	4.20	4.25	4.14	p **= 0.000 < 0.05**

Notes: Range scale where 1 means very often; 2 often; 3 rarely; 4 sometimes and 5 never. significant differences are when $p \leq 0.05$.

It is quite a big problem when people make unnecessary food, buy unplanned or order too much food in restaurants or cafés and do not consume it, and then they just throw it away. We asked respondents to evaluate some statements related with this. The research data reveal that the respondents are quite sustainable consumers, they disagree with most statements and the mean ranges from 2.57 to 4.10 (see Table 5). Significant differences between two groups of respondents are seen in four statements. Research results show that respondents from European countries are less likely to order too much food in cafes than respondents living in Lithuania, however, these differences are very small. The Lithuanian population is less likely to use food products that have an expiration date and are less likely to buy products at a discount, although they do not consume them later and discard food less often depending on the seasonality of the year.

Table 5. Food consumption habits.

Statements	Mean	Mean Lithuania	Mean European Countries	Significant Differences between Countries *p*-Value
I make too much food (and have to throw it away)	3.77	3.71	3.81	$p = 0.089 > 0.05$
I order too much food in a cafe/restaurant (which I do not eat and then leave)	4.10	4.09	4.12	p **= 0.020 < 0.05**
If I order too much food in a cafe/restaurant, I ask that I take it away (I eat it at home later)	2.57	2.58	2.56	$p = 0.058 > 0.05$
I consume products even if they have expired (e.g., groats, biscuits, etc.)	3.40	3.42	3.27	p **= 0.001 < 0.05**
I buy unplanned products at a discount, but I can no longer consume them	3.98	4.06	3.89	p **= 0.000 < 0.05**
I throw more food depending on the season (e.g., summer)	3.71	3.81	3.61	p **= 0.000 < 0.05**

Notes: Range scale where 1 means strongly agree, 2 agree, 3 partially agree, 4 disagree and 5 strongly disagree. Significant differences are when $p \leq 0.05$.

The results revealed that only about 17 percent of respondents compost food waste (Lithuania 17.5%; Europe 16.6%) (Figure 2). About 13 percent of respondents threw it together with other waste. A total of 4.4 percent of respondents would not think about it. A total of 17.4 percent of Lithuanian respondents and 13.5 percent of respondents from Europe would like to do it, but do not have a chance.

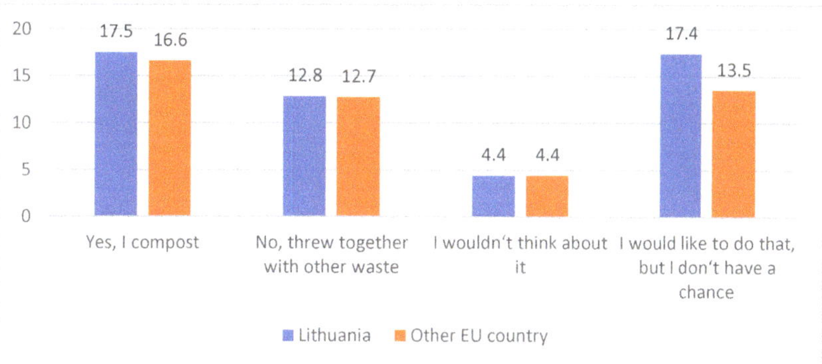

Figure 2. Results of food composting of respondents; Significant differences are when $p \leq 0.05$. Chi-square test $p = 0.467 > 0.05$.

Assessing the answers of both groups, it can be seen that a part of the respondents compost food; a large part of respondents would like to do it, but do not have the chance. In reality, only 34.3 percent of respondents are not worried about it and it is not relevant for them. We think the results are really promising. It is clear that it is important for the majority of respondents in the study not to waste food, while at the same time taking care of food waste disposal. It is likely that some live in apartments where composting is more difficult. Composting is an important element in sustainable waste management.

5. Discussion

Wasting food causes environmental and economic inefficiencies. It affects climate change, emissions, availability of natural resources, deterioration of land conditions, global hunger and can even be an underlying reason for an economic collapse [36]. The European Commission is taking the issue of tackling food waste very seriously. Reducing food waste has enormous potential for reducing the resources we use to produce the food we eat. Being more efficient will save food for human consumption, save money and lower the environmental impact of food production and consumption (EU actions against food waste).

This paper contributes to current debates on food waste management [10,14,37] by illustrating empirically what negative problems arise from unsustainable food waste.

To this aim, we conducted a study and assessed the food consumption habits of the European population. We compared the results with Lithuanian food consumption habits. This study allows to identify the respondents' shopping habits and the main problems of food consumption in the EU.

Excessive purchasing, over-preparation and unwillingness to consume leftovers are some of the main antecedents of food waste [11]. Part of the population does not even know that their actions are harmful to the environment and influence the economic circumstances negatively, which can be caused due to their cultural mindset, different traditions and certain everyday consumption routines. Therefore, having an educational intervention to increase consumers' awareness of the importance of green consumption enhances the general approach towards food management, its preparation and planning processes, which results in a remarkable decrease in food loss and waste levels [36,38]. Unreasonable food waste is promoted by improper planning of purchases and portions of food to be prepared, promotional shopping and so on [1]. Our study highlights that respondents are quite responsible, they do not make too much food and usually in a café/restaurant they order as much food as they can eat, but sometimes buy unplanned products at a discount.

In line with other research [11,14], we found that fruit and vegetables are the product group most commonly wasted. The fish and eggs food groups, which make up the smallest parts of the food supply chain, also generate the lowest quantities of food waste in absolute terms, despite the fact that much of these food groups (50% and 31%, respectively) go to waste [14,39]. The survey data also revealed that fish, eggs, grains and pasta are wasted less in the food supply chain.

Our findings also reveal that shopping habits of the respondents are quite different. Based on some research, customer shopping habits changed during the pandemic and unplanned shopping increased [34]. Our study revealed that the majority of the respondents go shopping for food two or three times per week, they plan purchases by making a list. It is very important to shop smartly and realistically [11].

Composting is an important element in sustainable waste management [40–42]. We find out that most of the respondents do not compost food, but would like to do that. In Lithuania, the amount of food waste in the mixed municipal waste is about 15%, with an average of 41 kg of food waste per person per year. Across the European Union, somewhere between 118 and 138 million tons of bio-waste arise annually, of which currently only about 40% (equivalent to 47.5 million tons per annum) is effectively recycled into high-quality compost and digestate [43].

6. Conclusions

The survey found that the majority of respondents state that food is not often wasted. This makes it a little more optimistic that global food waste and sorting problems will be addressed through people's awareness and real action efforts. The results of our survey show that the population of Lithuania emits slightly less food than the population of the European countries participating in the survey. Clearly, food wastage is not just a problem of family-specific intolerance, it is a global food security problem. This problem is directly linked to climate change, waste sorting and recycling, and other global ecological and economic or social problems. It is possible to notice the crucial role of educating and informing people. This should be the responsibility of national governments when allocating funds to educational programs. These programs should cover all age groups, from kindergarten to advertisements, flyers and conversations with adults. Adults would probably best understand and stimulate economic interest, with an emphasis on saving food and then composting food waste, because, unfortunately, not all adults are able to adequately assess the effects of climate change and their food supply needs. Consumers should also purchase food avoiding shopping routines and try to plan their food basket more so that they do not end up wasting edible food. On the other hand, generally, there are not many messages towards sustainable consumption in the majority of retailers; the most famous food retailers arrange communication strategies starting from their commercial goals rather than toward a zero-waste responsible behavior [44].

However, in line with recent research [25,26], it is possible to notice that the COVID-19 pandemic has affected consumer food habits, their management and consumption that, in turn, has reduced the household waste. Nowadays, most consumers try to define a meal list before shopping [27]. This increased awareness about food waste and its impacts on the environment helps reduce the quantity of spoiled and thrown products even if purchases have increased during the COVID-19 pandemic [28–31].

In line with [22], results highlight that adequate tools for reducing food loss and waste can become crucial to make green food supply chains. If, on one hand, it is relevant to implement infrastructures for recycling disposed products, on the other hand, training and education can shift habits towards responsible consumption as well as ethical consumption (i.e., by means of donations and food banks).

One of the best uses of discarded food is feeding livestock, saving precious resources that would have otherwise been used for producing commercial feed. If the food cannot be reused at all, we should at least try to recycle or compost it in a responsible manner instead of sending it to the landfills where it continues to rot [11]. The draft State Waste Prevention and Management Plan 2021–2027 by the Ministry of Environment of Lithuania defines tasks and goals for implementing separate collection of food and kitchen waste by 31 December 2023. High-quality compost used in agriculture has to be made from the separately collected bio-waste, and the restoration of areas for the preparation of energy plant media has to be vulnerable [45]. However, some findings show that people do not try to compost food and throw it away with other waste. The benefits of composting are significant: through composting, the quantity of garbage direct to the landfill is reduced, the organic matter is reused rather than dumped and it is recycled into a useful soil.

Composting can be defined as natural processes of recycling organic products such as leaves and food scraps into fertilizers that can enrich soil and plants. Recycling food and other organic waste into compost provides a range of environmental benefits, including improving soil health, decreasing greenhouse gas emissions, recycling nutrients, and mitigating the impact of climate changes. Composting can appear as much as an art as a science. Recent research and policies about managing wastes and producing food in an environmental way highlight a new interest in small-scale backyard composting as well as an interest in developing large-scale commercial and municipal composting systems.

Regarding research limitations, it can be noticed that only those respondents who use an internet connection and in the network of authors could participate in the study due to COVID-19 restrictions. The study makes uses also of a random selection of the respondents, so in the future it would be useful to do research that would cover all age groups and other demographical characteristics.

Future research direction aims to repeat the survey with as many European respondents as possible, to assess and understand food consumption habits and knowledge in the food waste chain and to make the widest possible range of consumers aware of the consequences of irresponsible food waste.

Author Contributions: Conceptualization, M.F., J.P. and L.Š.; methodology, M.F., J.P. and L.Š.; software, J.P. and L.Š.; validation, J.P. and L.Š.; formal analysis, J.P. and L.Š.; investigation, M.F., J.P. and L.Š.; resources, M.F., J.P. and L.Š; data curation, J.P.; writing—original draft preparation, M.F., J.P. and L.Š.; writing—review and editing, M.F., J.P. and L.Š.; visualization, J.P.; supervision, M.F. and L.Š.; project administration, L.Š.; funding acquisition, M.F. All authors have read and agreed to the published version of the manuscript.

Funding: This research received no external funding.

Institutional Review Board Statement: The study was conducted in accordance with the Declaration of Helsinki.

Informed Consent Statement: Informed consent was obtained from all subjects involved in the study.

Data Availability Statement: The data that support the findings of this study are available on request from the co-author J.P.

Conflicts of Interest: The authors declare no conflict of interest.

References

1. State Food and Veterinary Service. Available online: https://vmvt.lt/maisto-sauga/maisto-sauga-ir-kokybe/maisto-svaistymas-ir-parama-maistu (accessed on 10 May 2022).
2. EEA. Bio-Waste in Europe—Turning Challenges into Opportunities. 2020. Available online: https://www.eea.europa.eu/publications/bio-waste-in-europe (accessed on 6 June 2022).
3. EIT Food. Available online: https://www.eitfood.eu/ (accessed on 6 June 2022).
4. Wells, H.F.; Hyman, J.; Buzby, J.C. *Estimated Amount, Value, and Calories of Postharvest Food Losses at Retail and Consumer Levels in the United States*; Economic Research Service: Washington, DC, USA, 2014.
5. Fiore, M.A.; Contò, F.; Pellegrini, G. Reducing Food Losses: A (Dis)-Opportunity Cost Model. *Riv. Studi Sulla Sostenibilità* **2015**, *2015*, 151–166.
6. FAO. *Food Wastage Footprint, Summary Report*; FAO: Rome, Italy, 2013.
7. FUSIONS. Available online: http://www.eu-fusions.org/index.php/about-foodwaste/280-food-waste-definition (accessed on 10 April 2022).
8. FAO. Food Loss and Waste Database. Available online: https://www.fao.org/platform-food-loss-waste/flw-data/en/ (accessed on 3 May 2022).
9. Kummu, M.; de Moel, H.; Porkka, M.; Siebert, S.; Varis, O.; Ward, P.J. Lost food, wasted resources: Global food supply chain losses and their impacts on freshwater, cropland, and fertiliser use. *Sci. Total Environ.* **2012**, *438*, 477–489. [CrossRef] [PubMed]

10. Cordova-Buiza, F.; Paucar-Caceres, A.; Quispe-Prieto, S.C.; Rivera-Garré, A.P.; Huerta-Tantalean, L.N.; Valle-Paucar, J.E.; Ponce de León-Panduro, C.V.; Burrowes-Cromwell, T. Strengthening Collaborative Food Waste Prevention in Peru: Towards Responsible Consumption and Production. *Sustainability* **2022**, *14*, 1050. [CrossRef]
11. Parameshwari, S. Impact of food waste and its effect on environment. *Int. J. Food Sci. Nutr.* **2017**, *2*, 184–187.
12. Predergast, G.P.; Tsang, A.S.L. Explaining socially responsible consumption. *J. Consum. Mark.* **2019**, *361*, 146–154. [CrossRef]
13. Schrader, U. The moral responsibility of consumers as citizens. *Int. J. Innov. Sustain. Dev.* **2007**, *2*, 79–96. [CrossRef]
14. Caldeira, C.; Laurentiisa, V.; Corradoa, S.; Holsteijnb, F.; Salaa, S. Quantification of food waste per product group along the food supply chain in the European Union: A mass flow analysis. *Resour. Conserv. Recycl.* **2019**, *149*, 479–488. [CrossRef]
15. Nagarajan, M.; Saha, R.; Kumar, R.; Sathasivam, D. Impact of peer influence and environmental knowledge on green consumption: Moderated by price premium. *Int. J. Soc. Ecol. Sustain. Dev.* **2022**, *13*, 16. [CrossRef]
16. Venghaus, S.; Henseleit, M.; Belka, M. The impact of climate change awareness on behavioral changes in Germany: Changing minds or changing behavior? *Energy Sustain. Soc.* **2022**, *12*, 8. [CrossRef]
17. Ganglmair-Wooliscroft, A.; Wooliscroft, B. An investigation of sustainable consumption behavior systems—Exploring personal and socio-structural characteristics in different national contexts. *J. Bus. Res.* **2022**, *148*, 161–173. [CrossRef]
18. Block, L.; Vallen, B.; Paul Austin, M. Food waste (mis)takes: The role of (mis)perception and (mis)estimation. *Curr. Opin. Psychol.* **2022**, *46*, 101327. [CrossRef]
19. Cudjoe, D.; Zhu, B.; Wang, H. Towards the realization of sustainable development goals: Benefits of hydrogen from biogas using food waste in China. *J. Clean. Prod.* **2022**, *360*, 132161. [CrossRef]
20. Parra-Orobio, B.A.; Correa-Betancourt, M.A.; Lozano-Messa, M.T.; Foronda-Zapata, K.Y.; Marmolejo-Rebellón, L.F.; Torres-Lozada, P. Influence of Storage Time of Food Waste on the Substrate Characteristics and Energetic Potential Through Anaerobic Digestion. *Water Air Soil Pollut.* **2022**, *233*, 16. [CrossRef]
21. Cudjoe, D.; Chen, W.; Zhu, B. Valorization of food waste into hydrogen: Energy potential, economic feasibility and environmental impact analysis. *Fuel* **2022**, *324*, 124476. [CrossRef]
22. Dong, W.; Armstrong, K.; Jin, M.; Nimbalkar, S.; Guo, W.; Zhuang, J.; Cresko, J. A framework to quantify mass flow and assess food loss and waste in the US food supply chain. *Commun. Earth Environ.* **2022**, *3*, 83. [CrossRef]
23. Sundin, N.; Osowski, C.P.; Strid, I.; Eriksson, M. Surplus food donation: Effectiveness, carbon footprint, and rebound effect. *Resour. Conserv. Recycl.* **2022**, *181*, 106271. [CrossRef]
24. Zakarya, I.A.; Rashidy, N.A.; Izhar, T.N.T.; Ngaa, M.H.; Laslo, L. A Comparative Study on Generation and Composition of Food Waste in Desa Pandan Kuala Lumpur During COVID-19 Outbreak. In Proceedings of the 3rd International Conference on Green Environmental Engineering and Technology, Virtually, 8–9 September 2021; Springer: Berlin/Heidelberg, Germany, 2022; pp. 59–68. [CrossRef]
25. Kumar, B.; Dholakia, N. Firms enabling responsible consumption: A netnographic approach. *Mark. Intell. Plan.* **2020**, *40*, 289–309. [CrossRef]
26. Scacchi, A.; Catozzi, D.; Boietti, E.; Bert, F.; Siliquini, R. COVID-19 Lockdown and Self-Perceived Changes of Food Choice, Waste, Impulse Buying and Their Determinants in Italy: QuarantEat, a Cross-Sectional Study. *Foods* **2021**, *10*, 306. [CrossRef]
27. Nicewicz, R.; Bilska, B. Analysis of Changes in Shopping Habits and Causes of Food Waste Among Consumers Before and During the COVID-19 Pandemic in Poland. *Environ. Prot. Nat. Resour.* **2021**, *32*, 8–19. [CrossRef]
28. Pappalardo, G.; Cerroni, S.; Nayga, R.M., Jr.; Yang, W. Impact of COVID-19 on Household Food Waste: The Case of Italy. *Front. Nutr.* **2020**, *7*, 585090. [CrossRef]
29. Marotta, G.; Nazzaro, C. Public goods production and value creation in wineries: A structural equation modelling. *Br. Food J.* **2020**, *122*, 1705–1724. [CrossRef]
30. Lerro, M.; Vecchio, R.; Nazzaro, C.; Pomarici, E. The growing (good) bubbles: Insights into US consumers of sparkling wine. *Br. Food J.* **2019**, *122*, 2371–2384. [CrossRef]
31. Šimanskienė, L.; Paužuolienė, J.; Fiore, M.; Župerkienė, E. Responsible Consumption: Society Habits in Time of Crisis. In *Palgrave Studies in Cross-Disciplinary Business Research, In Association with EuroMed Academy of Business*; Springer: Berlin/Heidelberg, Germany, 2022; pp. 281–304.
32. Fiore, M. Food loss and waste: The new buzzwords. Exploring an evocative holistic 4Es model for firms and consumers. *EuroMed J. Bus.* **2020**, *16*, 526–543. [CrossRef]
33. Menold, N.; Wolf, C.; Bogner, K. Design aspects of rating scales in questionnaires. *Math. Popul. Stud.* **2018**, *25*, 63–65. [CrossRef]
34. Alaimo, L.S.; Fiore, M.; Galati, A. Measuring consumers' level of satisfaction for online food shopping during COVID-19 in Italy using POSETs. *Socio-Econ. Plan. Sci.* **2021**, *82*, 101064. [CrossRef] [PubMed]
35. Alaimo, L.S.; Fiore, M.; Galati, A. How the COVID-19 Pandemic Is Changing Online Food Shopping Human Behaviour in Italy. *Sustainability* **2020**, *12*, 9594. [CrossRef]
36. Adamashvili, N.; Chiara, F.; Fiore, M. Food Loss and Waste, a global responsibility?! *Food Econ.* **2019**, *21*, 825–846. [CrossRef]
37. Directive (EU) 2018/851 of the European Parliament and of the Council of 30 May 2018 Amending Directive 2008/98/EC on Waste. Available online: https://eur-lex.europa.eu/legal-content/EN/TXT/PDF/?uri=CELEX:32018L0851&from=EN (accessed on 6 June 2022).
38. Romani, S.; Grappi, S.; Bagozzi, R.P.; Barone, A.M. Domestic food practices: A study of food management behaviors and the role of food preparation planning in reducing waste. *Appetite* **2018**, *121*, 215–227. [CrossRef]

39. European Commission's Knowledge Center for Bioeconomy. 2020. Available online: https://ec.europa.eu/knowledge4policy/bioeconomy (accessed on 15 May 2022).
40. Slater, R.A.; Frederikson, J. Composting municipal waste in the UK: Some lessons from Europe. *Resour. Conserv. Recycl.* **2001**, *32*, 359–374. [CrossRef]
41. EU Actions against Food Waste. Available online: https://ec.europa.eu/food/safety/food-waste/eu-actions-against-food-waste_en (accessed on 6 June 2022).
42. Bernstad, A.; Jansen, J. Review of comparative LCAs of food waste management systems—Current status and potential improvements. *Waste Manag.* **2012**, *32*, 2439–2455. [CrossRef]
43. Siebert, S.; Gilbert, J.; Jürgensen, M.R. Compost Production in Europe. Available online: https://www.compostnetwork.info/wordpress/wp-content/uploads/190823_ECN-Compost-Production-in-Europe_final_layout-ECN.pdf (accessed on 6 June 2022).
44. Fiore, M.; Conte, A.; Contò, F. Retailers towards zero-waste: A walkthrough survey in Italy, Special Issue. *Ital. J. Food Sci. SLIM* **2015**, *2016*, 92–97.
45. Ministry of Environment of the Republic of Lithuania. Available online: https://am.lrv.lt/en/ (accessed on 15 May 2022).

Article

Influence of the Total Consumption of Households on Municipal Waste Quantity in Romania

Irina-Elena Petrescu [1,*], Mariarosaria Lombardi [2], Georgiana-Raluca Lădaru [1], Răzvan Aurelian Munteanu [3], Mihai Istudor [3] and Georgiana Adriana Tărășilă [3]

[1] Department of Agrifood and Environmental Economics, The Bucharest University of Economic Studies, 010374 Bucharest, Romania; raluca.ladaru@eam.ase.ro
[2] Department of Economics, University of Foggia, 71122 Foggia, Italy; mariarosaria.lombardi@unifg.it
[3] Doctoral School, The Bucharest University of Economic Studies, 010374 Bucharest, Romania; munteanuaurelian19@stud.ase.ro (R.A.M.); istudormihai14@stud.ase.ro (M.I.); tarasilageorgiana16@stud.ase.ro (G.A.T.)
* Correspondence: irina.petrescu@eam.ase.ro

Abstract: Sustainable development is a worldwide concern. This paper presents an analysis of the influence of the final consumption expenditure on the total consumption of households in Romania. The regression function of the association between "the amount of municipal waste" and "the total consumption of households" has a direct linear relationship. The regression variable "total household consumption" (X) has a regression coefficient of −0.03031, which indicates that the amount of municipal waste decreases by one unit as household consumption increases by 30.31 units. Therefore, this regression coefficient indicates that the volume of municipal waste decreases by 30.31 tons to an increase in the final consumption expenditure of households of EUR 1 million. The influence of the final consumption expenditure of households by consumption purpose on the quantity of municipal waste is in the following order: health; housing, water, electricity, gas and other fuels; clothing and footwear; miscellaneous goods and services; recreation and culture; food and non-alcoholic goods; restaurants and hotels; furnishing and household equipment and routine household maintenance; alcoholic and tobacco goods; communications; and education. The value of the Significance F must be less than 0.05. In the case of the model, it is found that this value exceeds the threshold of 0.05 in the case of consumption generated by health services, recreation and culture, restaurants and hotels, alcohol and tobacco goods, and communications. Regarding the high value of Significance F in relation to consumption, we find the sectors that generate the least amount of waste (services). In the case of all of the independent variables, we can note that the relationship is a negative one, which proves that an increase in the quantity of any expenditure of the households generates a decrease of the municipal waste quantity.

Keywords: municipal waste; consumption of households; waste management

Citation: Petrescu, I.-E.; Lombardi, M.; Lădaru, G.-R.; Munteanu, R.A.; Istudor, M.; Tărășilă, G.A. Influence of the Total Consumption of Households on Municipal Waste Quantity in Romania. *Sustainability* **2022**, *14*, 8828. https://doi.org/10.3390/su14148828

Academic Editor: Ming-Lang Tseng

Received: 16 May 2022
Accepted: 11 July 2022
Published: 19 July 2022

1. Introduction

Currently, environmental, social and ethical events are occurring with increasing frequency around the world, resulting in environmental costs and consequences that are difficult to correct in the short-to-medium term [1].

Waste is an important problem worldwide, as it is generated by all human activities. Economic growth and technological development have increased the consumption of goods and services, and the amount of waste. Some studies have shown that sustainable consumption is closely related to the concept of sustainable development and have proposed three dimensions of the latter for discussion: social, economic, and ecological [2]. In this case, one of the major concern of states is how to achieve economic growth while decreasing the quantity of waste. The entrepreneur remains the main facilitator of regional growth, through the role of the agent of change [3].

According to the definition of the OECD/Eurostat joint questionnaire, municipal waste covers household waste and waste similar in nature and composition to household waste (European Commission, 2016) [4]. According to OECD, municipal solid waste accounts for only around 10% of total waste, but its management and disposal often account for more than a third of public sector financial efforts to reduce and control pollution [5]. According to a report realised by The World Bank (What a Waste 2.0), waste generation is expected to grow by 70% by 2050, while our global population is expected to grow at less than half of that rate [6]. The importance of research is given by the need to achieve the objectives of the Europe 2030 Strategy in terms of reducing the amount of waste, except for major mineral waste.

Several authors have studied the possibilities of achieving the goals by member states (SDG). Regarding Romania, Firoiu et al. (2019) analysed 107 indicators and estimated that only 40 are forecast to reach the EU average value in 2030 [7]. The evolution of the waste quantity by capita is an important indicator for SDG No. 12 "Ensure sustainable consumption and production patterns". Firoiu et al. (2019) mentioned that Romania has set the following goals for 2030: a gradual transition to a new development model based on the rational and responsible use of resources, halving food waste per capita at the retail and consumer levels and reducing food waste in production and supply. The food waste recycling rate of municipal waste should reach 55% by 2025, and 60% by 2030; the packaging waste recycling rate should reach 65% by 2025, and 70% by 2030; this should apply to hazardous household waste collection by 2022, hazardous household waste by 2023, and bio-waste and textiles by 2025; we should establish a binding wholesale producer responsibility scheme for all packaging by 2024, and implement sustainable green procurement practices in line with national and European policy priorities [7].

The Waste Framework Directive sets a waste hierarchy, represented as a pyramid. At the bottom of the pyramid is the prevention action, followed by preparing for re-use, recycling, recovery, and at the top of the pyramid is the disposal action. The Waste Framework Directive and the Europe 2030 Strategy set objectives for member states, including reducing the quantity of municipal waste and increasing the recycling rate of municipal waste. Thus, any scientific approach that can contribute to the analysis of the possibilities for achieving these objectives proves important. The results of the paper could be useful for the public authorities (central and local) in their actions for the prevention of waste generation, the analysis of the evolution of the quantity of municipal waste, and the determination of the best solutions for influencing the behavior of citizens regarding waste generation.

The novelty of the paper is given by the determination of the influence of the final consumption expenditure of households by consumption purpose (health, housing, clothing, restaurant, education etc.) on the quantity of municipal waste. The analysis of these influences brings benefits to the local public authorities in the elaboration of the management plans. The results could also help central public authorities to develop the National Strategy for the Circular Economy.

The following sections in the paper review the literature in the field of waste, especially from recent years, and present the study hypothesis and findings regarding the influence of the final consumption expenditure of households by consumption purpose (health, housing, clothing, restaurant, education etc.) on the quantity of municipal waste. At the end of the paper are presented the conclusions of the study, as well as the limitations and future research perspective.

2. Literature Review

Environmental problems threaten the health and economic prospects of many countries. One of the important factors causing global warming and environmental degradation is the generation of waste [8].

Waste is a relational concept: it is a collection of items that no longer have any value or utility for an individual in a given social context. Due to the ambiguity of these representations, different actors can see different "wastes" at the same time. For example, informal recyclers and libertarians (ideologically motivated people who feed on the waste of a consumer society) often recycle items from the waste stream for reuse or resale, such as household items and groceries (in some cases). These potentially usable/saleable items are initially thrown into the trash stream, indicating a different perception of the value of a particular item [9].

Waste is generated from a variety of sources, such as domestic, industrial, and commercial sources, and the rate of waste generation is steadily increasing: a common phenomenon observed in urban centers around the world. It is common to dump rubbish in and out of cities. Trash can clog sewers (causing flooding during monsoons), curbs and even the middle of streets, markets, commercial centers, residential buildings and open spaces, wherever it is found [10].

Countries around the world are working to improve their approach to solid household waste [11].

Many countries were to temporally postpone their plastic use reduction policies and plastic waste management strategies [12].

In terms of sustainability, waste management is an activity that shapes environmental protection. Sustainability has become an internationally adopted development model, with organizations and individuals acting in accordance with principles and changes [13].

Waste management discusses the "Waste Hierarchy", which is a concept according to which different waste management measures are grouped according to both their long-term impact on the environment and the type of waste category with the lowest impact on environment. As such, the waste management hierarchy involves: waste prevention, reducing (minimizing) the amount of waste produced, the reuse of materials, the recycling (recovery) of waste, conditions of economic efficiency, the energy recovery of waste, the controlled storage of waste, and the recovery of gases resulting from waste storage [14].

Currently, in the field of municipal waste management, there are various models that help to highlight deficiencies in the field. However, most of the models identified in the literature are decision support models, and for model improvement there is research in which the models are divided into three categories based on cost–benefit analysis, based on waste life cycle assessment, and based on decision-making criteria [15].

The most popular approaches to municipal waste management are reprocessing, composting, combustion, and landfilling/open dumping [16]. Waste management processes and standards rely heavily on the operations strategy [17].

According to Han et al. (2018) [18] poor domestic waste management in developing countries, including a lack of sanitation and inadequate waste disposal facilities, leads to severe environmental pollution, landscape damage, and even negative impacts on the health of local people; therefore, these considerations give a topical importance to the chosen research theme.

As an important part of modern cities, municipal waste management not only aims to improve the living environment of urban residents but also plays an important role in preventing resource waste and protecting the overall ecological environment [19].

Various sectors have been affected by COVID-19 and its consequences. The waste management system is one of the sectors affected by such unpredictable pandemics. The experience of COVID-19 proved that adaptability to such pandemics and the post-pandemic era has become a necessity in waste management systems, and this requires an accurate understanding of the challenges that have been arising [20].

Additionally, given the evidence that COVID-19 spreads through food, food containers or food packaging, the use of reusable bags is discouraged in order to minimize the risk of store workers contracting the virus on the surface of the bags [21].

According to the results of the research developed by Yoada et al. (2014) [22], who studied the domestic waste disposal practice and perceptions of private sector waste management, it turned out that "93.1% of households disposed of food debris as waste and 77.8% disposed of plastic materials as waste. The study also showed that 61.0% of the households disposed of their waste at community bins or had waste picked up at their homes by private contractors. The remaining 39.0% disposed of their waste in gutters, streets, holes and nearby bushes. Of those who paid for the services of private contractors, 62.9% were not satisfied with the services because of their cost and irregular collection."

Considering that waste management could represent an important aspect for public authorities in any country, Ahangar et al. (2021) designed a disposal for municipal solid waste using fuzzy programming, with an impact on the cost through the decrease of the use of manpower as well as the amount of polution [23].

Different authors have analysed the medical waste supply chain and the opportunity of investing in environmetal aspects. Some of them consider that the generators of medical waste would allocate more funds for the suppliers that incorporate environmetal aspects in their products [24]. In terms of consumption, this has changed recently, as Kearney (2010) [25] stated in his paper. Changes in agricultural practices in recent years have improved the world's ability to feed its population by increasing productivity, increasing food variety and reducing seasonal dependence. Food availability has also increased due to higher income levels and lower food prices. This has led to significant changes in food consumption in recent years.

The COVID-19 pandemic has changed the way people live by requiring people to work from home. This has increased the home delivery of food and groceries, leading to an increase in demand for single-use plastic bags and food packaging materials. The use of online shopping and delivery services has surged during the pandemic [26].

The demand for natural resources has increased enormously in recent decades in the world, as a result of population growth, welfare and consumption. The highest levels of wellbeing and consumption were recorded in developed countries. The identification factors for the higher demand for resources are the increase of the income level, which allowed the purchase of more products and the wider use of services, the increase in the number of smaller households, and the changes in lifestyles, which supported the more individualized purchasing model [27].

Given the significant increase in the population, which puts pressure on resources, obviously leading to an increase in consumption, there will be some pressure on waste management.

People consume an increasing variety of goods and services produced by industrial sectors that generate direct and indirect waste. Humans are the main driver of production, consumption, and the resulting waste [28]. Although waste generation can be directly attributed to increased consumption, the authors concluded that "75% of Australian household waste generation is related to indirect waste generation."

In developed countries, a large amount of food is wasted at the end of the food supply chain, and according to the existing literature, this is mainly due to consumer behavior, habits and attitudes [29].

The awareness of environmental issues associated with consumption habits has led to the more careful use of resources. The Eurobarometer Attitudes of European Citizens towards the environment shows a positive evolution of the behavior of European citizens. A larger share of Europeans reported taking measures to improve resource efficiency. From those presented in relation to the behavior towards the environment, the data of the Barometer show that in European Union countries, including Romania, there is currently a positive evolution [27].

3. Study Hypothesis

Given the growing concern of the European Union and, implicitly, of Romania regarding waste management, there is a need to determine the correlation between the amount of municipal waste and the final consumption expenditure of households in Romania. Considering the previous research regarding the influence of household consumption on municipal waste, the hypotheses tested in the paper are the followings:

- The amount of municipal waste and household consumption are interconnected, and as a society develops, it becomes more aware of the environmental impact of waste from consumption, and the amount of waste should decrease.

- The relationship between the final consumption expenditure of households for education and the municipal waste quantity presents an inversely proportional relationship.

In order to determine if the hypotheses of the paper are validated, were analyzed the dynamics of the total quantity of municipal waste from Romania in the period 2000–2020, as well as the evolution the final consumption of households in the period 2000–2019. Following this analysis, in order to measure the influence of the final consumption expenditure of households (total and by consumption purpose) on the quantity of municipal waste, was used the linear regression function. In the same time, the modern lifestyle can generate a greater waste quantity if governments do not implement a waste management strategy in order to increase the awareness of citizens regarding the importance of environmental protection. Lately, e-commerce has increased significantly, especially due to the restriction imposed by states for the pandemic of COVID-19, and this type of commerce requires more packaging.

4. Findings

According to OECD (2013), household final consumption expenditure is typically the largest component of the final uses of GDP, representing in general around 60% of GDP; it is therefore an essential variable for the economic analysis of demand [30]. According to the same source, household final consumption expenditure covers all purchases made by resident households (home or abroad) to meet their everyday needs: food, clothing, housing services (rents), energy, transport, durable goods (notably cars), health, leisure, and miscellaneous services [30].

As presented in Figure 1, in 2020, the final consumption expenditure of households in the European Union was EUR 6,767,546.5 million in current prices, registering a decrease of 7.83% compared with 2019, when the consumption was EUR 7,342,501.2 million in current prices, the highest value from 2000. From 2000 until 2020, the consumption expenditure of households in the European Union increased constantly by EUR 873.190 million/year, with two exceptions: one in 2009, due to the financial crisis from 2008, when the consumption expenditure decreased by 3%, and the second in 2020, caused by the sanitary crisis generated by the COVID-19 pandemic, with a decrease of 7.83%. The data series has a standard deviation of ±0.872 million euro, which determines a variation of ±15%, with the standard error being 0.19.

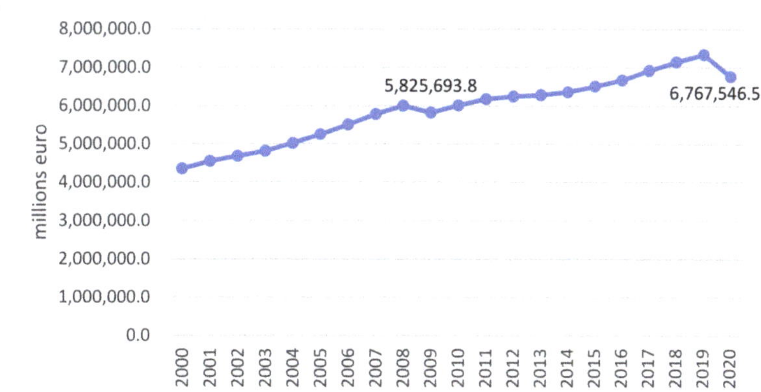

Figure 1. Evolution of the final consumption expenditure of households, in the European Union, in the period 2000–2020.

In 2020, all European Union countries decreased the final consumption expenditure of their households. Figure 2 presents the decrease of the final consumption expenditure of households in 2020 compared with 2019 for all European countries. The main key factors for household consumption are represented by income, inflation, wealth, and preferences, etc. The decreases are different from one country to another; the highest decrease compared to 2019 was registered in Malta, with a 21.13% decrease, while the minimum decrease was registered in Slovakia, with only 0.30%. According to Broom (World Economic Forum, 2021) and Eurostat, household consumption fell the most in countries with the severest lockdowns, while lockdowns, social distancing and restrictions on non-essential business activity reduced household consumption across the EU by 7.83%—an impact not seen since the 2008 financial crisis [31].

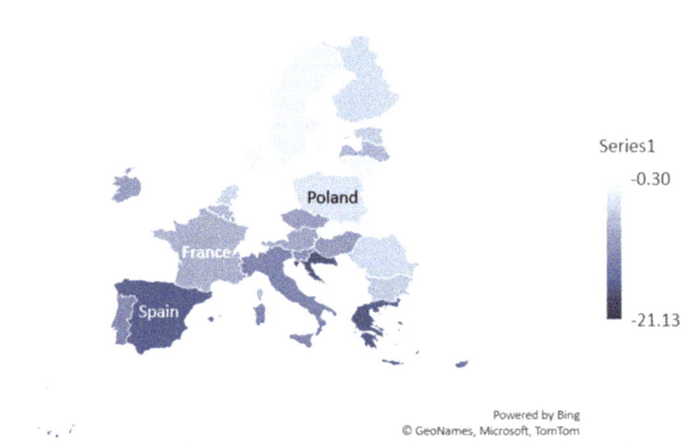

Figure 2. Decrease of the final consumption expenditure of households in 2020 compared with 2019.

The year 2020 was marked by the onset of the crisis caused by the COVID-19 pandemic. Starting in March 2020, household consumption has been declining massively. Figure 3 presents the decreases of household consumption by category in Romania in 2020 compared with 2019 and the most affected categories are restaurants and hotels, miscellaneous goods and services, recreation and culture, and transportation. The first case declared to be COVID-19 was registered in Europe on 24 January 2020, in France. According to Al-Salem et al. (2021), the Maastricht Treaty (1993) guaranteed free movement for the citizens of Iceland, Norway, Switzerland, the UK and all European signatory countries, and this freedom of movement led to an acceleration in the transmission of the disease throughout Europe [32]. As of 17 March 2020, all EU member states had reported cases of COVID-19, and Italy was the first country to declare a lockdown due to the alarming increase in cases. Thus, on 18 March 2020, most states in the European Union were in lockdown [33].

In Romania, on 16 March 2020, a state of emergency was declared, and on 25 March authorities announced the following lockdown measures: individuals were allowed to leave their homes to buy groceries, seek urgent medical attention, and for work, with a signed note from their employer; all shopping malls were closed, except for businesses selling food, veterinary services, and pharmaceutical products; additionally, elderly people who were aged 65 and above were only permitted to leave their homes between 11:00 to 13:00 for essential reasons, such as purchasing necessities etc. [34].

Imposing restrictions on social distance, the movement of people, the closure of restaurants and hotels, and the cancellation of social events have led to a reduction in household consumption, especially for the HoReCa, recreation and culture, and transportation categories. The reduction of consumption for this sector contributed to the registration of massive losses for the hospitality industry not only in Romania but also worldwide. In Romania, despite the potential value of our country for tourism development, the industry is not a significant source of contribution to the Gross Domestic Products, only 2.8%, but tourism has a larger sense when focusing on its multiplicative effect [35]. According to García-Madurga et al. (2021), the HoReCa channel is the set of commercial catering food establishments whose main activity is the production and sale of the direct out-of-home consumption of food [36].

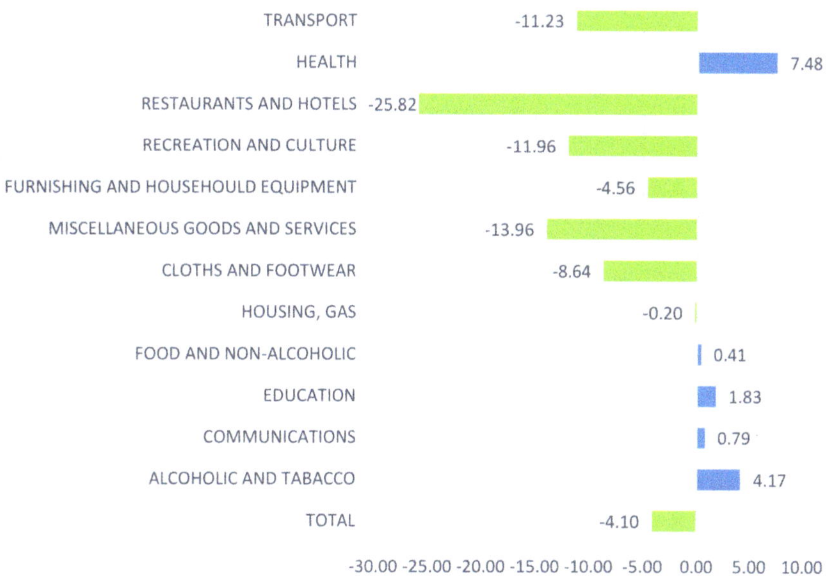

Figure 3. Decreases of household consumption by category in Romania in 2020 compared with 2019.

The categories of the final consumption expenditure of households that increased are the following: health, food and non-alcoholic, education, communication, and alcoholic and tobacco. The total consumption in Romania decreased by 4.1% in 2020 compared to 2019.

Figure 4 presents the evolution of municipal waste in Romania, in the period 2000–2019. Between 2000 and 2019, the average amount of municipal waste was 6631 thousand tons/year, and the variation was 1443.98 thousand tons/year. The maximum value of the amount of municipal waste was registered in 2008, at 8439 thousand tons/year, and the minimum value was registered in 2015, at 4904 thousand tons/year. Starting in 2015, there was an increase in the quantity of waste, with an average of 205.19 thousand tons/year. The data series has a standard deviation of ±1.45 thousand tons/year, which determined a variation of ±22%, with the standard error being 0.32.

Given the oscillating evolution of the total quantity of municipal waste, there is a need to study its influencing factors, especially because the concerns related to ensuring the sustainable development of society and the transition to a circular economy are growing, both in the European Union and worldwide. According to Hasan (2004), a critical component in any waste management program is public awareness and participation, in addition to appropriate legislation, strong technical support, and adequate funding, as waste represents the result of human activities and everyone needs to have a proper understanding of waste management issues, without which the success of even the best-conceived waste management plan becomes questionable [37].

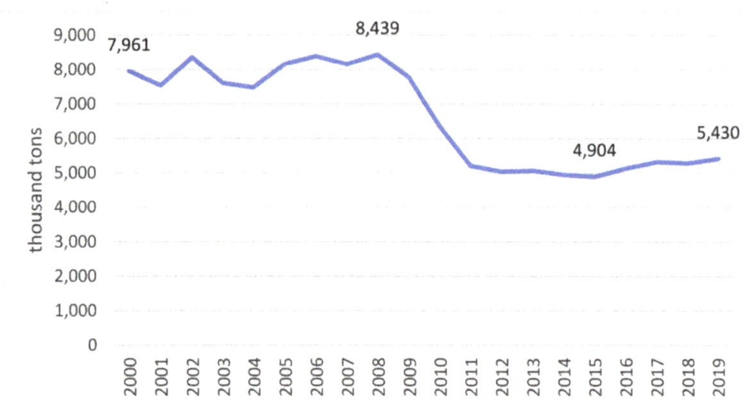

Figure 4. Evolution of the municipal waste in Romania, in the period 2000–2019, in terms of thousands of tons.

Figure 5 presents the evolution of the final consumption expenditure of households, in Romania, in the period 2000–2020. The data series has a standard deviation of $\pm$ 33.78 million euros, which determines a variation of $\pm$43%, with the standard error being 0.75.

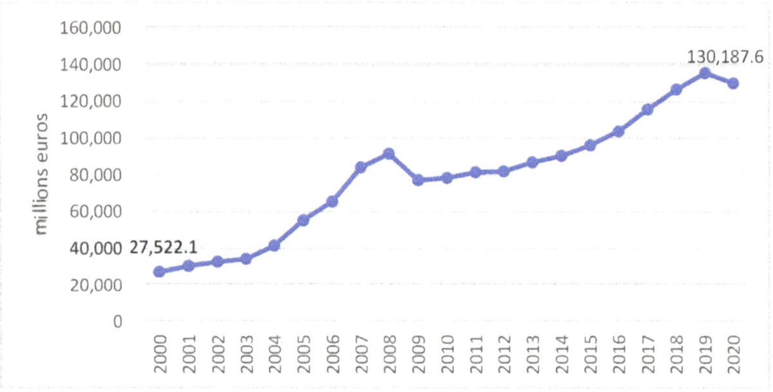

Figure 5. Evolution of the final consumption expenditure of households in Romania, in the period 2000–2020, in terms of millions of Euro (EUR).

5. Results

Regarding the correlation between the amount of waste and household consumption, the regression function of the association between "the amount of municipal waste" and "the total consumption of households" has a direct linear relationship. The regression variable "total household consumption" (X) has a regression coefficient of −0.03031, which indicates that the amount of municipal waste decreases by one unit as household consumption increases by 30.31 units.

The free term, i.e., the ascertainable variable of the regression function (Y), presents a positive value of 8966 thousand units (representing the value of Y regardless of the variation of X, or even in the absence of (X). The coefficient of determination R^2 expresses the fact that 44.75% of the variation of the quantity of municipal waste is determined by the variation of the total consumption of the households.

From the analysis of the obtained parameters presented in Table 1a and Figure 6, the following hypotheses are attested:

1. The correlation coefficient is quite high, at 0.66, which denotes a close relationship in terms of intensity. The high value of the correlation ratio shows that there is a strong dependence between the two series (Romer, 1997) [38].
2. By testing the correlation ratio, the regression model is statistically validated; the value of Significance F $\leq$ 5% (0.05). As such, the model adjusts well the data in the sample.
3. By testing the parameters of the regression model (Y and X) for a significance threshold of α = 5%, the existence of an association between the two variables is attested by statistically significant coefficients (p value $\leq$ 5%; the parameter is significant).
4. The confidence intervals of the parameters are statistically significant, as the limits have the same sign. For the estimated evaluation of X, the confidence interval is (−0.04)–(−0.01), and for the free term Y, the confidence interval is 7580.9–10,351.4. In both cases, the confidence interval does not include the value 0.

Table 1. Linear regression results.

(a)

Regression Statistics	Value
Multiple R	0.66893939
R Square	0.447479908
Adjusted R Square	0.416784347
Standard Error	1102.752513
Observations	20

(b)

ANOVA					
	df	*SS*	*MS*	*F*	*Significance F*
Regression	1	17,727,768.9	17,727,768.9	14.57800078	0.001259506
Residual	18	21,889,135.9	1,216,063.105		
Total	19	39,616,904.8			

(c)

	Coefficients	Standard Error	t Stat	*p*-Value	Lower 95%	Upper 95%	Lower 95.0%	Upper 95.0%
Intercept	8966.21007	659.3526598	13.59850444	6.57501×10^{-11}	7580.961535	10,351.4586	7580.961535	10,351.4586
CONSUMPTION	−0.030316478	0.007940169	−3.818114821	0.001259506	−0.046998155	−0.013634801	−0.046998155	−0.013634801

Source: Data processing, Eurostat.

Therefore, the regression equation between the dependent variable (municipal waste) and the independent variable (the final consumption expenditure of households) is as follows:

$$\text{Municipal waste} = -0.03 \times \text{Consumption} + 8966.21$$

The value of the Significance F must be less than 0.05. In the case of the model, as presented in Table 2, it was found that this value exceeds the threshold of 0.05 in the case of consumption generated by health services, recreation and culture, restaurants and hotels, alcohol and tobacco goods, and communications. For the high value of Significance F in relation to consumption, we found that the sectors that generate the lowest amount of waste (services). In the case of all of the independent variables, we can see that the relationship is a negative one, which proves that an increase in the quantity of any expenditure of the households generates a decrease of the municipal waste quantity.

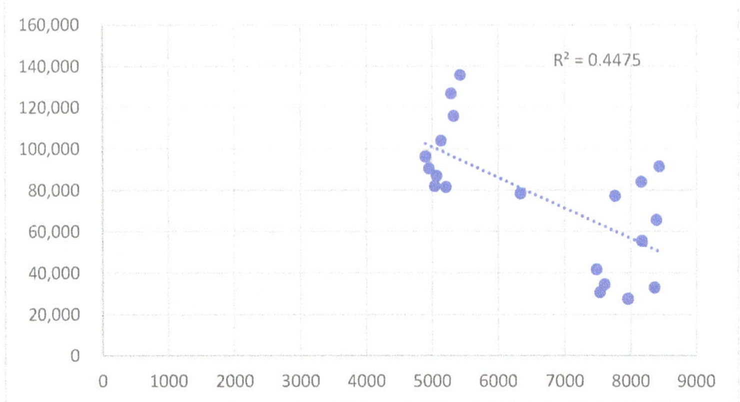

Figure 6. The regression function of the association between "the amount of municipal waste" and "the total consumption of households".

Table 2. The influence of the final consumption expenditure of households by consumption purpose on the quantity of municipal waste.

Independent Variable	Multiple R	R Square	Intercept	Regression Function Coefficients	Sig.
Health	0.87	0.77	8706.29	−0.54	3.24
Housing, water, electricity, gas and other fuels	0.60	0.36	8983.85	−0.14	0.005
Clothing and footwear	0.58	0.34	7901.19	−0.37	0.006
Miscellaneous goods and services	0.73	0.53	8485.13	−0.64	0.000
Recreation and culture	0.78	0.61	8724.91	−0.45	4.17
Food and non-alcoholic goods	0.66	0.44	9505.33	−0.13	0.001
Restaurants and hotels	0.14	0.02	7079.98	−0.14	0.54
Furnishing and household equipment and routine household maintenance	0.60	0.36	8200.36	−0.37	0.004
Alcoholic and tobacco goods	0.78	0.61	9070.44	−0.59	3.81
Communications	0.89	0.80	9019.20	−0.95	8.99
Education	0.65	0.42	8470.23	−1.31	0.001

Source: Data processing, Eurostat.

In the case of the final consumption expenditure of households for housing, gas, cloths and footwear, miscellaneous goods and services, food and non-alcoholic goods, furnishing and household equipment, and education, the Significance F value is less than 0.05, which proves that the models are statistically valid. The value of the R Square coefficient is relatively small (below 0.5) except for the independent variable miscellaneous goods and services (R square = 0.53), which means that the amount of municipal waste is influenced by other factors. The coefficient of determination R Square, for correlations whose Significance F is less than 0.05, states that:

1. 36% of the variation of the amount of municipal waste is determined by the variation of the final consumption expenditure of households for housing, water, electricity, gas and other fuels;
2. 34% of the variation of the amount of municipal waste is determined by the variation of the final consumption expenditure of households for clothing and footwear;
3. 53% of the variation of the municipal waste quantity is determined by the variation of the final consumption expenditure of households for miscellaneous goods and services;

4. 44% of the variation of the quantity of municipal waste is determined by the variation of the final consumption expenditure of households for food and non-alcoholic goods;
5. 36% of the variation of the quantity of municipal waste is determined by the variation of the final consumption expenditure of households for furnishing and household equipment, and routine household maintenance;
6. 42% of the variation of the municipal waste quantity is determined by the variation of the final consumption expenditure of households for education.

Worldwide, more than three billion people depend on solid fuels, including biomass (wood, dung and agricultural residues) and coal, to meet their most basic energy needs: cooking, boiling water, and heating [39]. Figure 7 presents the correlogram indicating the correlation between the amount of municipal waste and final consumption expenditure of households by consumption purpose. The R Square determination coefficient states that 36% of the variation in the amount of municipal waste is determined by the variation in the final consumption expenditure of households for housing, water, electricity, gas and other fuels.

The relationship between the amount of municipal waste and final consumption expenditure of households for housing, water, electricity, gas and other fuels shows that changing the consumption expenditure of households for housing, water, electricity, gas and other fuels by 1000 monetary units generates a decrease of municipal waste of 0.14 tons. The increase in the consumption expenditure of households for housing, water, electricity, gas and other fuels can be explained either by an increase in prices, which our country is currently facing, or by an increase in energy or gas consumption. In either case, in order to meet basic needs, households tend to reduce the consumption of other services or products that can generate large amounts of waste. This shows that there is a negative relationship between the two variables analyzed. The coefficient of Significance F is 0.005, which expresses a significant probability and determines a valid model.

According to the European Parliament (2020), clothes, footwear and household textiles are responsible for water pollution, greenhouse gas emissions and landfill; likewise, fast fashion—the constant provision of new styles at very low prices—has led to a big increase in the quantity of clothes produced and thrown away [40]. In order to decrease the impact upon the environment of the textile industry, the European Commission adopted, in March 2020, the New Circular Economy Action Plan, showing actions for changing the way we produce and consume in order to achieve a climate-neutral, competitive economy of empowered consumers [19]. The measures presented in the New Circular Economy Plan complete the EU Industrial Strategy, and include the following: making sustainable products the norm in the EU, empowering consumers, focusing on the sectors that use the most resources and where the potential for circularity is high, and ensuring less waste [41].

According to the model, the R Square determination coefficient states that 34% of the variation in the amount of municipal waste is determined by the variation in the final consumption expenditure of households for clothing and footwear. The relationship between the amount of municipal waste and the final consumption expenditure of households for clothing and footwear shows that changing the consumption expenditure of households for clothing and footwear by 1000 monetary units generates a decrease of municipal waste of 0.37 tons. The increase in the final consumption expenditure for clothing and footwear does not necessarily represent an increase in the quantity purchased of clothes and footwear, this being determined more by an increase in prices for these product categories. This means that there is a negative relationship between the two variables analyzed. The coefficient of Significance F is 0.001, which expresses a significant probability and determines a valid model.

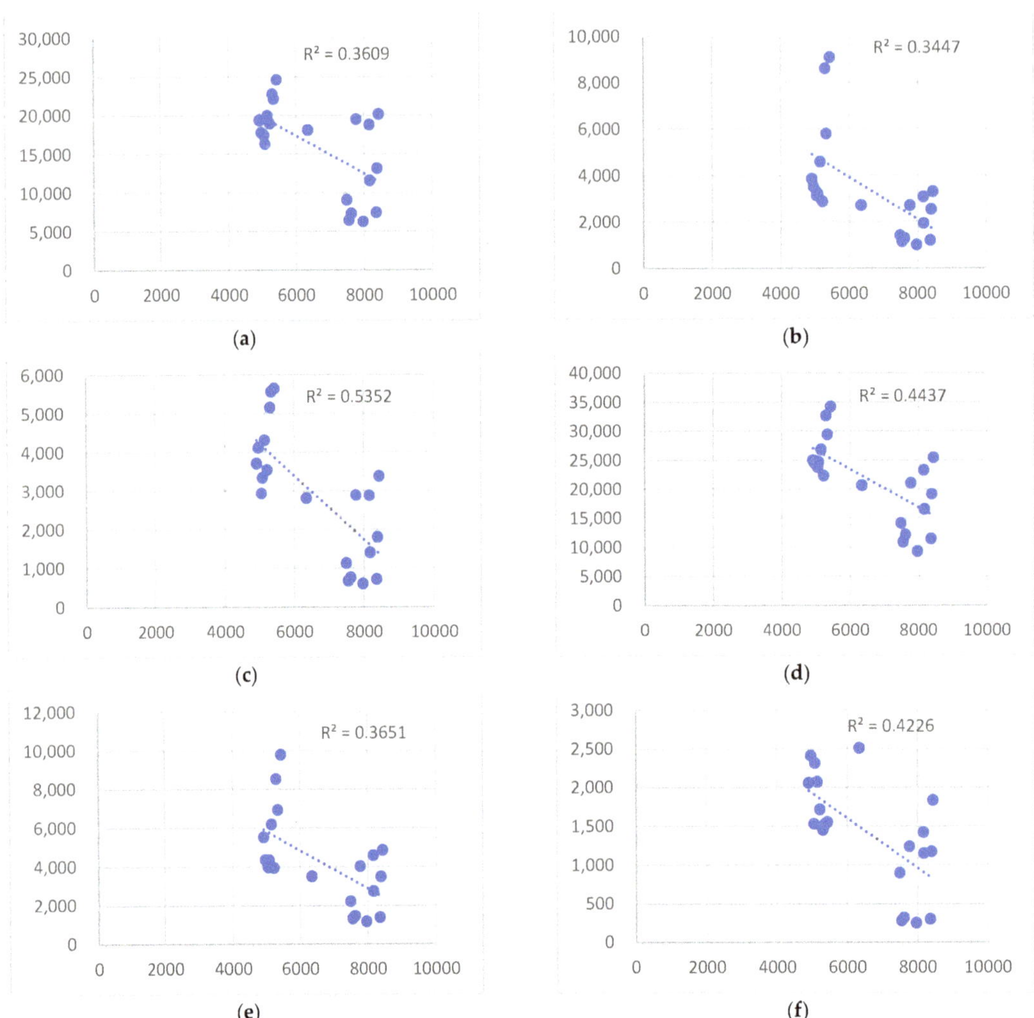

Figure 7. Correlogram indicating the correlation between the amount of municipal waste and the final consumption expenditure of households by consumption purpose, depending on the components. (**a**) Water, electricity, gas and other fuels; (**b**) Clothing and footwear; (**c**) Miscellaneous goods and services; (**d**) Food and non-alcoholic goods; (**e**) Furnishing and household equipment and routine household maintenance; (**f**) Education.

According to the model, the R Square determination coefficient states that 53% of the variation in the amount of municipal waste is determined by the variation in the final consumption expenditure of households for miscellaneous goods and services. Regarding the relationship between the amount of municipal waste and the final consumption expenditure of households for miscellaneous goods and services, a change in the consumption expenditure of households for clothing and footwear by 1000 monetary units generates a decrease in municipal waste of 0.64 tons. The increase in the final consumption expenditure of households for miscellaneous goods and services does not necessarily represent an increase in the quantity purchased, with this being more determined by an increase in prices for these product categories. This determines that there is a negative relationship between the two variables analyzed. The Significance Coefficient F is 0.000, which is a significant probability and means a valid model.

According to Food and Agriculture Organization, food loss and waste has indeed become an issue of great public concern, and The 2030 Agenda for Sustainable Development reflects the increased global awareness of the problem [42]. Food and non-alcoholic consumption has concentrated in the last years to a healthier and more ecological model. This model includes less processed food, in a smaller quantity, but at higher prices for consumers. This explains the increase of the consumption expenditure of households for food and non-alcoholic goods, which does not necessarily represent an increase of the quantity of the food.

The correlation between the two variables, the municipal waste quantity and the final consumption expenditure of households for food and non-alcoholic goods proves that consumers have spent more and generated a smaller quantity of waste. Additionally, the concerns for a healthier environment have focused consumers from our country on decreasing food waste.

Analyzing the coefficient of determination R Square, we can notice that 44% of the variation of the quantity of municipal waste is determined by the variation of the final consumption expenditure of households for food and non-alcoholic goods. The relationship between the amount of municipal waste and the final consumption expenditure of households for food and non-alcoholic goods shows that changing the consumption expenditure of households for food and non-alcoholic goods by 1000 monetary units generates a decrease of municipal waste of 0.13 tons. The coefficient of Significance F is 0.005, which is a significant probability and means a valid model.

An important part of the municipal waste is represented by furniture and household equipment, as mentioned by Article 3 of Directive (EU) 2018/851 of the European Parliament and of the Council of 30 May 2018 amending Directive 2008/98/EC on waste. Additionally, the same directive mentions in Article 9 that member states should prevent the generation of municipal waste by encouraging the re-use and repair of products, including furniture and household equipment [43].

Analyzing the coefficient of determination R Square, we can notice that 36% of the variation of the amount of municipal waste is determined by the variation of consumption expenditure of households for furnishing and household equipment and routine household maintenance. The relationship between the amount of municipal waste and the final consumption expenditure of household consumption for furnishing and household equipment and routine household maintenance demonstrates that changing the consumption expenditure of households for furnishing and household equipment and routine household maintenance by 1000 monetary units results in a decrease in municipal waste of 0.37 tons. The coefficient of Significance F is 0.004, which is a significant probability and means a valid model.

Ozturk (2008) mentioned that education is indispensable and fundamental for economic development, and there is a positive feedback from improved education to greater income equality, which in turn is likely to favor higher rates of growth [44]. The effects of investing in education not only contribute to the expansion of human knowledge, abilities and competences but also to the improvement of values and even better decision-making [45]. This is the reason why sustainable economic development requires important investments in the education system.

The determination coefficient R Square shows the fact that 42% of the variation of the amount of municipal waste is determined by the variation of the final consumption expenditure of households for education. At an increase of consumption expenditure of households for education by 1000 units of consumption, this will determine a decrease of 1.31 tons, representing the largest decrease of all of the variables analyzed.

6. Conclusions

In September 2015, Romania adopted the objectives of Agenda 2030 for sustainable development. Agenda 2030 involves achieving 17 objectives for sustainable development. As mentioned by Firoiu et al. (2019), achieving the targets of the 2030 Agenda depends exclusively on an effective approach for the implementation of objectives by ensuring a continuous dialogue between all of the entities directly and indirectly involved [7].

The quantity of municipal waste increased in Romania until 2009. Due to approaches to the environment, Romania developed programs through its local authorities for decreasing this quantity. Still, economic growth has increased the consumption of goods and services.

The correlation between the quantity of municipal waste and household consumption shows a close relationship in terms of intensity. The high value of the correlation ratio shows that there is a strong dependence between the two series. The regression variable "total household consumption" (X) has a regression coefficient of −0.03031, which indicates that the amount of municipal waste decreases by one unit as household consumption increases by 30.31 units.

The policies of the business environment for waste management influence the quantity of municipal waste. Additionally, with proper plans for recyciling, the business environment can contribute to increasing the recycling rate. Presenting the evolution of the household consumption expenditure by category in Romania in 2020 compared with 2019 could bring useful information for business decisions with strong managerial and practical implications.

Analysing the evolution of consumption brings economical implications, as consumption represents an important factor for economic growth. Moreover, the correlation with the municipal waste quantity is useful, with social implications for political decision factors. In order to reduce the quantity of municipal waste while increasing the recyciling rate, public athorities make efforts to develop consumer education programs for sustainable consumption.

One of the main concerns of Romanian local authorities should be to develop programs to prevent waste generation. Additionally, Romania should increase the rate of recycling of municipal waste by 55% until 2025, and by 60% until 2030. Selective waste collection generates raw material for reuse, recycling and energetic recovery, with a positive impact for the environment.

It is our desire that the analysis will present support for local public authorities in the elaboration of the management plans, and for the central public authorities in developing the National Strategy for the Circular Economy and achiveing the Sustainable Development Goals (SDG). All European countries must meet the Sustainable Development Goals (SDG). Municipal waste is an important issue for the achievement of SDG 11 regarding Sustainable Cities and Communities, and monitoring the evolution of the waste quantity is an essential condition for fulfilling this objective. The originality of the paper in the determination of the consumption expenditure's influence on municipal waste could be extended to other countries or regions, especially because the level of income, as main key factor that influences the consumption of households, is different in different European countries and could bring important conclusions for the elaboration of waste management policies.

The limitations of the paper are caused by data availability. A future research perspective of the present study is to analyse the influence of the final consumption of households by purpose, but at the regional level (NUTS 2), especially after March 2020, a period in which the household consumption was influenced by the COVID-19, as soon as the data are available.

Author Contributions: The authors worked together for this research, but, per structure: conceptualization I.-E.P., M.L. and G.-R.L.; methodology, software validation and resources, I.-E.P., R.A.M. and M.L.; formal analysis, G.A.T., M.L., R.A.M. and M.I.; writing—original draft preparation, and writing—review and editing, I.-E.P., M.L., G.-R.L., R.A.M. All authors have read and agreed to the published version of the manuscript.

Funding: This research received no external funding.

Institutional Review Board Statement: Not applicable.

Informed Consent Statement: Not applicable.

Data Availability Statement: The data presented in this study are available on Eurostat (https://ec.europa.eu/eurostat (accessed on 28 November 2021)) and National Institute of Statistics (https://insse.ro/cms/en (accessed on 28 November 2021)).

Conflicts of Interest: The authors declare no conflict of interest.

References

1. Pătărlăgeanu, S.R.; Negrei, C.; Dinu, M.; Chiocaru, R. Reducing the carbon footprint of the bucharest university of economic studies through green facades in an economically efficient manner. *Sustainability* **2020**, *12*, 3779. [CrossRef]
2. Boboc, D.; Ariciu, A.L.; Ion, R.A. Sustainable consumption: Analysis of consumers' perceptions about using private brands in food retail. *Sustainability* **2015**, *7*, 9293–9309. [CrossRef]
3. Ignat, R.; Constantin, M. Multidimensional facets of entrepreneurial resilience during the COVID-19 crisis through the lens of the wealthiest Romanian counties. *Sustainability* **2020**, *12*, 10220. [CrossRef]
4. Guidance on Municipal Waste Data Collection. September 2016. Eurostat—Unit E2—Environmental Statistics and Accounts; Sustainable Development. Available online: https://ec.europa.eu/eurostat/documents/342366/351758/Guidance+on+municipal+waste/3106067c-6ad6-4208-bbed-49c08f7c47f2 (accessed on 5 January 2022).
5. OECD iLirary. Municipal Waste. Available online: https://www.oecd-ilibrary.org/environment/municipal-waste/indicator/english_89d5679a-en (accessed on 11 February 2022).
6. Slipa, K.; Lisa, Y.; Perinaz, B.-T.; Woerden, V. *What a Waste 2.0., A Global Snapshot of Solid Waste Management to 2050*; World Bank Group, International Bank for Reconstruction and Development: Washington, DC, USA, 2018; Available online: https://blogs.worldbank.org/sustainablecities/new-phenomenon-realizing-economic-growth-while-cutting-waste-how (accessed on 24 January 2022).
7. Firoiu, D.; Ionescu, G.H.; Băndoi, A.; Florea, N.M.; Jianu, E. Achieving sustainable development goals (SDG): Implementation of the 2030 agenda in Romania. *Sustainability* **2019**, *11*, 2156. [CrossRef]
8. Tan, X.; Zhu, K.; Meng, X.; Gu, B.; Wang, Y.; Meng, F.; Liu, G.; Tu, T.; Li, H. Research on the status and priority needs of developing countries to address climate change. *J. Clean. Prod.* **2021**, *289*, 125669. [CrossRef]
9. Parizeau, K.; Lepawsky, J. Legal orderings of waste in built spaces. *Int. J. Law Built Environ.* **2015**, *7*, 21–38. [CrossRef]
10. Zia, H.; Devadas, V. Municipal solid waste management in Kanpur, India: Obstacles and prospects. *Manag. Environ. Qual.* **2007**, *18*, 89–108. [CrossRef]
11. Azevedo, B.D.; Scavarda, L.F.; Caiado, R.G.G.; Fuss, M. Improving urban household solid waste management in developing countries based on the German experience. *Waste Manag.* **2021**, *120*, 772–783. [CrossRef]
12. Izvercian, M.; Ivascu, L. Waste management in the context of sustainable development: Case study in Romania. *Procedia Econ. Financ.* **2015**, *26*, 717–721. [CrossRef]
13. da Costa, J.P. The 2019 global pandemic and plastic pollution prevention measures: Playing catch-up. *Sci. Total Environ.* **2021**, *774*, 145806. [CrossRef]
14. McDougall, F.; White, P.; Franke, M.; Hindle, P. *Integrated Solid Waste Managementa Life Cycle Inventory*; Blackwell Publishing: Hoboken, NJ, USA, 2001.
15. Morar, F.; Maior, P. General considerations regarding waste and its risks upon the environment. In *Descrierea CIP a Bibliotecii Naţionale a României Globalization and Intercultural Dialogue: Multidisciplinary Perspectives*; Arhipelag XXI: Tîrgu-Mureş, Romania, 2014; pp. 874–882, ISBN 978-606-93691-3-5.
16. Muhammad, A.; Hussein, M.Z.S.M.; Veera, M.H.Z.; Sundram, P.K. Reverse logistics activities for household E-waste management: A review. *Int. J. Sup. Chain* **2020**, *9*, 312.
17. Mostafa, N. Logistics of waste management with perspectives from Egypt. In *Waste Management in MENA Regions*; Springer: Cham, Switzerland, 2020; pp. 171–191.
18. Han, Z.; Liu, Y.; Zhong, M.; Shi, G.; Li, Q.; Zeng, D.; Zhang, Y.; Fei, Y.; Xie, Y. Influencing factors of domestic waste characteristics in rural areas of developing countries. *Waste Manag.* **2018**, *72*, 45–54. [CrossRef] [PubMed]
19. Wang, C.; Qin, J.; Qu, C.; Ran, X.; Liu, C.; Chen, B. A smart municipal waste management system based on deep-learning and Internet of Things. *Waste Manag.* **2021**, *135*, 20–29. [CrossRef]
20. Mahyari, K.F.; Sun, Q.; Klemeš, J.J.; Aghbashlo, M.; Tabatabaei, M.; Khoshnevisan, B.; Birkved, M. To what extent do waste management strategies need adaptation to post-COVID-19? *Sci. Total Environ.* **2022**, *837*, 155829. [CrossRef]
21. Hale, R.C.; Song, B. Single-use plastics and COVID-19: Scientific evidence and environmental regulations. *Environ. Sci. Technol.* **2020**, *54*, 7034–7036. [CrossRef] [PubMed]
22. Yoada, R.M.; Chirawurah, D.; Adongo, P.B. Domestic waste disposal practice and perceptions of private sector waste management in urban Accra. *BMC Public Health* **2014**, *14*, 697. [CrossRef] [PubMed]
23. Ahangar, S.S.; Sadati, A.; Rabbani, M. Sustainable design of a municipal solid waste management system in an integrated closed-loop supply chain network using a fuzzy approach: A case study. *J. Ind. Prod. Eng.* **2021**, *38*, 323–334. [CrossRef]

24. Maihami, R.; Ghalehkhondabi, I. Pricing problem in a medical waste supply chain under environmental investment: A game theory approach. *J. Ind. Prod. Eng.* **2022**, 1–17. [CrossRef]
25. Kearney, J. Food consumption trends and drivers. *Phil. Trans. R. Soc.* **2010**, *365*, 2793–2807. [CrossRef]
26. Parashar, N.; Hait, S. Plastics in the time of COVID-19 pandemic: Protector or polluter? *Sci. Total Environ.* **2021**, *759*, 144274. [CrossRef]
27. Humă, C. Modificarea comportamentului de consum al populaṫiei, din perspectivă ecologistă, în țările uniunii europene. *Qual. Life Calitatea Vietii* **2013**, *24*, 353–372.
28. He, H.; Reynolds, C.J.; Hadjikakou, M.; Holyoak, N.; Boland, J. Quantification of indirect waste generation and treatment arising from Australian household consumption: A waste input-output analysis. *J. Clean. Prod.* **2020**, *258*, 120935. [CrossRef]
29. Principato, L.; Mattia, G.; Di Leo, A.; Pratesi, C.A. The household wasteful behaviour framework: A systematic review of consumer food waste. *Ind. Mark. Manag.* **2020**, *93*, 641–649. [CrossRef]
30. OECD. Household consumption. In *National Accounts at a Glance*; OECD Publishing: Paris, France, 2013; Available online: https://www.oecd-ilibrary.org/docserver/na_glance-2013-12-en.pdf?expires=1641467357&id=id&accname=guest&checksum=7D6355B535069A27AD384783D84C64A2 (accessed on 14 January 2022). [CrossRef]
31. Broom, D. This Is How COVID-19 Hit Household Expenditure in Europe. 2021. Available online: https://www.weforum.org/agenda/2021/12/pandemic-impact-europe-consumer-spending/ (accessed on 14 January 2022).
32. Al-Salem, W.; Moraga, P.; Ghazi, H.; Madad, S.; Hotez, P.J. The emergence and transmission of COVID-19 in European countries, 2019–2020: A comprehensive review of timelines, cases and containment. *Int. Health* **2021**, *13*, 383–398. [CrossRef] [PubMed]
33. COVID-19 Pandemic in Europe. Available online: https://en.wikipedia.org/wiki/COVID-19_pandemic_in_Europe (accessed on 17 January 2022).
34. Romania: Government Announces lockdown Measures on March 25/Update 2. Available online: https://www.garda.com/fr/crisis24/alertes-de-securite/326626/romania-government-announces-lockdown-measures-on-march-25-update-2 (accessed on 17 January 2022).
35. Tigu, G.; Ciora, C.; Petcu, M.A.; Boboc, D.; Crismariu, O.D.; Curteanu, A.B. *Restart the Hotel, Restaurant and Ttravel Industry in Romania after the COVID-19 Pandemic. Economic Recovery after COVID-19*; Dima, A., Anghel, I., Dobrea, R.C., Eds.; Springer: Berlin/Heidelberg, Germany, 2021; p. 87.
36. García-Madurga, M.-Á.; Esteban-Navarro, M.-Á.; Morte-Nadal, T. COVID key figures and new challenges in the HoReCa sector: The way towards a new supply-chain. *Sustainability* **2021**, *13*, 6884. [CrossRef]
37. Hasan, S.E. Public awareness is key to successful waste management. *J. Environ. Sci. Health A Tox Hazard Subst. Environ. Eng.* **2004**, *39*, 483–492. [CrossRef]
38. Romer, D. *Advanced Macroeconomics*; McGraw-Hill, Inc.: New York, NY, USA, 1996.
39. World Health Organization. *Fuel for Life—Household Energy and Health*; World Health Organization: Geneva, Switzerland, 2006; p. 8.
40. European Parliament. The Impact of Textile Production and Waste on the Environment (Infographic). 2020. Available online: https://www.europarl.europa.eu/news/en/headlines/society/20201208STO93327/the-impact-of-textile-production-and-waste-on-the-environment-infographic (accessed on 15 February 2022).
41. European Commission. *Changing How We Produce and Consume: New Circular Economy Action Plan Shows the Way to a Climate-Neutral, Competitive Economy of Empowered Consumers*; European Commission: Brussels, Belgium, 2020.
42. Food Loss and Waste Database. Available online: https://www.fao.org/food-loss-and-food-waste/flw-data (accessed on 15 February 2022).
43. Official Journal of the European Union. Directive (Eu) 2018/851 of The European Parliament and of The Council of 30 May 2018 Amending Directive 2008/98/EC on Waste. Available online: https://eur-lex.europa.eu/legal-content/EN/TXT/PDF/?uri=CELEX:32018L0851&rid=5 (accessed on 17 February 2022).
44. Ozturk, I. The Role of Education in Economic Development: A Theoretical Perspective. 2001. Available online: https://papers.ssrn.com/sol3/papers.cfm?abstract_id=1137541 (accessed on 2 March 2022).
45. Colin Power Junior Sophister. Education Development: Importance, Challenges and Solutions. The Student Economic Review Vol. XXVIII. Available online: https://www.tcd.ie/Economics/assets/pdf/SER/2014/Colin_Power.pdf (accessed on 3 March 2022).

Review

How to Monitor the Transition to Sustainable Food Services and Lodging Accommodation Activities: A Bibliometric Approach

Christian Bux [1,*], Alina Cerasela Aluculesei [2] and Simona Moagăr-Poladian [2]

1 Department of Economics, Management and Business Law, University of Bari Aldo Moro, 70124 Bari, Italy
2 Institute for World Economy, Romanian Academy, 050711 Bucharest, Romania; alina.cerasela@iem.ro (A.C.A.); smpoladian1@gmail.com (S.M.-P.)
* Correspondence: christian.bux@uniba.it

Abstract: The transition to sustainable food systems is one of the main challenges facing national and international action plans. It is estimated that food services and lodging accommodation activities are under pressure in terms of resource consumption and waste generation, and several tools are required to monitor their ecological transition. The present research adopts a semi-systematic and critical review of the current trends in the food service and lodging accommodation industries on a global scale and investigates the real current environmental indicators adopted internationally that can help to assess ecological transition. This research tries to answer the subsequent questions: (i) how has the ecological transition in the food service industry been monitored? and (ii) how has the ecological transition in the lodging accommodation industry been monitored? Our study reviews 66 peer-reviewed articles and conference proceedings included in Web of Science between 2015 and 2021. The results were analyzed according to content analysis and co-word analysis. Additionally, we provide a multidimensional measurement dashboard of empirical and theoretical indicators and distinguish between air, water, energy, waste, health, and economic scopes. In light of the co-word analysis, five research clusters were identified in the literature: "food cluster", "water cluster", "consumers cluster", "corporate cluster", and "energy cluster". Overall, it emerges that food, water, and energy are the most impacted natural resources in tourism, and users and managers are the stakeholders who must be involved in active monitoring.

Keywords: ecological transition; tourism; environmental indicators; circular economy; sustainability; monitoring framework; food services; hotels; hospitality

Citation: Bux, C.; Aluculesei, A.C.; Moagăr-Poladian, S. How to Monitor the Transition to Sustainable Food Services and Lodging Accommodation Activities: A Bibliometric Approach. *Sustainability* **2022**, *14*, 9102. https://doi.org/10.3390/su14159102

Academic Editor: Mario D'Amico

Received: 7 July 2022
Accepted: 22 July 2022
Published: 25 July 2022

1. Introduction

Ecological transition, defined as an implementation of the sustainable development concept, aims at ensuring resilience soon after economic crises and ecological disasters [1] and represents one of the main challenges of national and international action plans [2,3]. Among others, its main pillars are represented by sustainability and efficient waste management paradigms, the development of renewable energy sources, and more sustainable agriculture [4,5]. Therefore, to achieve the 17 sustainable development goals (SDGs) enacted by the United States [6], several proper strategies and policies have been promoted on a global scale. At the European level, on the one hand, the circular economy action plan is based on depicting production and consumption systems which rely on recycling, re-use, and repairing and remanufacturing products, as well as on green consumption patterns [7]. On the other hand, there is a need to put the 'farm to fork' practice into effect, which aims at ensuring fair, healthy, and environmental-friendly food systems. The concept of the circular economy has a high impact at the international level due to its ambition to redefine a sense of a resource's value. Even though companies and public authorities have the most powerful role in accelerating the transition from the linear to the circular economy, customer preferences are key to making this change [8].

Moreover, shifting production from one based on primary resources to one based on recycling resources generates the premise of applying a sustainable economic model that could change the world into a better place for future generations. However, many questions arise when putting into practice the circular economy model. The most well-known issues are circular economy rebound [9] and the high costs of implementing barriers like technological costs, market readiness, institutional resistance, or cultural aspects [10]. Also, even though the circular economy is based on the four '*R*'s (i.e., reduce, reuse, recycle, and recover), research shows that most companies focus their activity on recycling and neglect the other three pillars of circularity [11]. Another critique of the circular economy is that it represents just a concept that is hard to transform into an economic reality [12].

Theoretically, ecological transition is possible and affordable to achieve [13] but it is still difficult to evaluate and monitor. Therefore, to assess Member States' progress towards ecological transitions, several indicators have been implemented at the European level, such as production and consumption indicators, waste management indicators, secondary raw materials indicators, and competitiveness and innovation indicators [14,15], in addition to environmental performance indicators [16].

The present paper investigates ecological transition within different sub-sectors of so-called hospitality management, considering an expected revival in tourism activity after the recovery from the COVID-19 pandemic [17]. The authors explore the food services and food and beverage activities within restaurants, fast-food chains, takeaways, catering, bars, and pubs on the one side, and lodging accommodation activities, from luxury hotels to campgrounds, on the other. The hospitality industry, encompassing establishments such as canteens, elderly care hospitals, hotels, schools, restaurants, and universities [18], represents one of the major incomes and sources of labor in Europe, as well as one of the most impactful sectors on the economy, environment, and society [19]. Besides, food services and lodging accommodation activities are under pressure in terms of resource consumption and waste generation, requiring a conversion towards circular economic systems focused on savings and recovering resources. Hotels and restaurants must be considered resource-intensive activities considering that their processes are orientated toward space conditioning (i.e., heating and cooling systems, ventilation, air conditioning), lighting, hot water, and electricity use, as well as cooking activities and washing activities, among others. However, although resource and waste management represent a topical and current concern, little research has been conducted on the environmental impacts of tourism, and additional investigations must be addressed to enhance environmental sustainability.

Several authors have considered these burdens in the field of transportation, analyzing aviation [20], train [21,22], or cruise ship [23] impacts, whereas increasing environmental loads should be refocused toward energy consumption (i.e., heating, air conditioning, and lighting) and greenhouse gases emissions [24], as well as to waste (and food waste) management [25]. Since the eruption of the COVID-19 pandemic, pollution and ecological burdens have significantly decreased by 30% on a global scale, highlighting the need to find and understand the unsustainable drivers and gradually switch them towards investment opportunities [17]. At present, the environmental and economic performance of food services and lodging accommodations are included in eco-effective strategies promoted at the EU level [26,27]. Still, the monitoring and measurement of specific policies need to be implemented either at the local or global level, representing a challenge toward ecological transition [28]. Monitoring represents a challenge to reaching sustainable development and economic growth. In light of the uncommon measurement and monitoring tools used in the hospitality industry, it seems essential to develop shared, common, and homogeneous systems to reach such targets, filling in empirical and theoretical gaps, which highlight the lack of harmonized indicators for measuring ecological transition.

In light of these premises, the present research carries out a semi-systematic literature review on the environmental indicators adopted by international realities to assess the sustainable transition of food services and lodging accommodation activities. Besides, by answering to two research questions (Section 2), the authors have provided a multidimensional measurement dashboard of empirical and theoretical indicators, which distinguishes between air, water, energy, waste, health, and economic scope (Section 3). The main purpose of this research is to identify the current state and the future prospects of sustainability in the food service and the lodging accommodation industries. Although several authors and practitioners have experienced, explored, and tested circular economy and sustainability strategies in the hospitality industry, a comprehensive and critical review on this topic still needs to be conducted. Hence, the present research contributes to the empirical studies dealing with sustainable tourism and environmental best practices in the hospitality sector, providing theoretical and managerial recommendations for supporting either academics or practitioners.

2. Theoretical Background and Research Questions Development

Tourism and hospitality activities have been the preferred subject of several research studies from environmental, social, or economic perspectives. Ecological transition, based on the sustainable development concept, encompasses transversal and multidimensional areas, covering not only environmental and green issues but revising the entire concept of work and enterprise and opening modern and innovative paths for boosting competitiveness while ensuring environmental protection [1–3].

In recent years, with the exception of the pandemic period, there has been an increase in the number of tourists. International tourism has continued to sustain global economic development [29]. Latest international statistics have estimated that more than 1.5 billion international tourist arrivals were recorded before the pandemic, representing a substantial investment opportunity for communities all around the world. In Europe, over 740 million tourists have been registered, accounting for more than 50% of the total tourists in the global market [30]. According to Eurostat, more than one in ten EU enterprises, excluding the financial sector, was based on tourism in 2019 [12]. However, considering that tourism represents a holistic industry and encompasses several businesses endowed with divergent characteristics and management operations, the present research focuses especially on food services and accommodation lodging activity. Food services, defined as the operations related to preparing, transporting, or selling foods in restaurants, cafeterias, or catering services [31,32], play an important social and economic role in modern societies [25]. In western Europe, food services revenue rose to EUR 427 billion, while in eastern Europe, it rose to over EUR 45 billion before the outbreak of the pandemic. The United Kingdom and France have the most developed markets for food services, whereas France and Italy are ranked first in the number of food service companies. France accounts for 161,466 restaurants and mobile food service activities, with 155,875 for Italy [33]. Still, several major issues in this area are under-researched, such as their related environmental externalities and issues, such as food waste or packaging waste, which represent an increasing share on a global scale. It is estimated that food services amount to 12% of global food waste [34]. Hence, the first research question (RQ) is proposed to investigate the ecological transition in the food service industry as follows:

RQ1: How has the ecological transition in the food service industry been monitored?

On the other hand, accommodation lodging activities that include hotels, motels, resorts, and bed and breakfast units, are supposed to constitute the most energy-intensive buildings due to their multi-usage functions, such as food and beverage production and consumption, recreation, and hygiene procedures [35]. Overall, over 73.2% of worldwide CO_2 emissions are derived from energy consumption, of which 24.2% come from industrial use (e.g., iron and steel production, food and tobacco production, and chemical and petrochemical production) and over 17.5% come from energy use in buildings [36]. It emerges that approximately 6.6% of CO_2 emissions are generated by commercial buildings, such as restaurants or hotels, to produce electricity for lighting, appliances, and heating. The amount of energy consumed in non-residential buildings amounts to over 700 million tons of the equivalent oil (Mtoe), with approximately 233 Mtoe needed for space heating,

116 Mtoe needed for lighting and 86 Mtoe needed for water heating. Besides, over 232 Mtoe is spent on other end-uses, such as IT equipment [37]. Among others, several strategies and action plans have been addressed to enhance energy and waste management [38], promoting guests' awareness of sustainable behaviors [39]. It is estimated that hotels produce more than 289,000 t of waste each year, of which approximately 80 t comes from food waste. More specifically, an amount of approximately 0.8–1.2 kg of waste per guest daily, which doubles on checkout days, has been assessed. Hotels and restaurants generate over 160–200 kg of CO_2/m^2. In terms of waste composition, it is estimated that about 44% comes from organic matter, 16% from glass, 13% from plastic, 11% from paper, and 9% from cardboard, whereas only 7% comes from unsorted waste [40,41]. Therefore, the second research question investigates the monitoring of the ecological transition in the lodging accommodation industry as follows:

RQ2. How has the ecological transition in the lodging accommodation industry been monitored?

The circular economy is presented as a possible solution to actual economic challenges, such as the increasing global demand for natural resources, delays in supply chains, climate change, and industrial pollution effects. In this context, the circular economy can also be considered a potential solution for the hospitality industry, known to be a consumer of resources [42]. However, the impact of the circularity concept in the hospitality industry is still new and under-researched. Moreover, few studies have been written in the field of circular economics for tourism. According to da Silva et al. [43], most of the studies in these areas began to be written and published in 2019, with most of them being focused primarily on a theoretical approach. Despite the desire of the stakeholders in the hospitality sector to focus more on sustainability and resource waste, the COVID-19 pandemic proved that the principles of circular economics and the signs of progress with the adoption of its rules are extremely sensitive to unpredicted events. An eloquent example is the increased use of plastic packages in the hospitality sector. On the one hand, due to the prevention rules applied for preventing the spread of COVID-19, and especially the distancing rules, the restaurants developed delivery services to continue their activity and to assure food security conditions. Customers preferred to order food at home to meet the travel conditions of the pandemic [44]. In this context, increasing numbers of home orders led to increasing use of plastic packaging. However, the desire of the customers to reduce single-use packaging plastics has not disappeared yet. According to Kitz et al. [45], in Canada, there is an increasing desire in consumers to use biodegradable single-use packaging.

The circular economy concept endorses either policy aimed at diminishing environmental burdens or those elaborating strategies oriented toward fostering economic growth, which is often difficult to define [14]. Environmental and sustainable strategies are usually part of a circular economy-monitoring framework [46]. This includes ten indicators used to capture the main features of a circular economy in a synthetic and suggestive way [15], as follows: (a) production and consumption (four indicators); (b) waste management (two indicators); (c) secondary raw materials (two indicators); and (d) competitiveness and innovation (two indicators). In addition, environmental performance indicators should be taken into account, measuring the environmental impacts related to climate change, consumption of natural resources, and waste generation [16]. Environmental performance indicators are used for comparing performance over time, communicating results in a transparent way, and supporting national and international policy for monitoring goals [47]. These are related to air, water, or land emissions, as well as resource consumption. Among other sustainability and environmental indicators, environmental footprint (i.e., water and carbon footprint), being an assessment based on inventories such as material flow analysis or environmental tools and life-cycle assessment [48–50], has emerged as a popular concept. Therefore, several studies have been devoted to the assessment of energy, carbon, and water consumption [51].

3. Research Methodology

3.1. Research Strategy and Review Criteria

Different approaches could be adopted to conduct a literature review, such as systematic, semi-systematic, and integrative reviews [52], each having unique characteristics and research strategies. Among them, the present research paper applies a semi-systematic method, being orientated toward providing an overview of the research area and tracking its development over time, contributing to the state of contextualized knowledge. In accordance with Wong et al. [53] and McColl-Kennedy et al. [54], the semi-systematic approach offers a comprehensive understanding of complex and transversal areas (e.g., environmental sciences, social sciences, and business and management), covering broad topics and different types of studies, contributing, at the same time, to detecting themes, theoretical perspectives, or common threads of research [55].

To perform a clear and replicable strategy, achieving a double purpose of methodological objectivity and efficacy of this review itself [56–58], there are four different steps as follows: (a) defining the background of the analysis and identification of the review criteria (i.e., inclusion and exclusion criteria); (b) selecting suitable research strings and creating a preliminary global database; (c) in-depth screening of selected items and improvement of the opening database; and (d) data synthesis and data analysis through co-word and content analysis. Such an approach is the so-called SALSA (search, appraisal, synthesis, and analysis) framework [59]. Considering the theoretical background of the environmental indicators mentioned, this research paper reviews academic articles based on ecological transition indicators by answering the identified research questions. Ecological transition is investigated under the so-called "environmental approach", which is considered a concept-oriented term that encompasses pollution control, cleaner production, green chemistry, eco-design, life-cycle assessment, waste minimization, zero waste, and social welfare [60].

As discussed by Snyder [52], the research questions have been formulated to investigate broader topics, which have been conceptualized differently and analyzed within transversal disciplines. Such research queries aim at mapping the theoretical and practical approaches related to the food service and lodging accommodation industries, exploring the collective evidence in diverse research areas. To pursue reliability and dependability, the present research explores academic peer-reviewed journals and conference proceedings written in English, whereas it does not include book chapters. Regarding the publication timeline, articles published between 2015 and 2021 have been considered from the SDG and the circular economy action plan developed and implemented starting from 2015, whereas, under the geographical perspective, either global or European experiences have been taken into account. Figure 1 illustrates the research strategy according to the PRISMA (Preferred Reporting Items for Systematic Reviews and Meta-Analyses) model.

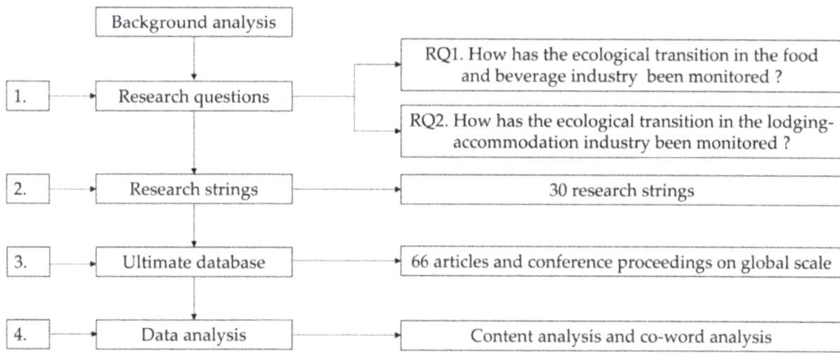

Figure 1. Research strategy overview. Source: Personal elaboration by the authors.

3.2. Data Collection and Research Strings

Data collection began with a TITLE-ABS-KEY query and was conducted using Web of Science (WoS), which represents a collector of standardized, reputable, and high-quality research. Web of Science is one of the world's leading databases and is increasingly used for academic papers [61], being the leading scientific citation search and analytical information platform [62]. The authors have considered all the indexes included in the WOS database and used them for export selection using the integrated function on the platform, integrated with the VOS viewer software, from 2015 to 2021. The research considers a timespan from 2015–2021 since the SDGs [6], as well as the circular economy action plan [2], were implemented in 2015 by the United Nations and the European Union, respectively. Although the main purpose of the research is to focus on monitoring the overall ecological transition of food services and lodging accommodation activities as a starting point for future research, the authors have tried to identify the possible strategies that could be adopted soon after the COVID-19 pandemic, highlighting the implications of these in Section 5.4.

Considering the research questions, designed to address and investigate food services and lodging accommodation activities as well as environmental indicators (also defined as circular indicators or sustainability indicators), the present research has identified several keywords to cover as many relevant aspects as possible.

Regarding RQ1, the subsequent (truncated) keyword combinations have been selected: "indicator" or "index" or "monitoring" AND "environment" AND "restaurant", or "fast food" or "takeaway" or "catering" or "pub", for an amount of 15 research strings. Regarding RQ2 the subsequent combinations were investigated: "indicator" or "index" or "monitoring" AND "environment" AND "hotel" or "motel", "resort" or "bed and breakfast" or "spa", for an amount of 15 research strings. Overall, 30 research strings were investigated within titles, abstracts, and keywords. Figure 2 illustrates the research string strategy.

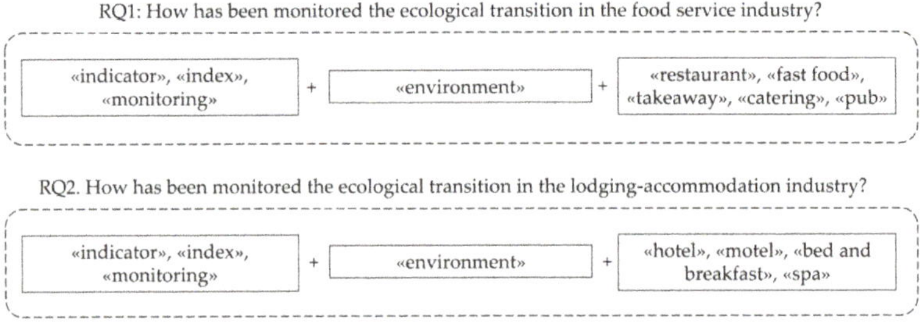

Figure 2. Research Strings. Source: Personal Elaboration by the Authors.

Research strings helped to create a preliminary database, which included the article authors' name, title, publishing journal, year of publication, geographical area, paper type, and DOI (digital object identifier). Then, a further in-depth selection of the most relevant studies was conducted to avoid duplication. Collected data were catalogued in Microsoft Excel sheets.

3.3. Data Synthesis and Data Analysis

The collected data were synthesized according to a co-word analysis [63]. Such an approach investigates the co-occurrences of keywords and also identifies the relationship and interactions between different research topics [64], highlighting research trends, as well as the most influential authors in the field of environmental transition in tourism. It provides a systematic overview of research evolution, creating suggestive maps of keywords and textual data [65] and represents a rigorous method for investigating huge amounts of data [66,67]. Furthermore, using the VOSviewer software, clusters of keywords used in the metadata of the different articles were identified. VOSviewer is an open-source software and was successfully applied to co-word analysis [68,69] to investigate the links between research topics and to identify the most relevant publishing journals. It was developed by Leiden University.

In addition, to integrate the meta-analysis and better evaluate and interpret the results, a content analysis was carried out. This helped to identify "not-so-obvious" perspectives, creating potentially groundbreaking perspectives for either academics or practitioners, playing a critical role in creating new theories [37]. Furthermore, it has previously been successfully applied in tourism and hospitality research [70,71], guaranteeing its suitability for pursuing the aims and scope of the present research. Considering that environmental performance indicators are essential for comparing hotel and restaurant performances over time and communicating transparent and comparable results, selected indicators from the content analysis were synthesized into a dashboard, which distinguished between the scopes of the indicators, as follows: (a) air scope, including the indicators which monitor the activities affecting climate change and causing risks to human health; (b) water scope, including indicators which evaluate human impact on freshwater systems, either from single products or entire processes; (c) energy scope, including indicators which assess the use of energy, energy performance, and resource consumption rates; (d) waste scope, including indicators which measure waste generation from a economic, environmental, or social perspective; (e) health scope, including indicators which monitor the impact of unhealthy or unsafe foods on human health, as well as suitable human conditions on an employee's well-being; and (f) economic scope, including indicators which monitor economic performance and service quality [47,48,50,58].

4. Results

4.1. Publishing Journals, Geographical Areas, and Research Timeline

To better contextualize the selected papers ($n = 56$) and conference proceedings ($n = 10$), Table 1 includes the publishing journals, geographical areas, and timelines investigated and also highlights those journals and countries that account for two or more contributions. The limited number of the selected papers represents this topical yet still under-researched (in the literature) niche subject. Considering the high heterogeneity of the hospitality sector, the present research took into sole consideration the food service and lodging accommodation industries to obtain as accurate results as possible. Although the dataset is composed of 66 papers, such results are in line with other scientific research papers in the tourism and hospitality field [72,73].

It has emerged that the vast majority of the selected contributions were published on Sustainability (9 articles) and in both the Journal of Hospitality Management (4 articles) and the International Journal of Environmental Research and Publish Health (2 articles). Indeed, among others, such journals aim at publishing research which provides solutions toward tackling climate change, pursuing sustainable development, and guaranteeing either economic growth or environmental protection. The selection by geographical area shows the majority of authors have explored environmental and social indicators in the United States (14 articles), followed by China (10 articles) and then Brazil (13 articles), whereas in the European Union, only several manuscripts have investigated Poland, The Czech Republic, Italy, Portugal, and Romania. In terms of the timeline of publication, the highest number of contributions were published in 2019 (15 articles).

To comprehend the main research trends, we searched for the most cited articles in the WoS database. The most cited contribution (38 citations) identified possible opportunities to enhance restaurants' competitiveness through online reviews, based on the competitive index and dissimilarity index [73]. The second most cited contribution (37 citations) dates back to 2015 and investigates fast-food consumption on healthy diets and obesity reduction through the body mass index z-score [74]. The same number of citations was achieved by Charlebois et al. [75], in which food service procurement, kitchen practices, cost management, menu design, and technical literacy in the field of food waste minimization through "performance indicators" was investigated. Therefore, using these preliminary insights, a transversal and holistic concept of "ecological transition" has emerged, which encompasses managerial strategies, human health, food safety, and food security toward the wider aim of sustainable development.

Table 1. Contributions per publishing journal. (a) Geographical areas and (b) timeline (c).

a. Publishing Journal	**N.**
Sustainability	9
Journal of Hospitality Management	4
International Journal of Environmental Research and Public Health	3
Anatolia	2
International Journal of Contemporary Hospitality Management	2
International Journal of Culture Tourism and Hospitality Research	2
Journal of Hospitality and Tourism Management	2
b. Geographical Areas	**N.**
United States	14
Brazil	4
Poland	3
Russia	3
Australia	2
Canada	2
Czech Republic	2
Italy	2
Portugal	2
Romania	2
Taiwan	2
Turkey	2
c. Timeline	**N.**
2015	6
2016	3
2017	9
2018	10
2019	15
2020	13
2021	10

Source: Personal elaboration by the authors.

4.2. *VOSviewer Analysis Results*

According to the VOSviewer analysis, which investigated the co-occurrences of keywords and interactions between research topics within article titles and abstracts, a total of five clusters emerged and are depicted in Figure 3. In the first cluster, defined as the "food cluster" and identified in red in Figure 4, 17 keywords were found for a total of 568 links and 84 co-occurrences. The most relevant keywords are represented by "fast food", "body mass index", "obesity", "home", and "restaurant", whereas specific countries have emerged, such as "Canada" and "U.S.A.". In the second cluster, defined as the "water cluster" and represented in green in Figure 4, 14 keywords were found for a total of 73 co-occurrences. Such a cluster highlights research topics related to "climate", "recreational water", "pools", and "water use", focusing on water consumption in health resorts. In the third cluster, the so-called "consumers cluster", illustrated in blue in Figure 4, 12 keywords were found for 51 co-occurrences. This reveals connections and links between words such as "consumer", "government", "market", and "hospitality". The fourth cluster, defined as the "corporate cluster" and represented in yellow in Figure 4, is composed of 11 keywords and reveals 48 co-occurrences. Among other factors, this cluster focuses on keywords like "companies", "hotels", "tasks", "work", and "work environment", revealing a novel attitude toward ecological transition in the field of corporate social responsibility, customer satisfaction, and employee wellbeing. In the last cluster, which we have called the "energy cluster", represented in violet in Figure 3, 9 keywords were found for 38 co-occurrences. Its keywords are related to "energy", "investment", "natural environment", and "sustainable development", highlighting a focus on energy transition towards sustainability. Furthermore, among its keywords, it presented the country "China". Table A1 illustrates the five clusters and their items.

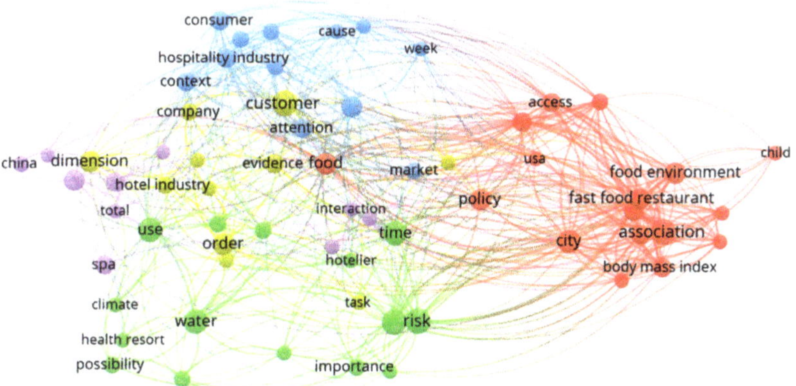

Figure 3. VOSviewer results based on text data. Source: personal elaboration by the authors.

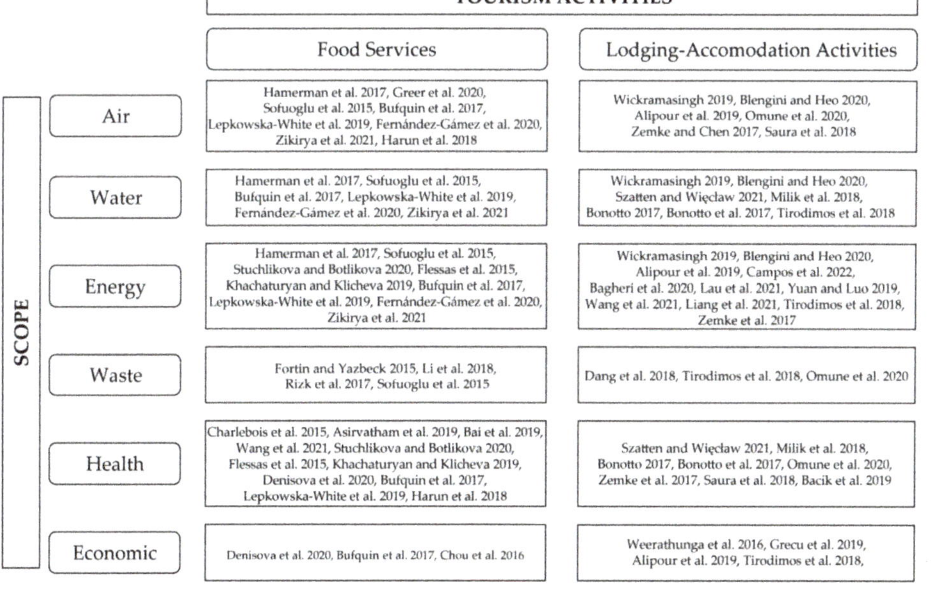

Figure 4. Multidimensional monitoring dashboard by selected indicators [75–115]. Source: personal elaboration by the authors.

5. Discussion

5.1. Monitoring the Ecological Transition of Food Services (RQ1)

Regarding RQ1, a total of 11 contributions were each been identified within the "food cluster" and the "corporate cluster", with only 9 contributions for the "customer cluster".

In the "food cluster", several authors have investigated the sustainability of fast foods through the analysis of childhood body mass index [76] or adolescent weight gain [77]. Wang et al. [78] have adopted the healthy eating index to measure diet quality and to estimate how specific food commodities align with the key recommendations of the dietary guidelines for Americans for healthy and nutritional diets. In addition, Li et al. [79] have explored the healthfulness of restaurants through the adoption of the Nutrition Environment Measures Survey-Restaurant, which represents an agent-based model, examining family dining patterns, individual and community sociodemographic characteristics, restaurant location, size, and food healthfulness. These authors focused on studying the food habits of school and university students, highlighting a correlation between fast-food influence on a child's body mass index and the geographical distance from school [76]. Food waste has been largely explored by different authors as a second pillar of the "food cluster". Considering that food waste is one of the most topical challenges from a social, environmental, and economic perspective [28], Hamerman et al. [80] have identified the service of offering to wrap leftovers as a positive indicator of restaurants' quality and environmental sustainability, considering it useful for reducing food waste and increasing future home consumption. In addition, Charlebois et al. [75] have highlighted the amount of food waste as an index of food service quality and economic performance. In the field of food waste disposal, Rizk et al. [81] have explored the environmental and public health consequences of organic waste treatment, monitoring the physical-chemical parameters of the composting process of food waste toward sustainable organic composting production. Lastly, as suggested by Greer et al. [82], successful indicators (i.e., key performance indicators) for measuring circular food services could be identified in the reduction and reporting of food waste. The third pillar of the "food cluster" was identified as indoor air quality during cooking and frying activities [83]. Among others, it seems essential to measure the CO_2 pollution, and volatile organic compound concentrations generated during deep-frying, which could cause negative health effects for chefs and consumers.

As regards the "corporate cluster", Stuchlikova and Botlikova [84] have identified transportation, climate change, waste management, water management and wastewater management as the most used indicators for calculating environmental impact analysis. However, as highlighted by several authors [85–87], additional financial and non-financial measures, as well as layout performance indicators, should be considered for food service sustainability. As proposed by Flessas et al. [85], two main indicators should be included: productivity and work-in-process. Productivity value shows the degree of human and natural resources used, whereas work-in-progress represents the proportion in which the inventory has been stored throughout the production process due to inefficiencies in the system. Furthermore, human-centered indicators for ecological transition processes include the number of work-related accidents and customer satisfaction. Similar results have been depicted by Bufquin et al. [88]. Regarding the improvement of food service warmth, competence, and competitiveness in global markets, the authors investigated customer satisfaction, changes in restaurant sales, and employee turnover rates, highlighting the interconnection between social responsibility and human development. Further, as suggested by Chou et al. [89], the ecological transition of food services should be explored in the light of five major indicators: sustainable service innovation, food service technology, organizational learning, adoption of innovation, and organizational environment. Among the social media marketing indicators, which show the success of small and medium enterprises [90], the authors highlighted the importance of evaluating performance indicators, such as likes, shares, and followers, as well as the restaurant's social media index.

Lastly, the "customer cluster" encompasses variables such as environmental and health perceptions [91], vitality [92], and corporate social responsibility [93] toward food services' economic, social, and environmental transitions. Considering that consumer review websites and consumer perception of sustainable practices have increased in importance in recent years, Fernandez-Gamez et al. [91] investigated the relationship between health and environmental conditions and a restaurant's corporate reputation at the country level through the use of the healthiest country index. Furthermore, regarding the measurement of the quality of urban life and the environment, Zikirya et al. [92] estimated the vitality of urban takeaways through building footprints, whereas Dang et al. [94] explored customers' perceptions, attitudes, and practices towards experienced secondhand smoke in restaurants, stressing the need to enhance the monitoring system of smokers. In the field of fast-food competitiveness, Harun et al. [93] have estimated that young generations are greatly involved in fast-food restaurants' responsible behavior and their concern for the environment. Therefore, economic development and purchasing behavior strategies should be grounded in environmental indicator measurement.

5.2. Monitoring the Ecological Transition of Lodging accommodation Activities (RQ2)

Regarding RQ2, all five clusters (i.e., food, water, customer, corporate, and energy) emerged as important. The largest number of contributions was identified in the "corporate cluster" (11 contributions), followed by the "energy cluster" (10 contributions), the "water cluster" (8 contributions), the "customer cluster" (4 contributions) and the "food cluster" (2 contributions). Such a variety of topics is possible because lodging accommodation activities are highly energy-intensive and cover multi-usage functions, from food and beverage consumption to recreation, resort, and hygiene procedures [35].

The "corporate cluster" encompasses the analysis of hotels' environmental orientation through composite indicators [95], as well as the investigation of their sustainability performance [96] or economic and financial performance through macroeconomic indicators [97,98]. Some authors address employees' environmental behavior and perception of their hotels' sustainability practices [99], whereas others link environmental and economic performance to a holistic perspective [100]. From an environmental perspective, several indicators were depicted: environmental records, environment policy, environment management systems, involvement in relevant projects/programs, and receipt of environmental awards and certifications. Furthermore, some authors [99] have applied the global sustainable tourism council hotel criteria indicators, as well as the indicators of sustainable development for tourism destinations and the European Union's sustainability framework for nearly zero-energy hotels. An additional contribution, as suggested by Weerathunga et al. [96], regards the development of a sustainable performance evaluation index, which encompasses either economic, social, government, or environmental sustainability. Regarding the environmental aspect, the authors assessed and evaluated the methods and processes adopted to reuse and recycle effluents and waste, as well as the awards and the certifications in recognition of environmentally friendly operations, as already proposed by Wickramasingh [95].

In the "energy cluster", Bagheri et al. [101] estimated that the hotel's energy dimension represents the most impactful dimension among all others, with the most significant indicators related to the supply and efficient use of energy, as well as the use of renewable energies. Further, Lau et al. [102] assessed chiller-power usage, highlighting the need to enhance energy control in deluxe waterfront hotels. Among the possible solutions to reduce the energy consumption of existing hotel buildings, some authors [103,104] discussed the importance of energy-saving retrofitting, including the application of photovoltaic panels and energy-efficient LED lights [105]. The indicators dedicated to estimating energy consumption include, for example, self-sufficiency ratio, energy production diversification, per-capita energy production, energy intensity, energy consumption, electricity consumption, and energy industry investment. The use of clean energy and solar power still represents a challenge for measuring the environmental sustainability of hotels [106].

In terms of the "water cluster", this dimension has been investigated in the fields of water quality, water savings, and water security. Milik et al. [107] investigated the physical and chemical parameters of fountains in health resorts, highlighting a need to introduce supervision over the quality of drinking water. Similar research was conducted by Bonotto [108], measuring alpha-emitting radionuclides, temperature, pH, Eh, electrical conductivity, dissolved gases, and major constituents in water, whereas other authors have investigated bacteria such as Legionella [109] or fecal water contamination [110]. As to water savings, Omune et al. [111] explored water conservation practices, paying attention to taps that are open unnecessarily.

The "customer cluster" includes research studies which explore the relationship between online comments and reviews and hotels' environmental management [112,113]. Among others, tourists evaluate garbage or fumes that harm and pollute nature, solar panels, self-sufficiency zones, sustainable maintenance policies, noise pollution, traditional foods, and sustainable energy. Furthermore, tourists seem to have a positive sentiment towards hotel ecosystems, paying attention to the surrounding environment, the abundance of nature and plants, and the pure air in the facilities. Such indicators could enhance either environmental or economic performance [114].

The "food cluster" includes research studies related to a hotel's readiness to offer local cuisines [115] and food-waste management practice [111]. Among other factors, it seems that serving local food could reduce food production costs, increase food profit margin, improve the attractiveness of menus, and make menu prices cheaper.

Overall, the five clusters highlight an evident link between specific subjects in the literature, with food, water, customer, corporate, and energy clusters able to monitor the transition to sustainable food services and lodging accommodation activities. Although the number of the selected articles is limited to 66 scientific contributions, it emerged that these studies are homogeneously focused on a few (but relevant) topics, which is in line with other research in the tourism and hospitality field concerning food services and lodging accommodation activities [116].

5.3. Theoretical and Managerial Implications

Figure 4 summarizes the selected indicators and provides a multidimensional measurement dashboard of empirical and theoretical tools useful for boosting decision-making processes and strategies in the hospitality industry. In the light of Eurostat [15], which provides a list of indicators used to monitor the progress towards the circular economy and ecological transition, Figure 4 distinguishes between air, water, energy, waste, health, and economic scopes [47,48,50,58]. Table A2 provides details related to the multidimensional monitoring dashboard.

Regarding food services, several articles have investigated the scope of air and energy, as well as health, highlighting the nexus between food safety, service quality, and economic performance. However, waste (and food waste) still requires further research, considering the low number of contributions exploring this issue. The same trends are detected in lodging accommodation-industry research, which appears marginally interested in waste management. Although restaurants and hotels have the potential to quantify food waste and packaging waste, which are mainly produced in the final consumption stage and especially outdoors, several authors have not addressed their efforts towards this issue, instead focusing on air emissions related to energy consumption or water pollution, yet neglecting all the hidden costs associated with waste generation, from cradle to grave. This means that the circular economy approach, which highlights the need to reuse or recycle waste and avoid unnecessary withdrawals of virgin raw materials from nature, has not yet been fully implemented. Several strategies can be adopted to convert waste (food and non-food) into secondary raw materials, useful both in terms of closed-loop-recycling and open-loop-recycling. As a consequence, tourism operators must be encouraged by authorities to measure waste since alternative pathways could be walked, such as (a) waste reduction through awareness-raising practices and improvements in customer behavior and (b) energy recovery or compost practices. Both alternatives would help improve environmental performance and reduce the burden on the environment.

Our research confirms a lack of harmonized indicators for measuring ecological transition, as well as a lack of data. In light of the uncommon measurement and monitoring tools used in the hospitality industry, public authorities should suggest practices, together with a set of indicators to monitor the environmental transition, with also a set of homogeneous and standardized measurement methodologies, which could boost comparability and replicability among food services and lodging accommodation activities. Although the European Union, through the Commission Delegated Decision (EU) 2019/1597 on common methodologies and minimum quality requirements for the homogeneous assessment of food waste quantities and composition [117], has suggested suitable tools to measure food waste, hotels and restaurants have not yet considered the importance of measurement and monitoring practices. Whether public authorities should make measurement mandatory rather than leave it voluntary practice remains a fundamental question. Nevertheless, the need for the development of a life-cycle inventory database, as well as the need for measurement programming, replicability, and comparability, remains a fixed point.

5.4. The Ecological Transition after the COVID-19 Pandemic

It has been proved that circular economy principles are sensitive to unpredictable events, such as the COVID-19 pandemic. Since the eruption of COVID-19, pollution and ecological burdens have decreased by 30% [17], which means that several economic activities, such as tourism, must switch from linear to circular practices through suitable investment opportunities. Although no selected articles in the present review deal with the COVID-19 issue, meaning that no authors have monitored ecological transition during the pandemic, the authors have identified some possible strategies to rebuild tourism soon after the pandemic. Hotels and restaurants have adopted new strategies in terms of hygiene protocols, and several units have proceeded to renovate their businesses during the stoppage of their activities due to the pandemic, increasing their attractiveness to customers. The COVID-19 pandemic has offered the food service and lodging accommodation industries an opportunity to more authentically practice corporate social responsibility, focusing their awareness on environmental and social challenges [118,119].

In light of the sudden business revival soon after the pandemic, the adoption of integrated management systems, which take care of environmental, social, health and economic issues, has emerged. Integration based on industrial symbiosis and circular economy paradigms could represent an essential step to switch from linear to circular activities, boosting the interactions between suppliers, companies, and customers. Besides, considering the nexus between economic growth and environmental protection, activities in food service and lodging accommodation could benefit from the adoption of environmental strategies since this could lead to cost reductions while increasing efficiency and clients′ trust. Monitoring ecological transition requires quantitative and qualitative data; on one hand, indicators are essential to gain as much knowledge as possible and compare data over time, and on the other hand, companies should engage in sustainable practices to enhance consumers′ loyalty, stakeholders′ interest to invest, and local-supplier networking, considering such variables as key factors to boost competitiveness (and sustainability) in the hospitality sector.

6. Conclusions

The ecological transition in the hospitality sector remains one of the main challenges towards sustainable development. The traditional way to do business, which involves a lack of sustainable practices, is beginning to be no longer acceptable by the decision makers and public authorities. As a consequence, representatives from the field should reconsider their "way of doing business". The present research, through a semi-systematic literature review of 66 peer-reviewed articles and conference proceedings included in Web of Science between 2015 and 2021, brings new details about the monitoring activities of the transition to sustainable food services and lodging accommodation activities. Overall, the current review has identified five intervention clusters, namely (a) "food cluster", (b) "water cluster", (c) "consumers cluster", (d) "corporate cluster", and (e) "energy cluster". Each cluster refers to different areas of intervention and specific indicators, either of sustainability or circularity, in order to identify strategies for pursuing the SDGs and ecological transition. One of the conclusions of the study is that the natural resources that are most impacted within the tourism sector are food, water, and energy, and the main stakeholders to be actively monitored are the users (or consumers) and the managers. Besides, companies and consumers should develop a positive attitude toward the implementation of economic and environmentally friendly-blended performance, whereas researchers should investigate each stage of the supply chain, from raw material supply to consumer behavior. Although the selected papers represent a small sample for inference, such scientific research still provides useful quantitative data for understanding the main sustainability trends in the food service and lodging accommodation industries.

As regards the limitations of the present research, one refers to the small number of peer-reviewed articles and proceedings that were investigated, explained by restricting the approach of the research to the food service industry and the lodging accommodation industries. Another limitation was the language criteria for inclusion. The present research contains only articles and proceedings published in English that are included in the WOS database while excluding books. To enlarge the research base, the authors intend to expand this research to other databases, like Scopus and Elsevier and also to include books and book chapters. Lastly, our research has a timespan between 2015–2021 and does not make a distinction between articles published before or during the COVID-19 pandemic, owing to the main purpose of the paper as being focused on monitoring the overall ecological transition of food services and lodging accommodation activities, setting a starting point for future research.

Future research directions will broaden the time span of the current study, as well as the research strings selected for screening, providing a comparative analysis among geographical areas and distinguishing between before, during, and after the COVID-19 pandemic. If it is true that an ecological transition ensures resilience soon after economic crises and natural disasters, a future extension of the present research based on the use of either quantitative or quantitative indicators, will confirm this assumption.

Author Contributions: This paper should be considered the result of the joint efforts of the authors, as follows: Conceptualization, C.B. and A.C.A.; methodology, C.B. and A.C.A.; software, C.B. and A.C.A.; investigation, C.B. and A.C.A.; resources, C.B. and A.C.A.; data curation, C.B. and A.C.A.; writing—original draft preparation, C.B. and A.C.A.; writing—review and editing, C.B., A.C.A. and S.M.-P.; visualization, S.M.-P.; supervision, S.M.-P. All authors have read and agreed to the published version of the manuscript.

Funding: This research received no external funding.

Institutional Review Board Statement: Not applicable.

Informed Consent Statement: Not applicable.

Data Availability Statement: Not applicable.

Conflicts of Interest: The authors declare no conflict of interest.

Appendix A

Table A1. VOSviewer clusters, items, links and occurrences.

"Food Cluster" (17 Items)			
Items	**Links**	**Total Link Strength**	**Occurences**
Access	32	43	5
Association	26	58	9
Body Mass Index	20	36	5
Canada	29	37	4
Child	8	11	3
City	26	36	7
Cycle	17	22	3
Fast Food	17	26	3
Fast Food Restaurant	26	41	6
Food	30	37	6
Food Environment	23	39	6
Home	31	44	5
Obesity	23	38	5
Policy	22	24	6
Population	24	40	5
Resident	16	19	3
U.S.	16	17	3
"Water Cluster" (15 Items)			
Items	**Links**	**Total Link Strength**	**Occurences**
Climate	10	11	3
Health Resort	9	10	3
Hotelier	9	10	3
Implementation	17	17	4
Pool	14	18	3
Possibility	10	13	4
Recreational Water	12	17	3
Requirement	17	18	4
Risk	30	46	9
Time	29	37	8
Type	34	44	9
Use	28	34	8
Water	24	35	8
Water sample	14	20	4
"Customer cluster" (12 Items)			
Items	**Links**	**Total Link Strength**	**Occurences**
Attention	27	33	5
Cause	19	23	3
Consumer	21	30	5
Context	26	38	5
Food waste	21	26	4
Government	26	32	4
Hospitality Industry	25	35	5
Interest	20	24	4
Market	15	15	4
Patron	19	24	3
Restaurant Industry	24	30	6
Week	25	27	3

Table A1. *Cont.*

"Corporate Cluster" (11 Items)			
Items	**Links**	**Total Link Strength**	**Occurences**
Company	14	16	3
Culture	11	11	3
Customer	30	44	9
Customer Satisfaction	8	10	3
Dimension	22	28	6
Evidence	17	18	5
Hotel Industry	14	18	4
Order	15	19	6
Task	7	7	3
Work	17	18	3
Work Environment	10	11	3
"Energy Cluster" (10 Items)			
Items	**Links**	**Total Link Strength**	**Occurences**
China	7	8	4
Energy	16	19	6
Interaction	19	20	3
Investment	12	12	3
Lack	11	11	3
Natural Environment	10	11	3
Spa	14	16	5
Star Hotel	11	12	3
Sustainable Development	12	13	5
Total	13	14	3

Appendix B

Table A2. Multidimensional monitoring dashboard using selected indicators.

RQ	Cluster	Selected Indicator	Scope	Reference
1	Food	Body Mass Index	Health	[76]
1	Food	Weight Gain Rate	Health	[77]
1	Food	Healthy Eating Index	Health	[78]
1	Food	Nutrition Environment Measures	Health	[79]
1	Food	Leftovers Rate	Waste	[80]
1	Food	Food Waste Rate	Waste	[75]
1	Food	Organic Waste Treatment	Air, Water, Energy	[81]
1	Food	Food Waste Reduction	Waste	[82]
1	Food	Indoor Air Quality	Air	[83]
1	Corporate	Environmental Impacts Indicators	Air, Water, Energy, Waste	[84]
1	Corporate	Productivity, Human Centered Indicators	Energy, Health	[85–87]
1	Corporate	Employees Turnover Rates	Health, Economic	[88]
1	Corporate	Sustainable Service Innovation	Air, Water, Energy	[89]
1	Corporate	Restaurant Social Media Index	Economic	[90]
1	Customer	Healthiest Country Index	Air, Water, Energy, Health	[91]
1	Customer	Urban Vitality/Building Footprints	Air, Water, Energy	[92]
1	Customer	Secondhand Smoke Rate	Air, Health	[94]
1	Customer	Fast Food Responsible Behavior	Air, Water, Energy	[93]

Table A2. *Cont.*

RQ	Cluster	Selected Indicator	Scope	Reference
2	Corporate	Waste Reuse/Recycling Rates	Waste	[95]
2	Corporate	Sustainability Performance Indicator	Air, Water, Energy	[96]
2	Corporate	Economic Performance	Economic	[97]
2	Corporate	Financial Performance	Economic	[98]
2	Corporate	GST Council Hotel Criteria Indicators	Air, Water, Energy	[99]
2	Corporate	Environmental/Economic Performances	Air, Energy, Economic	[100]
2	Energy	Energy Use/Renewable Energy Use	Energy	[101]
2	Energy	Chiller Power Use Rate	Energy	[102]
2	Energy	Building Energy Savings Retrofitting	Energy	[103–105]
2	Energy	Clean Energy/Solar Power Ratio	Energy	[106]
2	Water	Water Physical and Chemical Parameters	Water, Health	[107]
2	Water	Water Physical and Chemical Parameters	Water, Health	[108]
2	Water	Water Physical and Chemical Parameters	Water, Health	[109]
2	Water	Water Physical and Chemical Parameters	Water, Health	[110]
2	Water	Water Savings Rate	Water, Energy, Economic	[111]
2	Customer	Garbage Rate, Fumes Rate	Air, Waste, Health	[112]
2	Customer	Sustainable Maintenance, Energy Rate	Air, Energy, Health	[113]
2	Customer	Hotel Ecosystems Rate	Air, Health	[114]
2	Food	Local Cuisine Rate	Health	[115]
2	Food	Food Waste Rate	Waste	[111]

References

1. OECD. Building Back Better: A Sustainable, Resilient Recovery after COVID-19. Building Back Better, OECD. 2020. Available online: https://www.oecd.org/coronavirus/en/ (accessed on 30 August 2021).
2. European Commission. Circular Economy Action Plan. Available online: https://ec.europa.eu/environment/strategy/circular-economy-action-plan_it (accessed on 30 August 2021).
3. Friant, M.C.; Vermeulen, W.J.V.; Salomone, R. Analysing European Union circular economy policies: Words versus actions. *Sustain. Prod. Consum.* **2021**, *27*, 337–353. [CrossRef]
4. Jaroslaw, G.; Perzanowski, M.; Drejerska, N.; Fiore, M. From a concept to implementation of food chain within the Circular Economy paradigm: The case of Poland. *Riv. Studi Sulla Sostenibilità* **2018**, *8*, 71–86.
5. Fetting. The European Green Deal. ESDN Report, December 2020, ESDN Office, Vienna. Available online: https://www.esdn.eu/fileadmin/ESDN_Reports/ESDN_Report_2_2020.pdf (accessed on 30 August 2021).
6. United Nations. The 17 Goals. Available online: https://sdgs.un.org/goals (accessed on 30 August 2021).
7. Jaeger-Erben, M.; Jensen, C.; Hofmann, F.; Zwiers, J. There is no sustainable circular economy without a circular society. *Resour. Conserv. Recycl.* **2021**, *168*, 105476. [CrossRef]
8. Patwa, N.; Sivarajah, U.; Seetharaman, A.; Sarkar, S.; Maiti, K.; Hingorani, K. Towards a circular economy: An emerging economies context. *J. Bus. Res.* **2021**, *122*, 725–735. [CrossRef]
9. Zink, T.; Geyer, R. Circular Economy Rebound. *J. Ind. Ecol.* **2017**, *21*, 593–602. [CrossRef]
10. Grafstrom, J.; Aasma, S. Breaking circular economy barriers. *J. Clean. Prod.* **2021**, *292*, 126002. [CrossRef]
11. Corvellec, H.; Stowell, A.F.; Johansson, N. Critiques of the circular economy. *J. Ind. Ecol.* **2021**, *26*, 421–432. [CrossRef]
12. Eurostat. Tourism Industries—Economic Analysis. Statistics Explained. Available online: https://ec.europa.eu/eurostat/statistics-explained/index.php?title=Tourism_industries_-_economic_analysis (accessed on 26 April 2022).
13. Sauermann, H.; Vohland, K.; Antoniou, V.; Balázs, B.; Göbel, C.; Karatzas, K.; Mooney, P.; Perelló, J.; Ponti, M.; Samson, R.; et al. Citizen science and sustainability transitions. *Res. Policy* **2020**, *49*, 103978. [CrossRef]
14. Moraga, G.; Huysveld, S.; Mathieux, F.; Blengini, G.A.; Alaerts, L.; Van Acker, K.; de Meester, S.; Dewulf, J. Circular economy indicators: What do they measure? *Resour. Conserv. Recycl.* **2019**, *146*, 452–461. [CrossRef] [PubMed]
15. Eurostat. Which Indicators Are Used to Monitor the Progress Towards a Circular Economy? Available online: https://ec.europa.eu/eurostat/web/circular-economy/indicators (accessed on 13 October 2021).
16. Djekic, I.; Operta, S.; Djulancic, N.; Lorenzo, J.M.; Barba, F.J.; Djordjevic, V.; Tomasevic, I. Quantities, environmental footprints and beliefs associated with household food waste in Bosnia and Herzegovina. *Waste Manag. Res.* **2019**, *37*, 1250–1260. [CrossRef] [PubMed]
17. Casado-Aranda, L.A.; Sánchez-Fernández, J.; Bastidas-Manzano, A.B. Tourism research after the COVID-19 outbreak: Insights for more sustainable, local and smart cities. *Sustain. Cities Soc.* **2021**, *73*, 103126. [CrossRef]
18. Malefors, C.; Callewaert, P.; Hansson, P.A.; Hartikainen, H.; Pietiläinen, O.; Strid, I.; Strotmann, C.; Eriksson, M. Towards a baseline for Food-Waste quantification in the hospitality Sector-Quantities and data processing criteria. *Sustainability* **2019**, *11*, 3541. [CrossRef]

19. Styles, D.; Schonberger, H.; Galvez Martos, J.L. *Best Environmental Management Practice in the Tourism Sector;* European Commission—Joint Research Centre: Seville, UK; Publications Office of the European Union: Luxembourg, 2013.
20. Čokorilo, O. Environmental Issues for Aircraft Operations at Airports. *Transp. Res. Procedia* **2016**, *14*, 3713–3720. [CrossRef]
21. Givoni, M. Environmental Benefits from Mode Substitution: Comparison of the Environmental Impact from Aircraft and High-Speed Train Operations. *Int. J. Sustain. Transp.* **2007**, *1*, 209–230. [CrossRef]
22. Del Pero, F.; Delogu, M.; Pierini, M.; Bonaffini, D. Life Cycle Assessment of a heavy metro train. *J. Clean. Prod.* **2015**, *87*, 787–799. [CrossRef]
23. Johansen, B.H.; Aspen, D.M.; Sparrevik, M.; Æsøy, V. Applying System-Oriented Sustainability Scoring for Cruise Traffic Port Operators: A Case Study of Geiranger, Norway. *Sustainability* **2021**, *13*, 6046. [CrossRef]
24. Nagaj, R.; Žuromskaitė, B. Tourism in the Era of Covid-19 and Its Impact on the Environment. *Energies* **2021**, *14*, 2000. [CrossRef]
25. Filimonau, V.; Sulyok, J. 'Bin it and forget it!': The challenges of food waste management in restaurants of a mid-sized Hungarian city. *Tour. Manag. Perspect.* **2021**, *37*, 100759. [CrossRef]
26. Gossling, S.; Peeters, P.; Ceron, J.P.; Dubois, G.; Patterson, T.; Richardson, R.B. The eco-efficiency of tourism. *Ecol. Econ.* **2005**, *54*, 417–434. [CrossRef]
27. Calabro, G.; Vieri, S. The environmental certification of tourism: A tool to enhance the unicity of a territory. *Qual. Access Success* **2014**, *15*, 44–54.
28. Vermeir, I.; Weijters, B.; De Houwer, J.; Geuens, M.; Slabbinck, H.; Spruyt, A.; Van Kerckhove, A.; Van Lippevelde, W.; De Steur, H.; Verbeke, W. Environmentally Sustainable Food Consumption: A Review and Research Agenda from a Goal-Directed Perspective. *Front. Psychol.* **2020**, *10*, 1603. [CrossRef] [PubMed]
29. Li, K.X.; Jin, M.; Shi, W. Tourism as an important impetus to promoting economic growth: A critical review. *Tour. Manag. Perspect.* **2018**, *26*, 135–142. [CrossRef]
30. UNWTO. International Tourism Growth Continues to Outpace the Global Economy. Available online: https://www.unwto.org/international-tourism-growth-continues-to-outpace-the-economy (accessed on 13 October 2021).
31. Rodgers, S. Food service research: An integrated approach. *Int. J. Hosp. Manag.* **2011**, *30*, 477–483. [CrossRef]
32. Opolski Medeiros, C.; Salay, E. A Review of Food Service Selection Factors Important to the Consumer. *Food Public Health* **2013**, *3*, 176–190.
33. Statista. Restaurants and Food Services in Europe—Statistics & Facts. Available online: https://www.statista.com/topics/3966/restaurants-and-food-services-in-europe/#topicHeader__wrapper (accessed on 12 May 2022).
34. Aarnio, T.; Hämäläinen, T. Challenges in packaging waste management in the fast-food industry. *Resour. Conserv. Recycl.* **2008**, *52*, 612–621. [CrossRef]
35. Huang, Y.; Song, H.; Huang, G.Q.; Lou, J. A Comparative Study of Tourism Supply Chains with Quantity Competition. *J. Travel Res.* **2012**, *51*, 717–729. [CrossRef]
36. Ritchie, H.; Roser, M.; Rosado, P. CO_2 and Greenhouse Gas Emissions. 2020. Available online: https://ourworldindata.org/emissions-by-sector#citation (accessed on 14 June 2022).
37. Amicarelli, V.; Bux, C.; Lagioia, G.; Gallucci, T. Energy Efficiency Policies in Non-Residential Buildings: The Case of the University of Bari Aldo Moro. *Amfiteatru Econ.* **2019**, *21*, 845–860. [CrossRef]
38. Amicarelli, V.; Aluculusei, A.C.; Lagioia, G.; Pamfilie, R.; Bux, C. How to manage and minimize food waste in the hotel industry? An exploratory research. *Int. J. Cult. Tour. Hosp. Res.* **2021**, *16*, 152–167. [CrossRef]
39. Lee, S.; Oh, H. Effective Communication Strategies for Hotel Guests' Green Behavior. *Cornell Hosp. Q.* **2013**, *55*, 52–63. [CrossRef]
40. Nisa, C.; Varum, C.; Botelho, A. Promoting Sustainable Hotel Guest Behavior: A Systematic Review and Meta-Analysis. *Cornell Hosp. Q.* **2017**, *58*, 354–363. [CrossRef]
41. Hotel Energy Solutions. Analysis on Energy Use by European Hotels: Online Survey and Desk Research. Available online: https://www.e-unwto.org/doi/epdf/10.18111/9789284414970 (accessed on 12 May 2022).
42. Abdulredha, M.; Al Khaddar, R.; Jordan, D.; Kot, P.; Abdulridha, A.; Hashim, K. Estimating solid waste generation by the hospitality industry during major festivals: A quantification model based on multiple regression. *Waste Manag.* **2018**, *77*, 388–400. [CrossRef] [PubMed]
43. Rodriguez-Anton, J.M.; Alonso-Almeida, M.D. The Circular Economy Strategy in Hospitality: A Multicase Approach. *Sustainability* **2019**, *11*, 5665. [CrossRef]
44. da Silva, P.M.; da Silva, L.M.; Echeveste, S.S. Circular Economy in Tourism and Hospitality: Analysis of Scientific Production on the Theme. *Eur. J. Tour. Hosp. Recreat.* **2022**, *11*, 45–53. [CrossRef]
45. Nyide, C.J. Material flow cost accounting as a tool for improved resource efficiency in the hotel sector: A case of emerging market. *Risk Gov. Control Financ. Mark. Inst.* **2016**, *6*, 428–435. [CrossRef]
46. Kitz, R.; Walker, T.; Charlebois, S.; Music, J. Food packaging during the COVID-19 pandemic: Consumer perceptions. *Int. J. Consum. Stud.* **2022**, *46*, 434–448. [CrossRef]
47. Mayer, A.; Haas, W.; Wiedenhofer, D.; Krausmann, F.; Nuss, P.; Blengini, G.A. Measuring Progress towards a Circular Economy: A Monitoring Framework for Economy-wide Material Loop Closing in the EU28. *J. Ind. Ecol.* **2018**, *23*, 62–76. [CrossRef] [PubMed]
48. Jasch, C. Environmental performance evaluation and indicators. *J. Clean. Prod.* **2000**, *8*, 79–88. [CrossRef]
49. Van Den Bergh, J.; Grazi, F. On the Policy Relevance of Ecological Footprints. *Environ. Sci. Technol.* **2010**, *44*, 4843–4844. [CrossRef] [PubMed]

50. Amicarelli, V.; Bux, C.; Lagioia, G. How to measure food loss and waste? A Material Flow Analysis application. *Br. Food J.* **2020**, *123*, 67–85. [CrossRef]
51. Esposito, B.; Sessa, M.R.; Sica, D.; Malandrino, O. Towards Circular Economy in the Agri-Food Sector. A Systematic Literature Review. *Sustainability* **2020**, *12*, 7401. [CrossRef]
52. Lombardi, M.; Laiola, E.; Tricase, C.; Rana, R. Toward urban environmental sustainability: The carbon footprint of Foggia's municipality. *J. Clean. Prod.* **2018**, *186*, 534–543. [CrossRef]
53. Snyder, H. Literature review as a research methodology: An overview and guidelines. *J. Bus. Res.* **2019**, *2014*, 333–339.
54. Wong, G.; Greenhalgh, T.; Westhorp, G.; Buckingham, J.; Pawson, R. RAMESES publication standards: Meta-narrative reviews. *BMC Med.* **2013**, *11*, 20.
55. McColl-Kennedy, J.R.; Snyder, H.; Elg, M.; Witell, L.; Helkkula, A.; Hogan, S.J.; Anderson, L. The changing role of the health care customer: Review, synthesis and research agenda. *J. Serv. Manag.* **2017**, *28*, 2–33. [CrossRef]
56. Ward, V.; House, A.; Hamer, S. Developing a framework for transferring knowledge into action: A thematic analysis of the literature. *J. Health Serv. Res. Policy* **2009**, *14*, 156–164. [CrossRef] [PubMed]
57. D'Eusanio, M.; Zamagni, A.; Petti, L. Social sustainability and supply chain management: Methods and tools. *J. Clean. Prod.* **2019**, *235*, 178–189. [CrossRef]
58. Rana, R.L.; Tricase, C.; De Cesare, L. Blockchain technology for a sustainable agri-food supply chain. *Br. Food J.* **2021**, *123*, 3471–3485. [CrossRef]
59. Poponi, S.; Arcese, G.; Pacchera, F.; Martucci, O. Evaluating the transition to the circular economy in the agri-food sector: Selection of indicators. *Resour. Conserv. Recycl.* **2022**, *176*, 105916. [CrossRef]
60. Gunnarsdottir, I.; Davidsdottir, B.; Worrell, E.; Sigurgeirsdottir, S. Review of indicators for sustainable energy development. *Renew. Sustain. Energy Rev.* **2020**, *133*, 110294. [CrossRef]
61. Glavič, P.; Lukman, R. Review of sustainability terms and their definitions. *J. Clean. Prod.* **2007**, *15*, 1875–1885. [CrossRef]
62. Zhu, J.; Liu, W. A tale of two databases: The use of Web of Science and Scopus in academic papers. *Scientometrics* **2020**, *123*, 321–335. [CrossRef]
63. Li, K.; Rollins, J.; Yan, E. Web of Science use in published research and review papers 1997–2017: A selective, dynamic, cross-domain, content-based analysis. *Scientometrics* **2018**, *115*, 1–20. [CrossRef]
64. Aluculesei, A.C.; Nistoreanu, P.; Avram, D.; Nistoreanu, B.G. Past and Future Trends in Medical Spas: A Co-Word Analysis. *Sustainability* **2021**, *13*, 9646. [CrossRef]
65. Mulet-Forteza, C.; Genovart-Balaguer, J.; Mauleon-Mendeza, E.; Merigó, J.M. A bibliometric research in the tourism, leisure and hospitality fields. *J. Bus. Res.* **2019**, *101*, 819–827. [CrossRef]
66. Sigala, M.; Kumar, S.; Donthu, N.; Sureka, R.; Joshi, Y. A bibliometric overview of the Journal of Hospitality and Tourism Management: Research contributions and influence. *J. Hosp. Tour. Manag.* **2021**, *47*, 273–288. [CrossRef]
67. Niñerola, A.; Sánchez Rebull, V.; Hernández-Lara, A.B. Tourism Research on Sustainability: A Bibliometric Analysis. *Sustainability* **2019**, *11*, 1377. [CrossRef]
68. Vieira, E.S.; Madaleno, M.; Azevedo, G. Research on Corporate Governance: Bibliometric Analysis. In *Comparative Research on Earnings Management, Corporate Governance, and Economic Value*; IGI Global: Hershey, PA, USA, 2021.
69. Guo, Y.M.; Huang, Z.L.; Guo, J.; Li, H.; Guo, X.R.; Nkeli, M.J. Bibliometric Analysis on Smart Cities Research. *Sustainability* **2019**, *11*, 3606. [CrossRef]
70. Yu, Y.; Li, Y.; Zhang, Z.; Gu, Z.; Zhong, H.; Zha, Q.; Yang, L.; Zhu, C.; Chen, E. A bibliometric analysis using VOSviewer of publications on COVID-19. *Ann. Transl. Med.* **2020**, *8*, 816. [CrossRef]
71. Camprubì, R.; Coromina, L. Content analysis in tourism research. *Tour. Manag. Perspect.* **2016**, *18*, 134–140. [CrossRef]
72. Mehraliyev, F.; Cheng Chu Chan, I.; Choi, Y.; Koseoglu, M.A.; Law, R. A state-of-the-art review of smart tourism research. *J. Travel Tour. Mark.* **2020**, *37*, 78–91. [CrossRef]
73. Palácios, H.; Almeida, H.d.; Sousa, M.J. A Bibliometric Analysis of Service Climate as a Sustainable Competitive Advantage in Hospitality. *Sustainability* **2021**, *13*, 12214. [CrossRef]
74. Gao, S.; Tang, O.; Wang, H.; Yin, P. Identifying competitors through comparative relation mining of online reviews in the restaurant industry. *Int. J. Hosp. Manag.* **2018**, *71*, 19–32. [CrossRef]
75. Fortin, B.; Yazbeck, M. Peer effects, fast food consumption and adolescent weight gain. *J. Health Econ.* **2015**, *42*, 125–138. [CrossRef]
76. Charlebois, S.; Creedy, A.; von Massow, M. "Back of house"—focused study on food waste in fine dining: The case of Delish restaurants. *Int. J. Cult. Tour. Hosp. Res.* **2015**, *9*, 278–291. [CrossRef]
77. Asirvatham, J.; Thomsen, M.R.; Nayga, R.M.; Goudie, A. Do fast food restaurants surrounding schools affect childhood obesity? *Econ. Hum. Biol.* **2019**, *33*, 124–133. [CrossRef]
78. Bai, L.; Wang, M.; Yang, Y.; Gong, S. Food safety in restaurants: The consumer perspective. *Int. J. Cult. Tour. Hosp. Res.* **2019**, *77*, 139–146. [CrossRef]
79. Wang, S.D.; Nicolo, M.; Yi, L.; Dunton, G.F.; Mason, T.B. Interactions among Reward Sensitivity and Fast-Food Access on Healthy Eating Index Scores in Adolescents: A Cross-Sectional Study. *Int. J. Environ. Res. Public Health* **2021**, *18*, 5744. [CrossRef]
80. Li, Y.; Du, T.; Peng, J. Understanding Out-of-Home Food Environment, Family Restaurant Choices, and Childhood Obesity with an Agent-Based Huff Model. *Sustainability* **2018**, *10*, 1575. [CrossRef]

81. Hamerman, E.J.; Rudell, F.; Martins, C.M. Factors that predict taking restaurants leftovers: Strategies for reducing food waste. *J. Consum. Behav.* **2017**, *17*, 94–194. [CrossRef]
82. Rizk, M.C.; Bonalumi, I.P.; Almeida, T.S.; Camacho, F.P. Treatment of food waste from a university restaurant added to sugarcane bagasse. In *Waste—Solutions, Treatments and Opportunities*, 1st ed.; CRC Press: Boca Raton, FL, USA, 2017; pp. 1–6.
83. Greer, R.; von Wirth, T.; Loorbach, D. The diffusion of circular services: Transforming the Dutch catering sector. *J. Clean. Prod.* **2020**, *267*, 121906. [CrossRef]
84. Sofuoglu, S.C.; Toprak, M.; Inal, F.; Cimrin, A.H. Indoor air quality in a restaurant kitchen using margarine for deep-frying. *Environ. Sci. Pollut. Res.* **2015**, *22*, 15703–15711. [CrossRef] [PubMed]
85. Stuchlikova, J.; Botlikova, M. Selected aspects of sustainable tourism and their application in the operation of accommodation and catering facilities in the Moravian-Silesian Region. *SHS Web Conf.* **2020**, *74*, 05023. [CrossRef]
86. Flessas, M.; Rizzardi, V.; Tortorella, G.L.; Fettermann, D.; Marodin, G.A. Layout performance indicators and systematic planning: A case study in a southern Brazilian restaurant. *Br. Food J.* **2015**, *117*, 2098–2111. [CrossRef]
87. Khachaturyan, M.V.; Klicheva, E.V. The function of neural networks in strategic management of restaurant business. In Proceedings of the 34th IBIMA Conference, Madrid, Spain, 13–14 November 2019.
88. Denisova, T.; Gvozdev, M.; Markina, I.; Saveleva, I. Role of reserves in ensuring the continuity of public catering enterprises. In Proceedings of the 35th IBIMA Conference, Seville, Spain, 1–2 April 2020.
89. Bufquin, D.; DiPietro, R.; Park, J.-Y.; Partlow, C. Effects of Social Perceptions and Organizational Commitment on Restaurant Performance. *J. Hosp. Mark. Manag.* **2017**, *26*, 752–769. [CrossRef]
90. Chou, S.-F.; Horng, J.-S.; Liu, C.-H.; Huang, Y.-C.; Chung, Y.-C. Expert Concepts of Sustainable Service Innovation in Restaurants in Taiwan. *Sustainability* **2016**, *8*, 739. [CrossRef]
91. Lepkowska-White, E.; Parsons, A.; Berg, W. Social media marketing management: An application to small restaurants in the US. *Int. J. Cult. Tour. Hosp. Res.* **2019**, *13*, 321–345. [CrossRef]
92. Fernández-Gámez, M.A.; Santos, J.A.C.; Diéguez-Soto, J.; Campos-Soria, J.A. The Effect of Countries' Health and Environmental Conditions on Restaurant Reputation. *Sustainability* **2020**, *12*, 10101. [CrossRef]
93. Zikirya, B.; He, X.; Li, M.; Zhou, C. Urban Food Takeaway Vitality: A New Technique to Assess Urban Vitality. *Int. J. Environ. Res. Public Health* **2021**, *18*, 3578. [CrossRef]
94. Harun, A.; Prybutok, G.; Prybutok, V. Do the millennials in the USA care about the fast food industry's involvement in corporate social responsibility? *Young Consumers* **2018**, *19*, 358–381. [CrossRef]
95. Dang, A.K.; Tran, B.X.; Nguyen, L.H.; Do, H.T.; Nguyen, C.T.; Fleming, M.; Le, H.T.; Le, Q.; Latkin, C.A.; Zhang, M.; et al. Customers' Perceptions of Compliance with a Tobacco Control Law in Restaurants in Hanoi, Vietnam: A Cross-Sectional Study. *Int. J. Environ. Res. Public Health* **2018**, *15*, 1451. [CrossRef]
96. Wickramasingh, K. Measuring environmental orientation in hotels: Empirical evidence from Sri Lanka. *Anatolia* **2019**, *30*, 420–430. [CrossRef]
97. Weerathunga, P.R.; Xiaofang, C.; Samarathunga, W.; Kmmcb, K. Application of entropy based TOPSIS in analysis of sustainability performance of Sri Lankan hotels. In Proceedings of the 16th International Conference on Innovation and Management, University of Jaén, Jaén, Spain, 27–30 January 2016.
98. Grecu, A.; Gruia, A.K.; Marin, M.; Bănuță, M.; Olteanu, C.; Constantin, I.; Gadoiu, M.; Teodorescu, C.; Dobrea, R.C.; Drăghici, C.C. Specificity of Sustainable Structural Dynamics of Local Economy in Romanian Tourist Resorts. *Sustainability* **2019**, *11*, 7155. [CrossRef]
99. Blengini, I.; Heo, C.Y. How do hotels adapt their pricing strategies to macroeconomic factors? *Int. J. Hosp. Manag.* **2020**, *88*, 102522. [CrossRef]
100. Alipour, H.; Safaeimanesh, F.; Soosan, A. Investigating Sustainable Practices in Hotel Industry-from Employees' Perspective: Evidence from a Mediterranean Island. *Sustainability* **2019**, *11*, 6556. [CrossRef]
101. Campos, F.; Lima Santos, L.; Gomes, C.; Cardoso, L. Management Accounting Practices in the Hospitality Industry: A Systematic Review and Critical Approach. *Tour. Hosp.* **2022**, *3*, 243–264. [CrossRef]
102. Bagheri, M.; Shojaei, P.; Jahromi, S.A.; Kiani, M. Proposing a model for assessing green hotels based on ecological indicators. *Tour. Hosp. Res.* **2020**, *20*, 406–422. [CrossRef]
103. Lau, C.; Tang, I.L.F.; Chan, W. Waterfront Hotels' Chillers: Energy Benchmarking and ESG Reporting. *Sustainability* **2021**, *13*, 6242. [CrossRef]
104. Yuan, J.; Luo, X. Regional energy security performance evaluation in China using MTGS and SPA-TOPSIS. *Sci. Total Environ.* **2019**, *696*, 133817. [CrossRef] [PubMed]
105. Wang, D.; Meng, J.; Zhang, T. Post-evaluation on energy saving reconstruction for hotel buildings, a case study in Jiangsu, China. *Energy Build.* **2021**, *251*, 111316. [CrossRef]
106. Liang, Y.; Wang, H.; Hong, W.-C. Sustainable Development Evaluation of Innovation and Entrepreneurship Education of Clean Energy Major in Colleges and Universities Based on SPA-VFS and GRNN Optimized by Chaos Bat Algorithm. *Sustainability* **2021**, *13*, 5960. [CrossRef]
107. Szatten, D.; Więcław, M. Solar Climate Features Taking into Account the Morphometric Conditions of the Area and the Possibility of Using Them in Heliotherapy on the Example of the Cieplice and Kołobrzeg Health Resorts (Poland). *Atmosphere* **2021**, *12*, 383. [CrossRef]

108. Milik, J.; Pasela, R.; Budzinksa, K. Variability of physical and chemical parameters of water from fountains in health resorts. *E3S Web Conf.* **2018**, *44*, 00112. [CrossRef]
109. Bonotto, D.M. The dissolved uranium concentration and 234U/238U activity ratio in groundwaters from spas of southeastern Brazil. *J. Environ. Radioact.* **2017**, *166*, 142–151. [CrossRef] [PubMed]
110. Toyosada, K.; Otani, T.; Shimizu, Y.; Managi, S. Water Quality Study on the Hot and Cold Water Supply Systems at Vietnamese Hotels. *Water* **2017**, *9*, 251. [CrossRef]
111. Tirodimos, E.; Christoforidou, E.P.; Nikolaidou, S.; Arvanitidou, M. Bacteriological quality of swimming pool and spa water in northern Greece during 2011–2016: Is it time for Pseudomonas aeruginosa to be included in Greek regulation? *Water Supply* **2018**, *18*, 1937–1945. [CrossRef]
112. Omune, B.; Kambona, O.; Wadongo, B.; Wekesa, A. Environmental management practices implemented by the hotel sector in Kenya. *World Leis. J.* **2020**, *63*, 98–108. [CrossRef]
113. Zemke, D.M.V.; Chen, Y.-S. Hotel Design, Guest Satisfaction, and Behavioural Intentions. *Anatolia* **2017**, *28*, 338–350. [CrossRef]
114. Saura, J.R.; Reyes-Menendez, A.; Alvarez-Alonso, C. Do Online Comments Affect Environmental Management? Identifying Factors Related to Environmental Management and Sustainability of Hotels. *Sustainability* **2018**, *10*, 3016. [CrossRef]
115. Bacik, R.; Kmeco, L.; Richard, F.; Olearova, M.; Rigelsky, M. Marketing Instrument of Improving Hotel Management Service: Evidence of Visegrad Group Countries. *Mark. Manag. Innov.* **2019**, *1*, 208–220. [CrossRef]
116. Bondzi-Simpson, A.; Ayeh, J.K. Assessing hotel readiness to offer local cuisines: A clustering approach. *Int. J. Contemp. Hosp. Manag.* **2019**, *31*, 998–1020. [CrossRef]
117. Vong, C.; Rita, P.; António, N. Health-Related Crises in Tourism Destination Management: A Systematic Review. *Sustainability* **2021**, *13*, 13738. [CrossRef]
118. OJEU (Official Journal of the European Union). Commission Delegated Decision (EU) 2019/1597 of 3 May 2019. 2019. Available online: https://eur-lex.europa.eu/legal-content/EN/TXT/?uri=uriserv%3AOJ.L_.2019.248.01.0077.01.ENG (accessed on 15 June 2022).
119. He, H.; Harris, L. The impact of Covid-19 pandemic on corporate social responsibility and marketing philosophy. *J. Bus. Res.* **2020**, *116*, 176–182. [CrossRef] [PubMed]

Article

Economic and Environmental Assessment of Conventional versus Organic Durum Wheat Production in Southern Italy

Christian Bux [1], Mariarosaria Lombardi [2], Erica Varese [3] and Vera Amicarelli [1,*]

1 Department of Economics, Management and Business Law, University of Bari Aldo Moro, 70124 Bari, Italy; christian.bux@uniba.it
2 Department of Economics, University of Foggia, 71121 Foggia, Italy; mariarosaria.lombardi@unifg.it
3 Department of Management, University of Turin, 10132 Turin, Italy; erica.varese@unito.it
* Correspondence: vera.amicarelli@uniba.it

Abstract: Conventional and intensive agriculture systems represent an environmental challenge. This research aims at evaluating the economic and environmental implications of conventional and organic durum wheat production in Southern Italy by applying material flow analysis and the crop accounting method. The purpose is to evaluate and compare the natural resource consumption, waste generation and economic profitability of conventional and organic durum wheat farming, respectively. The functional unit is one hectare of cultivated land. System boundaries encompass all agronomic operations, from cradle to gate. The research applies a bottom-up approach and relies on either primary or secondary data. It emerges that organic durum wheat production reduces the use of synthetic chemical and phytosanitary products, as well as plastic waste, by up to 100%. Moreover, it decreases diesel use by 15%, with a consequent reduction in CO_2 emissions, and also avoids soil and groundwater pollution. From an economic perspective, gross income for conventionally farmed durum wheat is still 55% higher compared to organic production. Public authorities should boost environmental sustainability by supporting organic production from either an economic or a social perspective, by enhancing the sharing of best practices, by certification for farmers' groups, by research and innovation, and by incentives in taxation. Overall, this research represents a further step towards the adoption of sustainable agricultural practices.

Keywords: crop accounting; environmental sustainability; material flow analysis; resource management; organic farming; sustainable agriculture

Citation: Bux, C.; Lombardi, M.; Varese, E.; Amicarelli, V. Economic and Environmental Assessment of Conventional versus Organic Durum Wheat Production in Southern Italy. *Sustainability* **2022**, *14*, 9143. https://doi.org/10.3390/su14159143

Academic Editor: Michael S. Carolan

Received: 6 July 2022
Accepted: 24 July 2022
Published: 26 July 2022

1. Introduction

Conventional and intensive agriculture systems represent an environmental challenge [1]. Although the food system as a whole has enhanced agricultural yields through the adoption of monitoring crops' growth [2], accurate weather prediction technologies [3], and novel crop protection methods [4], to meet demand for food commodities and reduce hunger [5], such a rapid rise in productivity has had a detrimental effect on the environment. Among other issues, agricultural production is responsible for soil degradation, biodiversity losses, water pollution and climate change [6,7]. As reported by Ritchie and Roser [8], half of habitable land worldwide is used for agriculture, accounting for over 51 million km^2. For crop production (excluding animal feed), a figure of over 11 million km^2 of land has been estimated. Further, crop production for human consumption accounts for over 21% of food production emissions, equivalent to approximately 2.8 Gt of CO_2eq [9]. In Italy, agricultural greenhouse gas (GHG) emissions have been estimated at 418 kt of CO_2eq and represent the highest in Europe after Germany (810 kt of CO_2eq) and France (443 kt of CO_2eq) in 2019.

In the field of wheat production, Ritchie and Roser [8] have estimated that more than 3.8 m^2 of land is required to cultivate one kilogram of wheat, while the entire wheat chain generates more than 1.5 kg of CO_2 per kilogram of product (i.e., more than 50% is due to land use and farm management operations). Conventional crops are characterized by higher yields and profits compared to organic ones [10]. However, better economic performance is supplemented by negative externalities. From the environmental perspective, conventional crops cause soil depletion, groundwater pollution, and atmospheric contamination, as well as requiring extensive use of agrochemicals [11]. Further, from the economic and societal perspective, conventional crops are less likely to meet the increasing market demand for sustainable products [12], thereby frustrating international and national directives on sustainable production strategies, such as the Farm to Fork Strategy [13]. On the other hand, organic farming is defined as a system which relies on ecosystem management rather than external agricultural inputs and which eliminates the use of synthetic inputs such as synthetic fertilizers and pesticides, veterinary drugs, genetically modified seeds and breeds, preservatives, and additives [14]. It provides a sustainable alternative to conventional farming [15] since energy use per hectare of land. and levels of GHG emissions are both lower compared with conventional crops [16]. From the ecological perspective, organic farming does not revolutionize the soil structure, does not release polluting substances into nearby water bodies by leaching, and does not use chemicals which damage the ecosystem [17].

Recent studies [18] have found that although organic farming provides reliable environmental benefits and contributes towards food safety and food security goals, it also increases variability in crop yields, producing financial risks for farmers in terms of volatile profitability. This brings to light a critical issue in terms of environmental protection and economic growth: organic farms promote biotic abundance, soil carbon, and profitability, but conventional farms produce higher yields [18]. On the other hand, it has been argued that ecological and economic outcomes depend either on the adoption of different management systems (i.e., conventional or organic) or on crop types, underlining that each case must be treated individually.

In the quest for a fair trade-off between economic growth and environmental protection, the present research aims at evaluating the economic and environmental impacts of conventional and organic durum wheat (*Triticum turgidum* L. subsp. *durum*) production in the Apulia region (Southern Italy) by material flow analysis (MFA) and by crop accounting.

To the best of the authors' knowledge, there have been no studies which have combined these two methodologies to assess similarities and differences between these two types of farming systems in terms of natural resource consumption, waste generation and economic profitability. The originality of the present research thus relies on considering these factors in an attempt to fill in the academic data gaps. Specifically, the MFA produces transparent, comparable and replicable data, both quantitative and qualitative, and it identifies the hotspots of these two farming systems.

Finally, this research contributes to the integration of academic and practitioner knowledge in the field of durum wheat farming, representing a further step towards the adoption of sustainable agricultural practices.

2. Theoretical Background

Durum wheat is the tenth most important crop worldwide and is cultivated in three main areas, namely: the Mediterranean basin; the northern United States and Canada; and the desert areas of the southwest United States and northern Mexico [19]. In addition, durum wheat is the most cultivated cereal crop in the Mediterranean basin [19] and is essential in the Mediterranean diet, being the basis for the production of four different products: pasta, couscous, bulgur and bread.

On a global scale, durum wheat annual production has declined from 37 Mt in 2018 to 33.6 Mt in 2020 (−9%) [20] and represents approx. 4% of entire wheat production (895 Mt in 2020) as reported by FAOstat [21]. It is estimated that the European Union (EU) produced 7.3 Mt of durum wheat in 2020, with a cultivated land area of 2,199,000 ha and an average yield of 4 t/ha [22]. Italy is the largest EU producer of durum wheat, accounting for approx.

4 Mt cultivated on 1,210,415 ha of land. The Apulia region is the biggest producer of durum wheat in Italy, recording 0.99 Mt of durum wheat production in 2020 (24% of total Italian production) [23]. Moreover, the Apulia region has the largest area of Italy devoted to durum wheat farming, estimated at 344,400 ha in 2020, compared to 283,870 ha in 2010 (+21%) [24].

Several articles on durum wheat have been published in Italy in the fields of environmental sciences, social sciences, energy, and business management. In terms of organic farming, durum wheat is cultivated on more than 140,000 ha (10% of the total durum wheat cultivated land). Southern Italy, including Apulia, Sicily, Calabria and Molise, has contributed most to the organic conversion, covering over 50% of all organic cultivated land in Italy [25]. From the consumption perspective, of the rate of durum wheat self-sufficiency (apparent production/consumption) has been estimated at 56% [26].

Lately, researchers have applied the life cycle assessment (LCA) to improve the management of agri-food companies involved in whole-grain durum wheat pasta production, and to assess the energy and environmental impacts of durum wheat bread [27]. Such studies have estimated that the major environmental impacts along the entire wheat chain are generated during the cultivation stage, but no comparison has yet been made between conventional and organic farming. Furthermore, Todorović et al. [28] have investigated the different impacts of water and nitrogen on durum wheat eco-efficiency in the Mediterranean area, highlighting the need to adopt agronomic practices with low use of resources and higher eco-efficiency. Sustainable practices must address both precision agriculture and optimization of water and fertilizers, enhancing environmental, resources and economic performances at the same time. Similar results have been obtained by Alhajj Ali et al. [29], which have estimated wheat-cultivation-related GHG emissions and have evaluated lower carbon footprints associated with improved productivity and minimum inputs. Results like these suggest that the main contributors to negative emissions are farm inputs, as well as nitrogen fertilizers and pest management techniques [29,30]. As regards water footprints, several authors have stressed the importance of responsible water use [30,31] and have highlighted organic pest control and proper manure use as drivers to reduce water consumption towards sustainable practice levels [32,33].

As regards the comparison between conventional and organic wheat farming, several articles have investigated organic wheat quality and consumers' preferences [34,35] but few authors have compared the economic and environmental impacts associated with conventional and organic durum wheat farming, respectively. Montemurro and Maiorana [36] have estimated that conservative agricultural practices such as crop rotation, shallow tillage and organic fertilizers can reduce environmental impacts and contribute towards sustainable agriculture, whereas Tudisca et al. [37] highlighted a higher gross margin for organic durum wheat compared to the conventional crop, due to lower variable costs and higher production values. Further in-depth comparisons between conventional and organic crops have been conducted by Fagnano et al. [38], with the aim of evaluating the agronomic, technological, sensory and sanitary qualities of grains and pasta, but without assessing environmental consequences. In the field of water consumption analysis in Italy, relevant literature has considered the water footprint to evaluate the environmental and sustainable performance of companies. Ruini et al. [39] considered this indicator with respect to pasta production, highlighting its role in informing better decision-making regarding plant management, supplier collaboration and interaction between policy-makers and communities [40].

In addition to the environmental impacts of durum wheat production, and the economic savings arising from the adoption of innovative technologies [41], researchers have been interested in contractual arrangements within the Italian durum wheat sector [42]. It has emerged that Italian farmers are more likely to accept contractual clauses related to food quality than to adopt sustainable agronomic practices, highlighting the need to align economic incentives with environmental goals through measurable socio-environmental targets in contractual clauses [43]. It means that farmer preferences towards conventional or organic practices are not only financially driven but depend on several variables, such as social or financial ones [24].

3. Materials and Methods

The present research applies: (i) the MFA, to compare conventional and organic durum wheat production in terms of natural resource consumption and waste generation (Section 3.1); and (ii) the crop accounting method, to calculate economic indicators, such as gross income, total revenues and total costs (Section 3.2). The present research adopts a stepwise approach, as proposed by Bux and Amicarelli [44], and Hendriks et al. [45], as follows: (a) identification of the qualitative system, including functional unit, material flows and system boundaries definition; (b) assessment of the quantitative system along the entire supply chain, including energy and water use; (c) calculation of either the conventional or the organic durum wheat production level through an input–output table [46,47]; and (d) evaluation of the results by the crop accounting method. Section 3.3 describes the data collection process according to a bottom-up approach, which relies both on secondary data, taken from national and international reports, scientific research and official databases; and primary data, provided by a Southern Italian farm located in the Apulia region.

3.1. Material Flow Analysis

The MFA can be defined as "a systematic assessment of the state and change of materials flow and stock in space and time" [47] and has been applied with success at micro-, meso- and macro levels [48]. Some studies have explored local cereal supply chains from an economic, social and environmental perspective to aid decision-making [49], while others have evaluated energy use, GHG emissions, land use, use of pesticides, and blue water footprints associated with cereal production [50].

The authors selected one hectare of land as functional unit. Some authors have proposed 1 t of wheat produced as a functional unit [51,52], but such a unit is excessively influenced by the yield level, thereby compromising results and leading to incomparable outcomes [53]. Although the analysis of conventional durum wheat was conducted from October 2014 to June 2015, while the investigation of organic durum wheat was carried out between October 2019 and June 2020, the research relies on common characteristics in terms of structure and composition of the soil, organic endowment, crop in precession and water endowment.

As regards the system boundaries, the analysis encompasses all agronomic operations [54], from agricultural production to storage and warehouse operations (i.e., from cradle to farm gate). Conventional durum wheat production system boundaries include: (a) plowing, harrowing, and sowing; (b) fertilization; (c) chemical weeding and phytosanitary treatments; (d) combine harvesting (for third parties), straw harvesting (for third parties), and transport. Organic durum wheat system boundaries include: (a) light harrowing (or false sowing), sowing, tillage, and harrowing; (b) mechanical weeding; (c) combine harvesting (for third parties), straw shredding (for third parties), and transport. From a circular economy perspective, the straw on the field is subjected to shredding and sent to the fertilization phase together with the manure. The cradle-to-gate boundaries allow researchers to replicate the MFA in other geographical areas, and to compare trends and results obtained over time [55]. As regards the organic farming, durum wheat was cultivated according to a defined organic regime, as stated by the Council Regulation (EC) 824/2007 on organic production and labelling of organic products [56].

As regards the investigation of material flows, the authors consider material inputs such as seeds, fertilizers (i.e., urea, N, K, K_2O, O_2) and herbicides, as well as plastic nets for collecting and storing straw. Further, considering resource and energy inputs, the research takes water and diesel consumption into consideration. On the output side, the authors give an account of CO_2 emissions, plastic and paper waste (e.g., packaging), fertilizers and pollutants, wheat losses and straw. For an assessment of water consumption, the authors consider the average rainfall trends for the reference years (i.e., from October 2014 to June 2015; and from October 2019 to June 2020) recorded by Protezione Civile Puglia [57]. Rainfall trends have been considered comparable in both years [58,59]. For conventional farming, it is estimated that rainfall represents 90% of the entire water consumption by durum wheat, whereas for organic farming the figure is 100%. Overall, conventional farming requires 5225–5775 L/ha, whereas organic farming requires approx. 3610–3990 L/ha. As regards the organic method applied in the research, it assumes no use of synthetic fertilizers and chemicals, when ancient and indigenous grains are considered.

Figure 1 illustrates system boundaries and material flows for either conventional or organic durum wheat production. Cultivation and seeding include operations such as deep harrowing, ploughing, harrowing and seeding, followed by fertilization and chemical weeding for conventional durum wheat. In addition, mechanized harvesting and transport from field to warehouse are taken into consideration.

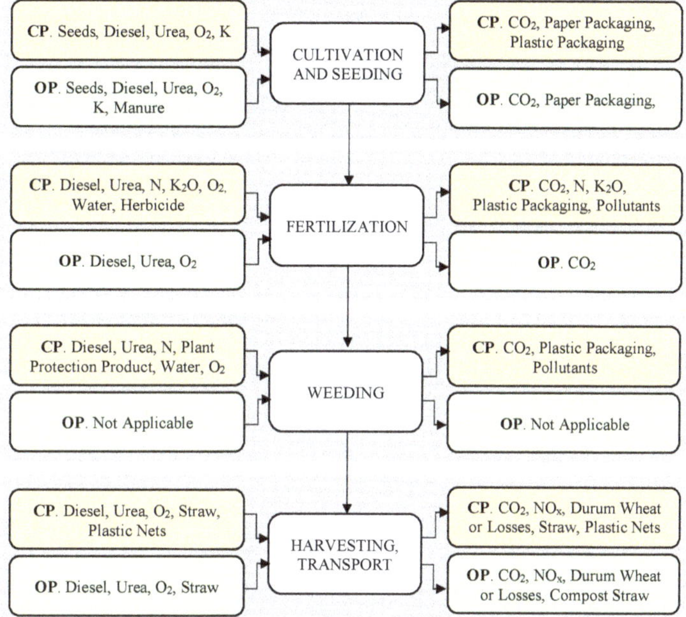

Figure 1. System Boundaries and Material Flows for the Durum Wheat Production. Notes: CP = conventional durum wheat production (yellow panels); OP = organic durum wheat production (green panels). Source: Personal elaboration by the authors.

3.2. *Crop Accounting Method*

As regards economic data, the authors consider a pre-pandemic scenario (i.e., baseline) and apply an estimate, budget, and costing tool defined as crop accounting. This represents a useful tool for crop enterprise management, since it considers all costs of growing crops until harvesting time [60]. In the case of conventional durum wheat, such an approach accounts for the costs of plowing, harrowing, sowing, fertilization, weeding, phytosanitary treatment, harvesting, and transport to the collection center. For organic production, the approach encompasses the costs of subsoiling, harrowing, sowing, fertilization (soil improver), mechanical weeding, harvesting, shredding, and transport to the collection center. This method calculates the gross income as the difference between total revenues and total costs associated with 1 t of durum wheat per ha (i.e., the functional unit). Either secondary data (i.e., national and international reports, scientific research, database), or primary data (i.e., site-specific data provided by a Mediterranean farm) are used to calculate the economic indicators.

3.3. Data Collection

Data collection represents a challenging step in MFA [61]. In Southern Italy, there is a lack of reliable, up-to-date data concerning water, nitrogen and carbon recycling, and the same applies in the case of carbon and nitrogen dioxide emissions [62]. Moreover, some criticalities are related to agricultural and transport operations on behalf of third parties. Bottom-up approaches provide more detailed information on material flows, offering a suitable empirical basis for practitioners and for academic research [63]. Furthermore, to acquire reliable data on conventional and organic durum wheat, the authors have adopted the research triangulation paradigm [64,65]. Such an approach combines data and observations at farm level and helps in boosting the credibility and validity of research findings [66]. As regards single material flows, data relating to seeds, fertilizers, diesel, urea, electricity and herbicides have been provided by an Apulian farm and compared with secondary data from scientific articles applying the life cycle assessment in Sicily [25], as well as the carbon and water footprints of Italian production [27–29]. Moreover, data related to nitrogen and water inputs have been compared to those provided by Todorović et al. [26] in the Mediterranean area. Straw, plastic and paper waste are primary data, whereas CO_2 emissions have been compared with Tedone et al. [17] and Alhajj et al. [27], which investigated GHG emissions from durum wheat production.

Table 1 illustrates the input–output table for either conventional or organic durum wheat production. As regards data uncertainties, the authors have determined a $\pm$ 5% error rate which encompasses measurement errors associated with the databases needed to conduct the MFA, data gaps due to confidentiality rules, errors due to assumptions or simplifications, and errors due to conversions into mass weight or the downscaling of data [67]. In addition, such an error rate takes into consideration the variability of agricultural activities. As regards rainfall trends, data have been taken from Protezione Civile Puglia [57]. Further, groundwater used by the crop has not been estimated, and rainwater has not been discounted by the crop coefficient of waste use [68].

Table 1. Input–Output Table for the Durum Wheat Production (Functional Unit: 1 ha).

Input–Output Table			**Conventional Production**		**Organic Production**	
Material Flows		**Unit**	**Min.**	**Max.**	**Min.**	**Max.**
Input	Seeds	kg/ha	142.5	157.5	142.5	157.5
	N	kg/ha	118 $^{\alpha}$	131 $^{\alpha}$	204 $^{\beta}$	226 $^{\beta}$
	P	kg/ha	57	63	28.5 $^{\gamma}$	31.5 $^{\gamma}$
	K	kg/ha	95	105	68.4	75.6
	Diesel	MJ/ha	148.2	163.8	127.3	140.7
	Urea	L/ha	11.4	12.6	8.6	9.5
	Electricity	MJ/ha	6070.5	6709.5	5198.4	5745.6
	Herbicides	kg/ha	4.7	5.2	0	0
	Water	L/ha	5225 $^{\delta}$	5775 $^{\delta}$	3610 $^{\varepsilon}$	3990 $^{\varepsilon}$
Output	Durum wheat	kg/ha	3087.5	3412.5	2185	2415
	Straw	kg/ha	2850	3150	1995	2205
	Paper waste	kg/ha	3.8	4.2	3.8	4.2
	Plastic waste	kg/ha	6.4	7	0	0
	CO_2 emissions	kg/ha	399	441	339.1	374.9

Notes: $^{\alpha}$ 20% from culture in precession, 80% of synthetic nitrogen; $^{\beta}$ 12% from culture in precession, 82% from soil conditioner, 6% from straw burial; $^{\gamma}$ 100% from soil conditioner and straw burial; $^{\delta}$ 10% from water treatments, 90% from rainfall; $^{\varepsilon}$ 100% from rainfall. Source: Personal elaboration by the authors.

4. Results and Discussion

4.1. MFA Results

Figure 2 illustrates the MFA for 1 ha of conventional durum wheat while Figure 3 shows the MFA for 1 ha of organic durum wheat. Organic farming does not rely on mineralized fertilization; it requires urea, K fertilizers and manure. From a circular perspective, natural fertilizers come from straw compost and residual straw elements. Additionally, no plastic packaging is generated during cultivation and seeding, fertilization or weeding. As regards diesel consumption, less fuel is required either for cultivation and seeding (−25%), or for fertilization (–55%). Considering that herbicides and plant protection products are not applied, and a weeding stage is not carried out, an additional 8.8–9.8 L/ha of diesel are saved. On the contrary, more diesel is required for the straw shedding stage (+43%). From a waste management perspective, plastic waste is totally avoided (from –7 to –6.4 kg/ha), and CO_2 emissions are reduced by approx. 15% (from 399–441 kg/ha to 339–374 kg/ha).

As regards the results from the water consumption analysis, it emerges that 5225–5775 L of water, of which 4702–5198 L is rainfall, are required to produce 3087–3413 kg of conventional durum wheat. The water consumption rate is estimated at 1.69 L/kg, of which 0.17 L/kg is the result of from irrigation. On the other hand, 3610–3990 L of water (100% rainfall) are required to produce 2185–2415 kg of organic durum wheat. The water consumption rate is assessed at 1.65 L/kg.

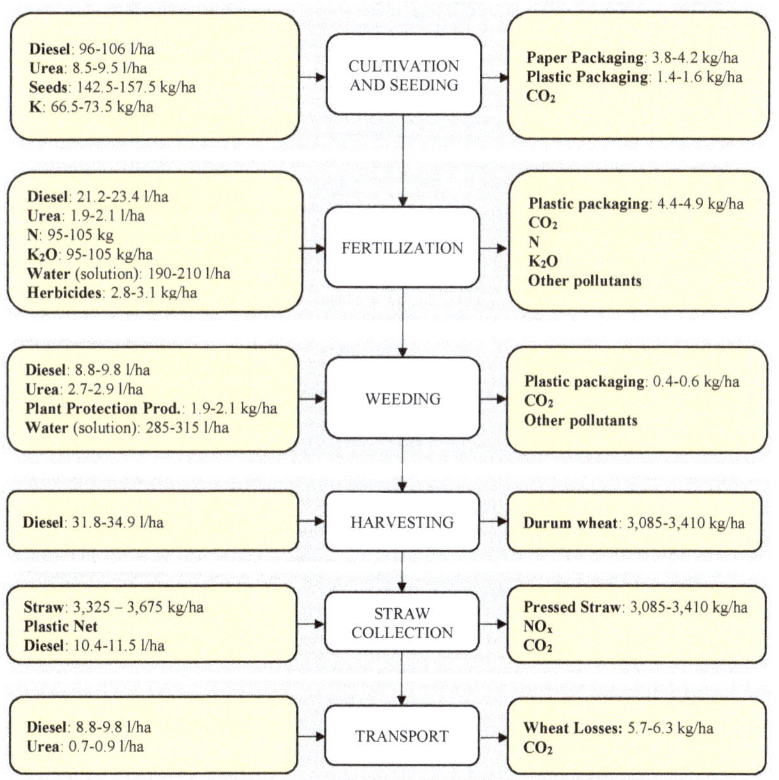

Figure 2. Material Flow Analysis for 1 ha of Conventional Durum Wheat. Source: Personal elaboration by the authors.

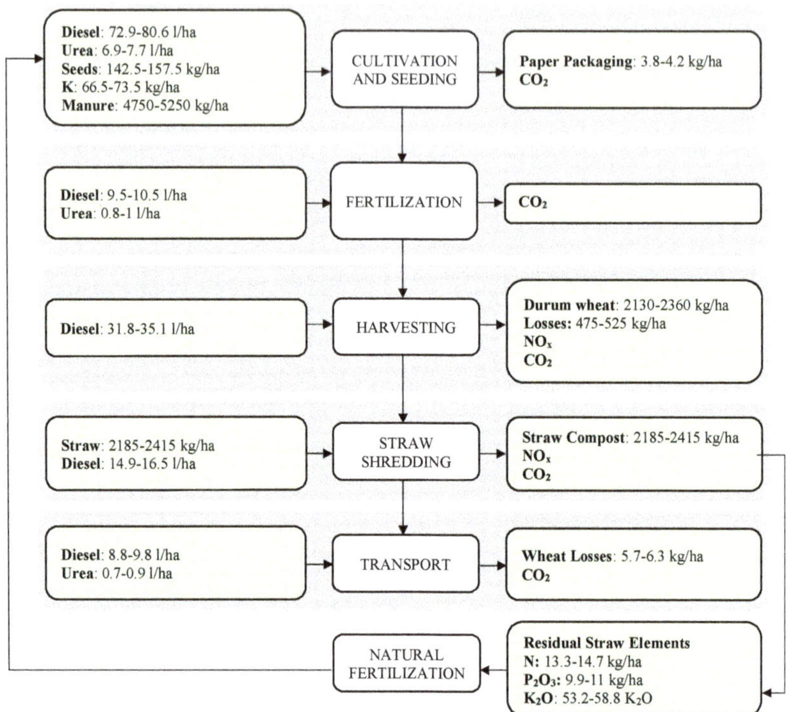

Figure 3. MFA for 1 ha of Organic Durum Wheat. Source: Personal elaboration by the authors.

Crop accounting for conventional and organic durum wheat considers average yields of 3.2 t/ha and 2.3 t/ha, respectively, while average prices are estimated at EUR 300/t for conventional durum wheat in 2014/2015 and EUR 420/t per organic durum wheat in 2019/2020. As regards straw, its price has been estimated at EUR 90/t and its yield at 3 t/ha. Table 2 illustrates the crop accounting for conventional and organic durum wheat production. At first glance, although conventional production costs are higher than organic production costs (+21%), it emerges that conventional durum wheat is more profitable (in terms of gross income) compared to organic durum wheat production. From an operational perspective, organic farming is less expensive considering that complex and environmentally impacting operations (i.e., plowing, cover fertilization, weeding, phytosanitary treatments) are absent. On the other hand, organic farming does not generate any revenues from straw selling, since straw is used as natural fertilizer for further cultivation and requires additional operations such as subsoiling (75–82 EUR/ha), basic fertilization (228–252 EUR/ha), mechanical weeding (57–63 EUR/ha), and collection and shredding (133–147 EUR/ha).

Table 2. Crop Accounting for Conventional and Organic Durum Wheat Production in EUR/ha.

Crop Accounting Method		CDW (Min.)	CDW (Max.)	ODW (Min.)	ODW (Max.)
Revenues	Revenues durum wheat	912	1008	917.7	1014.3
	Revenues straw	256.5	283.4	0	0
	Total revenues (a)	*1168.5*	*1291.5*	*917.7*	*1014.3*
Costs	Plowing	190	210	0	0
	Subsoiling	0	0	75.05	82.3
	Harrowing	123.5	136.5	47.5	52.5
	Sowing	190	210	190	210
	Cover fertilization	128.3	141.8	0	0
	Basic fertilization	0	0	228	252
	Weeding	59.9	66.2	0	0
	Mechanical weeding	0	0	57	63
	Phytosanitary treatments	80.8	89.3	0	0
	Collection	114	126	0	0
	Collection and shredding	0	0	133	147
	Transport to the collection center	39.9	44.1	31.4	34.7
	Total costs (b)	*926.5*	*1023.8*	*762*	*842*
	Gross income (a–b)	*242.3*	*267.8*	*156*	*172*

Notes: CDW = Conventional durum wheat; ODW = Organic durum wheat. Source: Personal elaboration by the authors.

4.2. Managerial Implications

Among its strengths, organic durum wheat farming enhances water retention and soil porosity through the presence of roots and soil microfauna. Moreover, from an environmental point of view, organic farming ensures lower environmental impacts by reducing the use of synthetic chemicals and phytosanitary products by up to 100% [69]. Similar benefits occur in terms of plastic waste, which is also reduced up to 100%, and the use of diesel, which is reduced by 15% (with consequent reduction in CO_2 emissions). In the same light, soil and aquifer pollution is avoided, as is the use of external materials, with associated damage to the surrounding environment (i.e., flora and fauna) [70]. From an economic perspective, organic production guarantees a positive, albeit reduced, margin. Finally, compliance with the Council Regulation EC/824/2007 on organic production and labeling of organic products allows for more attention, more analysis and the acquisition of experience in soil management by agricultural operators [71]. Considering future opportunities for organic durum wheat production, it is now possible to sign supply chain contracts cheaper than those signed for conventional production [72]. Moreover, genetic improvement and assisted evolution technologies, wheat grafting and the use of more efficient biostimulants could all be used in the future [73].

On the other hand, some weaknesses in organic farming have been evaluated, which may suggest a continuing preference for conventional rather than organic farming. Organic crops are affected by fungal diseases and erosion, and incur rather high production costs [74,75]. Furthermore, the required land use is higher than with conventional crops. It is estimated that, with a 25 to 64% increase in land use, there is a 20% to 45% lower yield compared to conventional crops (i.e., conventional yield is 3.2 t/ha, whereas organic yield is 2.3 t/ha). In terms of threats from climate change and the related consequences for crop variability, this price differential between organic and conventional products and market competitiveness in terms of price and/or consumer choice should not be underestimated.

From an economic perspective, organic farming provides lower gross incomes compared to conventional farming. Although farmers take care of the water–energy–food nexus in farming [76,77] and its related environmental consequences (i.e., higher waste generation, increased CO_2 emissions, higher water consumption), and even if resource-efficient, resilient and productive food systems are seen as fundamental to pursue sustainable development, still entrepreneurs pursue financial objectives [78]. In the light of the crop accounting results, it emerges that farmers are still more interested in cultivating conventional durum wheat compared to organic, since gross incomes are approx. 55% higher. Public authorities should boost environmental sustainability by supporting organic production from either an economic or social perspective. Key tools to enhance sustainable development could include networking for sharing best practices, certification for groups of farmers rather than for individuals, research and innovation, as well as providing economic benefits for organic producers, including incentives in taxation [79].

4.3. Limitations and Further Work

The present research provides environmental and socioeconomic data, addressing managerial concerns regarding the adoption of sustainable agriculture practices. Although this data concerns Southern Italian durum wheat crops, our research identifies broad hotspots of wheat production by comparing conventional and organic farming. Due to its replicability and comparability, it can be enlarged to other areas, as well as to other agricultural or processing practices. From the farmers' perspective, EU policies in the field of organic farming should encourage production and processing by stimulating conversion and reinforcing the entire value chain. Although the organic farming sector of the EU has increased by approx. 66% from 2010 to 2020 [80], farmers have reported insufficient access to stable markets for organic products, representing one of the largest barriers to economic viability, as well as a lack in information and technical assistance [74]. As a consequence, public authorities must boost consumer demand by preventing food frauds and strengthening consumer trust, improving traceability, reinforcing organic school schemes and facilitating the contribution of the private sector [80].

One means of boosting organic farming and increasing farmers' income is related to the diffusion of the organic certification, which could increase companies' visibility and consumers' trust. Starting from January 2022, the EU has activated a new organic legislation (Regulation EU 2018/848 on organic production and labelling of organic products), which ensures fair competition for farmers whilst preventing fraud and maintaining consumer reliance [80]. Among other things, the EU action plan aims to introduce simplified production rules, strengthen control systems along the entire supply chain, and implement an easier certification system for small farmers. In the light of the crop accounting analysis and considering the low incomes of organic durum wheat production, the adoption of group certification could represent a suitable instrument to both promote conversion to organic production methods, and maintain existing organic production, by reducing the costs of certification. Farmers could reduce either the cost of the control visit or the costs associated with bureaucratic requirements of organic certification, while maintaining quality assurance systems, and thereby counterbalance the lower incomes of organic production [81].

This research is limited by a lack of up-to-date data on water, nitrogen, carbon recycling and dioxide emissions by third parties. Moreover, results are influenced by meteorological and economic variables, such as market prices or inflation. In addition, the research is limited to a region of Southern Italy (Apulia) and does not allow the extension of its results to national or international realities. Although MFA provides transparent, comparable and replicable results under quantitative and qualitative perspective, and highlights hotspots in processes and stocks, more data are essential to guarantee reliability of results.

Future research directions might include the creation of a suitable inventory of durum wheat production in the Mediterranean area, by collecting data from as many farms as possible, and by calculating reliable eco-efficiency indicators [82]. The adoption of such indicators, which are based on the general concept of output maximization with resource consumption minimization, could be useful to identify the main environmental and economic criticalities in the organic and conventional durum wheat production, and they could also "capture the ecological efficiency of growth by measuring the efficiency of economic activities and its corresponding environmental impacts" [83]. Moreover, based on the present research, the authors are willing to apply the mass-balance approach to organic and conventional durum wheat production at the macro level (i.e., in Italy).

5. Conclusions

This research evaluated and compared natural resource consumption, waste generation and economic profitability in conventional and organic durum wheat farming by applying the MFA and the crop accounting method. The research focused on Southern Italy, specifically the Apulia region, which produces 0.99 Mt of durum wheat and represents 24% of the entire Italian production and over 13% of total EU production. Durum wheat represents a staple food, and it is the basis for the production of four essential products in the Mediterranean diet, namely, pasta, couscous, bulgur and bread.

It emerged that organic durum wheat production has lower environmental impacts, since the use of synthetic chemical and phytosanitary products, as well as the production of plastic waste, are reduced by up to 100% compared to conventional organic farming. Furthermore, such a sustainable agricultural practice allows for a decrease in diesel use of 15%, as well as related CO_2 emissions, which could be reduced from 399–441 kg/ha to 399–374 kg/ha. In addition, the adoption of organic farming practices enhances water retention and soil porosity through the presence of roots and soil microfauna. However, organic crops are subject to fungal diseases, erosion and rather high production costs.

From an economic perspective, although organic farming represents a more sustainable agricultural practice, its land use requirement is still higher, compared to conventional wheat production. It has been estimated that conventional yields are about 3.2 t/ha, whereas organic yields have been evaluated at 2.3 t/ha. Furthermore, the increase in land use is still associated with lower gross incomes, since the gross income for conventional durum wheat production is 55% higher, when compared to organic production.

Overall, public authorities should boost environmental sustainability by supporting organic production from either an economic or social perspective, and key tools to improve sustainable development and boost economic benefits while guaranteeing environmental protection must develop, including networking for sharing best practices among local farms, as well as enhanced certification for groups of farmers, research and innovation, and incentives in taxation.

Author Contributions: Conceptualization, C.B., M.L., E.V. and V.A.; methodology, C.B. and V.A.; software, C.B. and V.A.; validation, C.B. and V.A.; investigation, C.B. and V.A.; resources, C.B. and V.A.; data curation, C.B., M.L., E.V. and V.A.; writing—original draft preparation, C.B.; writing—review and editing, M.L., E.V. and V.A.; visualization, M.L., E.V. and V.A.; supervision, M.L., E.V. and V.A. All authors have read and agreed to the published version of the manuscript.

Funding: This research received no external funding.

Institutional Review Board Statement: Not applicable.

Informed Consent Statement: Not applicable.

Data Availability Statement: Not applicable.

Conflicts of Interest: The authors declare no conflict of interest.

References

1. Page, G.; Ridoutt, B.; Bellotti, B. Location and technology options to reduce environmental impacts from agriculture. *J. Clean. Prod.* **2014**, *81*, 130–136. [CrossRef]
2. Fritz, S.; See, L.; Bayas, J.C.L.; Waldner, F.; Jacques, D.; Becker-Reshef, I.; Whitcraft, A.; Baruth, B.; Bonifacio, R.; Crutchfield, J.; et al. A comparison of global agricultural monitoring systems and current gaps. *Agric. Syst.* **2019**, *168*, 258–272. [CrossRef]
3. Konduri, V.S.; Vandal, T.J.; Ganguly, S.; Ganguly, A.R. Data Science for Weather Impacts on Crop Yield. *Front. Sustain. Food Syst.* **2020**, *4*, 52. [CrossRef]
4. Hazra, D.K.; Purkait, A. Role of pesticide formulations for sustainable crop protection and environment management: A review. *J. Pharmacogn. Phytochem.* **2021**, *8*, 686–693.

5. Wiebe, K.; Sulser, T.B.; Dunston, S.; Rosegrant, M.W.; Fuglie, K.; Willenbockel, D.; Nelson, G.C. Modeling impacts of faster productivity growth to inform the CGIAR initiative on Crops to End Hunger. *PLoS ONE* **2021**, *16*, e0249994. [CrossRef] [PubMed]
6. Lo Piccolo, E.; Landi, M. Red-leafed species for urban "greening" in the age of global climate change. *J. For. Res.* **2021**, *32*, 151–159. [CrossRef]
7. Liao, C.; Tian, Q.; Liu, F. Nitrogen availability regulates deep soil priming effect by changing microbial metabolic efficiency in a subtropical forest. *J. For. Res.* **2021**, *32*, 713–723. [CrossRef]
8. Ritchie, H.; Roser, M. Environmental Impacts of Food Production. 2021. Available online: https://ourworldindata.org/environmental-impacts-of-food#citation (accessed on 17 February 2022).
9. Poore, J.; Nemecek, T. Reducing food's environmental impacts through producers and consumers. *Science* **2018**, *360*, 987–992. [CrossRef]
10. Froehlich, A.G.; Melo, A.S.S.A.; Sampaio, B. Comparing the Profitability of Organic and Conventional Production in Family Farming: Empirical Evidence from Brazil. *Ecol. Econ.* **2018**, *150*, 307–314. [CrossRef]
11. Ozturk, M.; Gul, A. *Climate Change and Food Security with Emphasis on Wheat*; Elsevier: Amsterdam, The Netherlands; Academic Press: Cambridge, MA, USA, 2020; pp. 1–29.
12. Eyinade, G.A.; Mushunje, A.; Yusuf, S.F.G. The willingness to consume organic food: A review. *Food Agric. Immunol.* **2021**, *32*, 78–104. [CrossRef]
13. European Commission. Farm to Fork Strategy, for a Fair, Healthy and Environmentally Friendly Food System. 2022. Available online: https://ec.europa.eu/food/farm2fork_en (accessed on 17 February 2022).
14. FAO. Organic Agriculture. 2022. Available online: https://www.fao.org/organicag/oa-faq/oa-faq1/en/ (accessed on 18 February 2022).
15. Halberg, N. Assessment of the environmental sustainability of organic farming: Definitions, indicators and the major challenges. *Can. J. Plant Sci.* **2012**, *92*, 981–999. [CrossRef]
16. Lynch, D.H.; Halberg, N.; Bhatta, G.D. Environmental impacts of organic agriculture in temperate regions. *CAB Rev.* **2012**, *7*, 1–17. [CrossRef]
17. Trydeman Knudsen, M.; Sillebak Kristensen, I.; Berntsen, J.; Molt Petersen, B.; Steen Kristensen, E. Estimated N leaching losses for organic and conventional farming in Denmark. *J. Agric. Sci.* **2006**, *144*, 135–149. [CrossRef]
18. Smith, O.M.; Cohen, A.L.; Rieser, C.J.; Davis, A.G.; Taylor, J.M.; Adesanya, A.W.; Jones, M.S.; Meier, A.R.; Reganold, J.P.; Orpet, R.J.; et al. Organic Farming Provides Reliable Environmental Benefits but Increases Variability in Crop Yields: A Global Meta-Analysis. *Front. Sustain. Food Syst.* **2019**, *3*, 82. [CrossRef]
19. Tedone, L.; Ali, S.A.; De Mastro, G. Optimization of Nitrogen in Durum Wheat in the Mediterranean Climate: The Agronomical Aspect and Greenhouse Gas (GHG) Emissions. In *Nitrogen in Agriculture—Updates*; Amanullah, A., Fahad, S., Eds.; IntechOpen: London, UK, 2019. [CrossRef]
20. Sicilian Wheat Bank. World Wheat Report 2020–2021. Available online: https://www.bancadelgrano.it/wp-content/uploads/2021/01/World-Wheat-Report-2020-2021.pdf (accessed on 20 June 2022).
21. FAOSTAT. Crops and Livestock Products. 2022. Available online: https://www.fao.org/faostat/en/#data/QCL (accessed on 28 June 2022).
22. European Commission. Cereals Market Situation. *Committee for the Common Organisation of Agricultural Markets.* 2022. Available online: https://circabc.europa.eu/sd/a/92653d37-7fff-40c1-8d5e-b6bb3625c04a/EU%20cereals%20market.pdf (accessed on 30 June 2022).
23. Istat. Coltivazioni: Cereali, Legumi, Radici Bulbi e Tuberi. 2022. Available online: http://dati.istat.it/Index.aspx?QueryId=33702 (accessed on 20 June 2022).
24. Istat. Statistiche Report. Coltivazioni Agricole. Annata Agraria 2019–2020 e Previsioni 2020–2021. 2021. Available online: https://www.istat.it/it/files//2021/04/Previsioni-coltivazioni-agricole.pdf (accessed on 30 June 2022).
25. Sinab. Bio in Cifre. 2020. Available online: http://www.sinab.it/sites/default/files/share/BIO%20IN%20CIFRE%202020.pdf (accessed on 23 February 2022).
26. Ismea. Cereali—Supply Balance Sheet. 2022. Available online: https://www.ismeamercati.it/flex/cm/pages/ServeBLOB.php/L/IT/IDPagina/4546 (accessed on 23 February 2022).
27. Zingale, S.; Guarnaccia, P.; Timpanaro, G.; Scuderi, A.; Matarazzo, A.; Bacenetti, J.; Ingrao, C. Environmental life cycle assessment for improved management of agri-food companies: The case of organic whole-grain durum wheat pasta in Sicily. *Int. J. Life Cycle Assess.* **2022**, *27*, 205–226. [CrossRef]
28. Todorović, M.; Mehmeti, A.; Cantore, V. Impact of different water and nitrogen inputs on the eco-efficiency of durum wheat cultivation in Mediterranean environments. *J. Clean. Prod.* **2018**, *183*, 1276–1288. [CrossRef]
29. Alhajj Ali, S.; Tedone, L.; Verdini, L.; De Mastro, G. Effect of different crop management systems on rainfed durum wheat greenhouse gas emissions and carbon footprint under Mediterranean conditions. *J. Clean. Prod.* **2017**, *140*, 608–621. [CrossRef]
30. Casolani, N.; Pattara, C.; Liberatore, L. Water and Carbon footprint perspective in Italian durum wheat production. *Land Use Policy* **2016**, *58*, 394–402. [CrossRef]
31. Ababaei, B.; Etedali, H.R. Estimation of water footprint components of Iran's wheat production: Comparison of global and national scale estimates. *Environ. Process.* **2014**, *1*, 193–205. [CrossRef]

32. Bouatrous, A.; Harbaoui, K.; Karmous, C.; Gargouri, S.; Souissi, A.; Belguesmi, K.; Cheikh Mhamed, H.; Gharbi, M.S.; Annabi, M. Effect of Wheat Monoculture on Durum Wheat Yield under Rainfed Sub-Humid Mediterranean Climate of Tunisia. *Agronomy* **2022**, *12*, 1453. [CrossRef]
33. Kourat, T.; Smadhi, D.; Madani, A. Modeling the Impact of Future Climate Change Impacts on Rainfed Durum Wheat Production in Algeria. *Climate* **2022**, *10*, 50. [CrossRef]
34. Drugova, T.; Curtis, K.R.; Akhundjanov, S.B. Organic wheat products and consumer choice: A market segmentation analysis. *Br. Food J.* **2020**, *122*, 2341–2358. [CrossRef]
35. Draghici, M.; Niculita, P.; Popa, M.; Duta, D. Organic Wheat Grains and Flour Quality versus Conventional Ones—Consumer versus Industry Expectations. *Rom. Biotechnol. Lett.* **2011**, *16*, 6572–6579.
36. Montemurro, F.; Maiorana, M. Agronomic Practices at Low Environmental Impacts for Durum Wheat in Mediterranean Conditions. *J. Plant Nutr.* **2016**, *38*, 624–638. [CrossRef]
37. Tudisca, S.; di Trapani, A.M.; Sgroi, F.; Testa, R. Organic farming and economic sustainability: The case of Sicilian durum wheat. *Qual.-Access Success* **2014**, *15*, 93–96.
38. Fagnano, M.; Fiorentino, N.; D'Egidio, M.G.; Quaranta, F.; Ritieni, A.; Ferracane, R.; Raimondi, G. Durum Wheat in Conventional and Organic Farming: Yield Amount and Pasta Quality in Southern Italy. *Sci. World J.* **2012**, *2012*, 973058. [CrossRef] [PubMed]
39. Ruini, L.; Marino, M.; Pignatelli, S.; Laio, F.; Ridolfi, L. Water footprint of a large-sized food company: The case of Barilla pasta production. *Water Resour. Ind.* **2013**, *1–2*, 7–24. [CrossRef]
40. Amicarelli, V.; Lagioia, G.; Gallucci, T.; Dimitrova, V. The water footprint as an indicator for managing water resources. The case of Italian olive oil. *Int. J. Sustain. Econ.* **2011**, *3*, 425–439. [CrossRef]
41. Finco, A.; Bucci, G.; Belletti, M.; Bentivoglio, D. The Economic Results of Investing in Precision Agriculture in Durum Wheat Production: A Case Study in Central Italy. *Agronomy* **2021**, *11*, 1520. [CrossRef]
42. Frascarelli, A.; Ciliberti, S.; Magalhães de Oliveira, G.; Chiodini, G.; Martino, G. Production Contracts and Food Quality: A Transaction Cost Analysis for the Italian Durum Wheat Sector. *Sustainability* **2021**, *13*, 2921. [CrossRef]
43. Ciliberti, S.; Del Sarto, S.; Frascarelli, A.; Pastorelli, G.; Martino, G. Contracts to Govern the Transition towards Sustainable Production: Evidence from a Discrete Choice Analysis in the Durum Wheat Sector in Italy. *Sustainability* **2020**, *12*, 9441. [CrossRef]
44. Bux, C.; Amicarelli, V. Separate collection and bio waste valorization in the Italian poultry sector by material flow analysis. *J. Mater. Cycles Waste Manag.* **2022**, *24*, 811–823. [CrossRef] [PubMed]
45. Hendriks, C.R.; Obernosterer, D.; Müller, S.; Kytzia, P.; Brunner, B.P.H. Material flow analysis: A tool to support environmental policy decision making Case-studies on the city of Vienna and the Swiss lowlands. *Int. J. Justice Sustain.* **2000**, *5*, 311–328. [CrossRef]
46. Yildiz, T. An Input-Output Energy Analysis of Wheat Production in Çarşamba District of Samsun Province. *J. Agric. Fac. Gaziosmanpasa Univ.* **2016**, *33*, 10–20. [CrossRef]
47. Brunner, P.H.; Rechberger, H. *Handbook of Material Flow Analysis. For Environmental, Resource and Waste Engineers*, 2nd ed.; CRC Press: Boca Raton, FL, USA; Taylor & Francis Group: London, UK; LLC: New York, NY, USA, 2017.
48. Camana, D.; Manzardo, A.; Toniolo, S.; Gallo, F.; Scipioni, A. Assessing environmental sustainability of local waste management policies in Italy from a circular economy perspective. An overview of existing tools. *Sustain. Prod. Consum.* **2021**, *27*, 613–629. [CrossRef]
49. Courtonne, J.-Y.; Alapetite, J.; Longaretti, P.-Y.; Dupré, D.; Prados, E. Downscaling material flow analysis: The case of the cereal supply chain in France. *Ecol. Econ.* **2015**, *118*, 67–80. [CrossRef]
50. Courtonne, J.-Y.; Longaretti, P.-Y.; Alapetite, J.-Y.; Dupré, D. Environmental Pressures Embodied in the French Cereals Supply Chain. *J. Ind. Ecol.* **2016**, *20*, 423–434. [CrossRef]
51. Brock, P.; Madden, P.; Schwenke, G.; Herridge, D. Greenhouse gas emissions profile for 1 tonne of wheat produced in Central Zone (East) New South Wales: A life cycle assessment approach. *Crop Pasture Sci.* **2012**, *63*, 319–329. [CrossRef]
52. Holka, M.; Jankowiak, J.; Bienkowski, J.F.; Dabrowicz, R. Life Cycle Assessment (LCA) of winter wheat in an intensive crop production system in Wielkopolska Region (Poland). *Appl. Ecol. Environ. Res.* **2016**, *14*, 535–545. [CrossRef]
53. McAuliffe, G.A.; Takahashi, T.; Lee, M. Applications of nutritional functional units in commodity-level life cycle assessment (LCA) of agri-food systems. *Int. J. Life Cycle Assess.* **2020**, *25*, 208–221. [CrossRef]
54. Tamburini, E.; Pedrini, P.; Marchetti, M.G.; Fano, E.A.; Castaldelli, G. Life Cycle Based Evaluation of Environmental and Economic Impacts of Agricultural Productions in the Mediterranean Area. *Sustainability* **2015**, *7*, 2915–2935. [CrossRef]
55. Fischer-Kowalski, M.; Krausmann, F.; Giljum, S.; Lutter, S.; Mayer, A.; Bringezu, S.; Moriguchi, Y.; Schutz, H.; Schandl, H.; Weisz, H. Methodology and Indicators of Economy-wide Material Flow Accounting. State of the Art and Reliability Across Sources. *J. Ind. Ecol.* **2011**, *15*, 855–876. [CrossRef]
56. Official Journal of the European Commission. Council Regulation (EC) No 834/2007 of 28 June 2007 on Organic Production and Labelling of Organic Products and Repealing Regulation (EEC) No 2092/91. 2007. Available online: https://eur-lex.europa.eu/legal-content/EN/ALL/?uri=CELEX%3A32007R0834 (accessed on 25 February 2022).
57. Protezione Civile Puglia. Annali Idrologici—Parte I—Download dal 1921 al 2020. 2022. Available online: https://protezionecivile.puglia.it/centro-funzionale-decentrato/rete-di-monitoraggio/annali-e-dati-idrologici-elaborati/annali-idrologici-parte-i-download/ (accessed on 17 February 2022).

58. Longobardi, A.; Buttafuoco, G.; Caloiero, T.; Coscarelli, R. Spatial and temporal distribution of precipitation in a Mediterranean area (southern Italy). *Environ. Earth Sci.* **2016**, *75*, 189. [CrossRef]
59. Caporali, E.; Lompi, M.; Pacetti, T.; Chiarello, V.; Fatichi, S. A review of studies on observed precipitation trends in Italy. *Int. J. Climatol.* **2020**, *41*, E1–E25. [CrossRef]
60. Golova, E.E.; Baranova, I.V.; Gapon, M.N. Crop Production Cost Accounting Audit. In *Land Economy and Rural Studies Essentials*; Nardin, D.S., Stepanova, O.V., Kuznetsova, V.V., Eds.; European Publisher: London, UK, 2021; Volume 113, pp. 72–78. [CrossRef]
61. Rahman, S.M.M.; Kim, J. Circular economy, proximity, and shipbreaking: A material flow and environmental impact analysis. *J. Clean. Prod.* **2020**, *259*, 120681. [CrossRef]
62. European Union. *Agriculture and Food Security in Climate Sensitive Areas in the Mediterranean*; Commission for Citizenship, Governance, Institutional and External Affairs: Bruxelles, Belgium, 2020. [CrossRef]
63. Schiller, G.; Gruhler, K.; Ortlepp, R. Continuous Material Flow Analysis Approach for Bulk Nonmetallic Mineral Construction Materials Applied to the German Building Sector. *J. Ind. Ecol.* **2017**, *21*, 673–688. [CrossRef]
64. Eisenhardt, K.M. Building theories from case study research. In *The Qualitative Researchers' Companion*; Huberman, A.M., Miles, M.B., Eds.; Sage Publications: Thousand Oaks, CA, USA, 2002.
65. Amicarelli, V.; Fiore, M.; Bux, C. Hidden flows assessment in the agri-food sector: Evidence from the Italian beef system. *Br. Food J.* **2021**, *123*, 384–403. [CrossRef]
66. Noble, H.; Heale, R. Triangulation in research, with examples. *Evid.-Based Nurs.* **2019**, *22*, 67–68. [CrossRef]
67. Patrício, J.; Kalmykova, Y.; Rosado, L.; Lisovskaja, V. Uncertainty in Material Flow Analysis Indicators at Different Spatial Levels. *J. Ind. Ecol.* **2015**, *19*, 837–852. [CrossRef]
68. da Silva, V.P.R.; da Silva, B.B.; Albuquerque, W.G.; Borges, C.J.R.; de Sousa, I.F.; Neto, J.D. Crop coefficient, water requirements, yield and water use efficiency of sugarcane growth in Brazil. *Agric. Water Manag.* **2013**, *128*, 102–109. [CrossRef]
69. Nielsen, K.M. Organic Farming. In *Encyclopedia of Ecology*, 2nd ed.; Fath, B., Ed.; Elsevier: Amsterdam, The Netherlands, 2019; pp. 550–558. [CrossRef]
70. Yadav, S.K.; Subhash Babu, M.K.; Yadav, K.S.; Yadav, G.S.; Pal, S. A Review of Organic Farming for Sustainable Agriculture in Northern India. *Int. J. Agron.* **2013**, *2013*, 718145. [CrossRef]
71. Liu, H.; Meng, J.; Bo, W.; Cheng, D.; Li, Y.; Guo, L.; Li, C.; Zheng, C.; Liu, M.; Ning, T.; et al. Biodiversity management of organic farming enhances agricultural sustainability. *Sci. Rep.* **2016**, *6*, 23816. [CrossRef] [PubMed]
72. Jouzi, Z.; Azadi, H.; Taheri, F.; Zarafshani, K.; Gebrehiw, K.; Van Passel, S.; Lebailly, P. Organic farming and small-scale farmers: Main opportunities and challenges. *Ecol. Econ.* **2017**, *132*, 144–154. [CrossRef]
73. Joshi, N.; Prasad Parewa, H.; Joshi, S.; Sharma, J.K.; Shukla, U.N.; Paliwal, A.; Gupta, V. Chapter 5—Use of microbial biostimulants in organic farming. In *Advances in Organic Farming*; Meena, V.S., Meena, S.K., Rakshit, A., Stanley, J., Srinivasarao, C., Eds.; Woodhead Publishing: Sawston, UK, 2021; pp. 59–73. [CrossRef]
74. Post, E.; Schahczenksi, J. Understanding Organic Pricing and Costs of Production. In *National Sustainable Agriculture Information Service*; ATTRA: Butte, MT, USA, 2012; pp. 1–12.
75. Durham, T.C.; Tamás, M. Comparative Economics of Conventional, Organic, and Alternative Agricultural Production Systems. *Economies* **2021**, *9*, 64. [CrossRef]
76. Slorach, P.C.; Jeswani, H.K.; Cuéllar-Franca, R.; Azapagic, A. Environmental sustainability in the food-energy-water-health nexus: A new methodology and an application to food waste in a circular economy. *Waste Manag.* **2020**, *113*, 359–368. [CrossRef]
77. Karamian, F.; Mirakzadeh, A.A.; Azari, A. The water-energy-food nexus in farming: Managerial insights for a more efficient consumption of agricultural inputs. *Sustain. Prod. Consum.* **2021**, *27*, 1357–1371. [CrossRef]
78. Sidhoum, A.A.; Dakpo, K.H.; Latruffe, L. Trade-offs between economic, environmental and social sustainability on farms using a latent class frontier efficiency model: Evidence for Spanish crop farms. *PLoS ONE* **2022**, *17*, e0261190. [CrossRef]
79. Viganò, E.; Maccaroni, M.; Righi, S. Finding the right price: Supply chain contracts as a tool to guarantee sustainable economic viability of organic farms. *Int. Food Agribus. Manag. Rev.* **2022**, *23*, 411–426. [CrossRef]
80. European Commission. Organic Action Plan. 2022. Available online: https://ec.europa.eu/info/food-farming-fisheries/farming/organic-farming/organic-action-plan_it (accessed on 29 June 2022).
81. Solfanelli, F.; Ozturk, E.; Pugliese, P.; Zanoli, R. Potential outcomes and impacts of organic certification in Italy: An evaluative case study. *Ecol. Econ.* **2021**, *187*, 107107. [CrossRef]
82. Saber, Z.; van Zelm, R.; Pirdashti, H.; Schipper, A.M.; Esmaeili, M.; Motevali, A.; Nabavi-Pelesaraei, A.; Huijbregts, M.A.J. Understanding farm-level differences in environmental impact and eco-efficiency: The case of rice production in Iran. *Sustain. Prod. Consum.* **2021**, *27*, 1021–1029. [CrossRef]
83. *ST/ESCAP/2561*; Eco-Efficiency Indicators: Measuring Resource-Use Efficiency and the Impact of Economic Activities on the Environment. United Nations: New York, NY, USA, 2009.

Article

Skill Needs for Sustainable Agri-Food and Forestry Sectors (I): Assessment through European and National Focus Groups

Luis Mayor [1,*], Line F. Lindner [1], Christoph F. Knöbl [1], Ana Ramalho [1,2], Remigio Berruto [3], Francesca Sanna [3], Daniele Rossi [4], Camilla Tomao [4], Billy Goodburn [5], Concha Avila [6], Marg Leijdens [7], Katharina Stollewerk [8], Michael Bregler [9], Christos Koidis [10], Alexandre Morin [11], Vesna Miličić [12], Giulia Fadini [13], Jonas Lazaro-Mojica [14] and Patrizia Busato [3]

1 ISEKI-Food Association, Impacthub Vienna, Lindengasse 56, 1070 Vienna, Austria
2 Research Centre for Natural Resources, Environment and Society (CERNAS), Coimbra Agriculture School, Bencanta, 3045-601 Coimbra, Portugal
3 Department of Agriculture, Forestry and Food Science (DISAFA), University of Turin Largo Paolo Braccini 2, 10095 Grugliasco, Italy
4 Confagricoltura, Corso Vittorio Emanuele II, 101, 00186 Roma, Italy
5 Irish Co-Operative Organisation Society, The Plunkett House, 84 Merrion Square, DO2 T882 Dublin, Ireland
6 Federación Española de Industrias de Alimentación y Bebidas, Velázquez 64, 28001 Madrid, Spain
7 AERES, Barnseweg 3, Postbus 331, 3770 AH Barneveld, The Netherlands
8 Lebensmittelversuchsanstalt, Zaunergasse 1-3, 1030 Vienna, Austria
9 Hohenheim Research Center for Bioeconomy, University of Hohenheim, Schloss Hohenheim 1, 70599 Stuttgart, Germany
10 Engineers for Business S.A., Doiranis 17, 54639 Thessaloniki, Greece
11 Association des Chambres d'Agriculture de l'Arc Atlantique, Maison de l'Agriculture, Rue P.A. Bobierre-La Géraudière, 44939 Nantes, France
12 Gospodarska Zbornica Slovenije, Dimičeva ulica 13, 1000 Ljubljana, Slovenia
13 Confederation of European Paper Industries, Avenue Louise, 250 Box 80, 1050 Brussels, Belgium
14 FoodDrinkEurope, Avenue des Nerviens 9-31, 1040 Brussels, Belgium
* Correspondence: luis.mayor@iseki-food.net

Citation: Mayor, L.; Lindner, L.F.; Knöbl, C.F.; Ramalho, A.; Berruto, R.; Sanna, F.; Rossi, D.; Tomao, C.; Goodburn, B.; Avila, C.; et al. Skill Needs for Sustainable Agri-Food and Forestry Sectors (I): Assessment through European and National Focus Groups. *Sustainability* **2022**, *14*, 9607. https://doi.org/10.3390/su14159607

Academic Editors: Mariarosaria Lombardi, Erica Varese and Vera Amicarelli

Received: 6 June 2022
Accepted: 1 August 2022
Published: 4 August 2022

Abstract: The agri-food and forestry sectors are under increasing pressure to adapt to climate change, consumer concern, technological and economic change, and complex global value chains. In turn, such challenges require that the necessary skills and competences are identified at various levels and within specific areas of the sectors. For that purpose, eleven focus groups in nine different EU-countries and two at EU-level were organized within the ERASMUS+ project "FIELDS" with the participation of farmers, cooperatives, agri-food companies, foresters, forest industries, advisors, and education providers to identify the skills needed in the agri-food and forestry sectors. The focus group participants identified business and strategic management skills, communication skills, and other skills related to sustainability, entrepreneurship, digital and soft skills to be most important for the agri-food and forestry sectors as a whole.

Keywords: education; training; skill needs; farmers; food industry; forestry; focus groups

1. Introduction

Climate change, the greening of products and processes, the reuse of side-stream products, the raised complexity of value chains, and the increased availability of information-driven novel challenges and opportunities in the agri-food and forestry sectors. Agri-food systems are highly dependent on climatic changes and integrally considered major world players in the fight for long term natural resource sustainability and a critical subsystem for the climate change challenge [1,2]. Furthermore, the agri-food and forestry sectors are more and more affected by enabling ICT technologies in practically all parts of the value chain. Such developments draw both sectors into a stream of global level innovation and lead to a readjustment of skills and job profiles.

To successfully address and react to these drivers, the agri-food and forestry workforce needs new skills and competences, which, in turn, require the identification of needed existing and emerging skills in the categories of bioeconomy, sustainability, and digital technology. This is necessary to develop a strategic approach keeping the European agri-food and forestry sectors competitive and sustainable in the long term [3,4].

Significant efforts have been invested in the assessment of skill needs in various working sectors in the last years [5], and in particular, research has been invested into the current and future agri-food and forestry workforce.

For the farming sector, literature reviews and bibliometric analyses have been performed on key trends and challenges for higher education regarding the development of a sustainable and resilient European economy [6]; specifically, farmers' entrepreneurship skills [7] and skill needs for professionals to engage in the transition towards sustainable agriculture [8]. Moreover, empirical research, such as surveys and focus groups, have been conducted examining skill needs in the precision agriculture workforce [9]; personal, communication, and leadership skills desired for agricultural and natural resources industry leaders [10]; and skill needs and competences for agronomists to promote sustainable agriculture [11].

Flynn et al. [12] organized workshops with food industry employers to identify the most desired knowledge, skills, and competences in the food industry workforce. Mayor et al. [13] compared the results by Flynn et al. [12] and carried out a survey to food industry professionals assessing training needs. Furthermore, Handayani et al. [14] performed surveys to identify green skill needs for food industry vocational graduates assessing the current skills in small and medium enterprises in the Thai food industry, and Akyazi et al. [15] developed a database on current and future skills emerging within Industry 4.0 for different food industry professional profiles.

The forestry skills forum, in its forestry workforce 2021 research report (forestry skills forum [16], highlighted the increased importance and lack of non-forestry skills, such as business and commercial skills, digital skills, and marketing/promotional skills. Forestry programs' performance in terms of provision of knowledge and skills for contemporary forestry professionals was assessed [17], while Blanc et al. [18] identified the forestry training sector stakeholders in the Western Italian Alps and described their characteristics and priorities in relation to training activities on entrepreneurial topics for forestry loggers.

The ERASMUS+ "FIELDS" project started in 2020 and aims at addressing the current and future skill needs for sustainability, digitalization, and bioeconomy in the agri-food and forestry sectors. When the project began in 2020, project partners encroached upon the assessment of skill needs through a set of activities that begun with the organization of focus groups (FGs), followed by a European survey and with the implementation of a future scenario analysis. All these activities were complementary and used later in the project to design training activities in different European countries.

FGs are small group discussions in which participants respond to a series of questions focused on a single topic. A skilled facilitator meets with five to twelve people to collect in-depth qualitative information about the group's ideas, perceptions, attitudes, or experiences on the defined topic [19]. FGs originate in marketing research as a method of collecting information about consumer perceptions and attitudes. Today, FGs are a common data collection technique in behavioral and social sciences, business, and in many other areas of knowledge production [20–22]. More specifically in education and training, FGs have been used in different activities, such as training needs assessment, the development of new training methodologies, curricula improvement, and marketing strategies for educational programs [23–25].

FG discussions typically involve face-to-face facilitation, although there is growing interest in utilizing digital technologies to conduct online FGs because of the many advantages: they are easier to attend, it is possible to recruit participants from different locations and the time and resources organizations spend are significantly reduced [26]. Another advantage is that by using online web-conferencing tools, FGs can be easily recorded, making the collection and processing of data easier and less time consuming.

In the FIELDS project, eleven FGs were carried out in nine countries during the period of May to July 2020 to assess the skills and training needs of current and future professionals of the agri-food and forestry sectors in the areas of sustainability, bioeconomy, digitalization, soft skills, and business entrepreneurship skills. The FGs, organized at national (all sectors) and at European Level (Forestry sector and Policy issues), tackled several aspects related to skill needs and best methodologies to carry out the required training. The results were later used to develop a European survey on skill needs [27] and to support a future scenario analysis on the same topic. Due to the COVID pandemic, FGs were held in online synchronous mode.

This work shows the findings of the first part of the FIELDS FGs related to skill needs, and also on the FG on policy aspects linked to those needs.

2. Materials and Methods

2.1. Focus Group Guidelines

FG guidelines were prepared by FIELDS project partners to plan, conduct, and report the findings of the FGs, making sure that a common methodology was followed in all of them. Due to the COVID-19 situation in spring of 2020, it was decided to conduct all FGs online using online web-conferencing tools and following guidelines providing specific online set-up recommendations. While the FGs were structured into two main sections on skill needs and training methods, this paper deals only with the results from the first section on skill needs.

2.2. Focus Group Organization

In the recruitment phase, FG organizers recruited between five and ten participants per FG from at least five of the following stakeholder profiles: farmers, cooperatives, agri-food companies, foresters, forest industries, advisors, education providers, and others (policy makers, market actors, consumers, etc.) and participants with proven experience and/or representing sectors at national level. For the European Forestry and Policy FGs, European associations representing the education, food, and forest industries, farmers and farm cooperatives, and trade unions were recruited. FG organizers sent invitations to participants including an information sheet about the project and an informed consent form to be filled in by participants beforehand.

2.3. Data Collection Previous to the Focus Groups

Skill categories on (A1) sustainability, (A2) digitalization, (A3) bioeconomy, (A4) soft skills, and (A5) business entrepreneurship were prepared by project partners allowing for input from different perspectives and backgrounds: companies, education providers (vocational education and training organizations, high schools and universities), chambers of commerce, national associations of cooperatives and agri-food companies, European representatives of the agri-food and forestry sectors, and others. Within the skill category of bioeconomy, it was decided to distinguish between agriculture (Table A3a), forestry (Table A3b), and food industry (Table A3c) skills. The resulting five FIELDS skills lists are outlined in Appendix A.

For the sake of simplicity, skill lists were shortened, skill concepts kept short, and the term "skills" was identified as a set of knowledge, skills, and competences related to a certain topic (as an example, the skill "communication" included the knowledge, skills, and competences related to communication, and the same applied to all the skills appearing in the lists).

The five skill lists were sent to all FG participants beforehand. Participants were asked to rank in order of importance (where one was most important and five least important), on each of the five skills lists, the five most important skills for the sector they represent (e.g., farmer, forester, food industry, etc.). Furthermore, participants were asked to look at their rankings on each of the five skills lists and select the 10 overall most important skills among all 25 skills and rank them in order of importance (where 1 was the most important and 10 the least important). Data was collected by email through excel sheets.

2.4. Running of the Focus Groups

In the period from May to July 2020, eleven online FGs were held, nine at national level (in Italy, Ireland, Spain/Portugal, Netherlands, Austria, Germany, Greece, France, and Slovenia) and two at a pan-European level (forestry sector training needs and EU policy issues). Table 1 shows the composition of the FGs.

Table 1. Composition of the FGs.

Country	Farmers	Cooperatives	Food Industry	Foresters	Forest Industry	Educators	Advisors	Other	TOT	Main Organizer
Italy	1	1	2	1		5		3	13	Confagricoltura [4]
Ireland	2	2	2	1		3	2	1	13	ICOS [5]
Spain/Portugal	1	2				2	3	2	10	FIAB [6]
Netherlands	1		1	1		3	2		8	AERES [7]
Austria	1	2	1	1	1	5	1		12	LVA [8]
Germany		1	1			1	1	3	7	UHOH [9]
Greece	1	1	1			2	1		6	EFB [10]
France	2		1			3		3	9	AC3A [11]
Slovenia	2	1	1			3	1		8	GZS [12]
EU-Forest		1		2	2	4	5		14	CEPI [13]
EU-Policy	3	1	2	1		4		2	13	FoodDrinkEurope [14]
TOT	14	12	12	7	3	35	16	14	113	

Superscripts correspond to affiliation information.

Each of the FGs followed the same procedure of conduction and the same questions (except for the FG Policy) were posed in all of them. All FGs included a facilitator and a rapporteur, both from the FG organizing institution, who both acted as observers and from whom no data was collected. In the first section of the FG, participants were asked to present the three most important skills on their top 10 list and explain why these skills were important for them, following a Round Robin format [19], where the facilitator asked each person to respond to the same question in turn. The FGs followed a list of questions related to skill needs and training methodologies, although this article focuses on the first *Round Robin* part.

On average, each FG lasted two hours. They were conducted in the national languages with different online web-conferencing tools (GoToMeeting, Zoom, etc.) and digitally recorded for the further processing of information.

2.5. Reporting

Each FG organizer was asked to prepare:

1. An executive summary of the FG.
2. A transcription of the audio file in English. The free YouTube transcription tool was used when available (depending on the language), and when not, transcription was carried out manually from the video recording. When quoting FG participants, data were anonymized by assigning 5-digit ID strings indicating FG country, stakeholder profile, participant number, gender, and role in the FG (participant or moderator/rapporteur).

3. Results and Discussion

3.1. Most Important Skills, All Countries, All Sectors

Ninety-five participants carried out the 10-ranking exercise. Figure 1 shows the most selected skills in the 10-rankings, considering working sectors and countries as a whole.

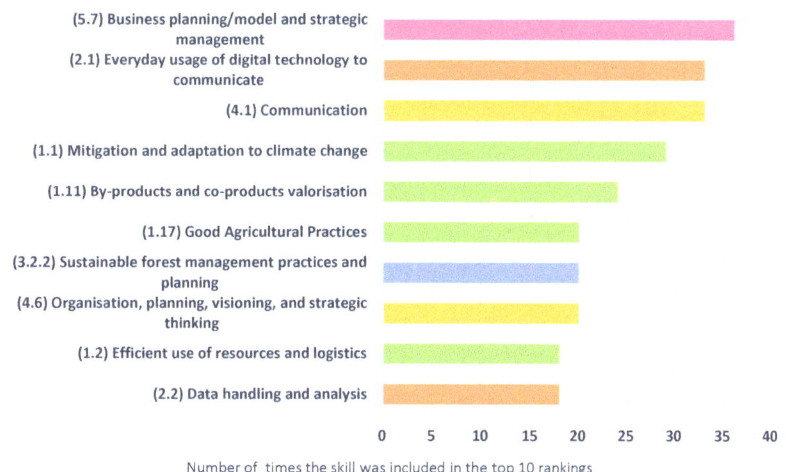

Figure 1. Most selected skills from the skills lists, 10-ranking exercise. In brackets skill numbers from Appendix A tables.

Viewed independently of the sector or country participants represent, skills related to *business planning/model and strategic management* were deemed very important, followed by communication skills. Skills related to doing business, including business management and planning and financial and economic understanding were relevant to all types of stakeholders represented in the FGs. As a cooperative advisor said in the Spanish/Portuguese FG: "All our farmers, forest producers, industries, are in a field of economic development and it must be very clear what their role is, how value is generated, how profit is generated. I have seen how farmers have been successful, even as entrepreneurs, many times in businesses that they did not know [...] But I think they started from a base of understanding the business". Additionally, the need for understanding business and where it is going so as to adapt to current and future challenges was mentioned: "we need to sort of reassess what we do on a day-to-day basis, we can't push additional costs onto the customer, and understand where business is going and be prepared for what's happening in the future" (agri-food industry, Irish FG). In this sense, Bröring and Vanaker [28] stressed the need for designing new business models from a bioeconomy perspective to enable the translation of new emerging technologies into value propositions and, thus, product–market applications.

Following skills related to *business planning/model and strategic management*, FG participants identified *digital skills* to be very important. While digitalization is here to stay, participants also recognized that digital skills entail communication skills and that the acquisition of more demanding digital skills is equally important. As an education provider said in the French FG: "[...] is something that is of fundamental importance today and we realize that codes and instructions for use are not necessarily mastered by everyone". In the same vein, the European Commission stated: "Information and communication technologies profoundly and irreversibly affect the ways of working, accessing knowledge, socializing, communicating, collaborating—and succeeding—in all areas of the professional, social, and personal life of European young people and citizens" [29].

FG participants addressed the importance of *communication skills* at different levels: among day-to-day collaborators, among different actors of the agri-food and forestry sectors, and with consumers and society in general; emphasizing not only one-way communication but as a tool for engagement and interaction with networks and for collaboration concerning higher-level topics, such as sustainability. As a cooperative participant in the Irish FG said: "there's a vast amount of stakeholders that we need to engage with whether that's our own shareholders, board employees, customers, so I think communication skills are critical in terms of creating a rationale for change and making sustainability or some of the complex concepts much more accessible and putting them into more layperson's terms". Additionally, the ability to inform and create awareness about processes and decisions to various stakeholders was highlighted as an important aspect of communication skills: "I think that everywhere we have attacks on agriculture [...] and that agriculture does not know how to communicate what it does, the food it produces daily for everyone and in the times we are living, is important the skill to communicate and learn to publish what we do" (farm advisor, Spanish/Portuguese FG). Communication and social cooperation were identified as skills strongly related to employability for the 21st century [30].

Skills related to sustainability were selected by participants and within this category skills for *mitigation and adaptation to climate change, by-products and coproducts valorization*, and *good agricultural practices* particularly. Skills for *mitigation and adaptation to climate change* were regarded as important to find solutions to exacerbate climate conditions and important for contributing to more sustainable and adaptable agri-food and forestry sectors. As a farmer of the Austrian FG commented: "[...] weather and the climate are constantly changing and extreme weather events must be taken into account, and we must find solutions to this and make adjustments, especially in the producing sector of agriculture". Skills for mitigating and adapting to climate change were also connected to communication skills and the ability to inform and create awareness in society about actions taken by key actors for environmental sustainability. As a cooperative participant from the Irish FG emphasized: "It's absolutely critical that we will be employing people that know this, that are able to talk about this and who can basically go one-to-one with de NGOs and others that are attacking our sector". Skills for valorizing *by products and co-products* were found important in the context of circular economy, resource efficiency, and conservation; their use as an energy source; for innovative product development; or as an opportunity for creating or relocating jobs. As a representative from an agri-food company in France said: "I think that a fundamental element for the bioeconomy is also the knowledge of potential resources, and more specifically the biomass field, there is training to be done in this area"; an Austrian farmer commented: "By-product use and economic usability, is again connected with the bioeconomy, that new value added chains can open up, and if you have something innovative that you can bring to market"; and a French education provider indicated: "When you need jobs that cannot be relocated, the bioeconomy is one of the major activities that allows that, the valorization of co-products is one of them". Skills for *good agricultural practices* from a sustainability perspective were also selected as important for solving day-to-day problems in farming operations, while the normative aspect of these practices was also noted as a market requirement. As a cooperative participant in the Greek FG said: "It is not enough just to harvest and transport to the premises of the cooperative, the product must also meet all these characteristics required by the market. The standards that our customers ask (GLOBAL GAP, ISO ...) are essentially the standards that we give priority to our producers". In this regard, viewing practitioners as change-agents in the move towards more sustainable agri-food chains was also maintained. As an advisor in the Iris FG commented: "[...] to have and to get really simple practices out there into practice on farms to try and give people the confidence, and I suppose an attitude change to how they can do an awful lot to contribute towards those challenges in terms of climate".

Skills for *sustainable forest management practices and planning* were also regarded as important, including forest regulations on sustainability. As a farmer participant in the Austrian FG put it: "[...] forestry already has many regulations on sustainability, and yet there is still more pressure coming from the NGOs in the direction of more and stricter sustainability requirements in the forestry sector". Additionally, innovative forestry practices and products were highlighted: "there are some sustainable products, what you can or cannot make out of wood, so there is also the term non-timber forest products. I believe that a great deal of knowledge can and should still be created here" (forest industry, Austrian FG).

Soft skills related to *organization, planning, visioning, and strategic thinking* were mentioned as important for looking ahead and making sound and long-term choices and decisions. As a cooperative participant of the Spanish/Portuguese FG said: "To do a strategic planning, your digitization needs, your logistics needs, your training of personnel. Not in all agri-food companies has the culture of strategic planning been acquired and it is fundamental". Along the same lines, an education provider in the Slovenian FG pointed out: "We focus too much on the present or on some short-term survival decision-making. Too little emphasis is given to strategic thinking". Bikse et al. [30] identified *perspective taking* as a core skill for employability in the 21st century and also as organizational skills [15].

Skills for the efficient use of resources and logistics were regarded as important for sustainability and for business reasons. Skills for *handling and analyzing data* were seen not only as technical skills but also related to issues such as GDPR and ethics. As an education provider in the Spanish/Portuguese FG said: "This is not only the matter of analyzing the information at a technical level, but it also has many things associated with it: using our personal data and privacy issues, ethical issues ... that is, using the information well". Furthermore, creating awareness about the strong potential that data handling and analysis may have in agriculture and forestry was emphasized by a farmer in the French FG: "Data processing and analysis aspect is only in its infancy. Farmers are big producers of data, but these are not fully exploited because of a lack of effective and competent treatment, certainly on the part of farmers. There is also a lack of visualization of the interest that this can have, particularly in pooling and massifying data". As reported by the European Commission [31], the transition towards Industry 4.0 will require workers to interact with digital interfaces and analyze larger amounts of data in their day-to-day decisions. The awareness of data security and protection will acquire importance as will trust in new technologies.

FG participants' ranking of the most important skills, considering working sectors and countries as a whole (Figure 2), show that the non-technical skills of business and entrepreneurship and soft skills (together accounting for 40% of all skills), followed by sustainability and digital skills, predominate the ranking, while technical bioeconomy skills in the agri-food and forestry sectors are less pre-eminent.

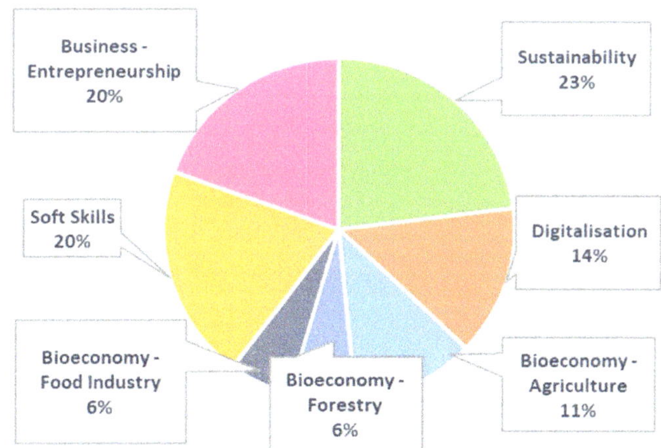

Figure 2. Distribution of skills in categories, 10-ranking exercise.

3.2. *Most Important Skills by Producing Sector*

3.2.1. Farmers

When seen independently, farmers also selected *business planning/model and strategic management* as the most important skill (Figure 3). This is in line with McElwee [32] asserting that farmers are businesspeople running businesses, but in practice they do not necessarily have well-defined business skills. Previous studies suggested that a minority of farmers have a vision or mission statements in written form, and many have no strategic mission at all [33]. Bailey et al. [34], after conducting focus groups with young farmers, found a need for more education relating to agriculture business management skills. Dias et al. [7] stated that farmers can be considered as entrepreneurs and decision-makers aiming to maximize profits to develop various typologies of agricultural diversification. In this sense the authors, based on previous literature, emphasized the importance of differentiating between entrepreneurial skills and management skills. Entrepreneurship is more than management as it is increasingly centered on innovation, risk-taking, and the discovery and exploitation of opportunities [35]. Although successful business creation also requires management skills, it is the entrepreneurial attitude which allows for perceiving opportunities [36]. In the FGs, recognizing business opportunities was included as part of *business planning/model and strategic management* (see Table A5, skill 5.7), inducing that FG participants may have perceived this skill as including both management and entrepreneurial skills.

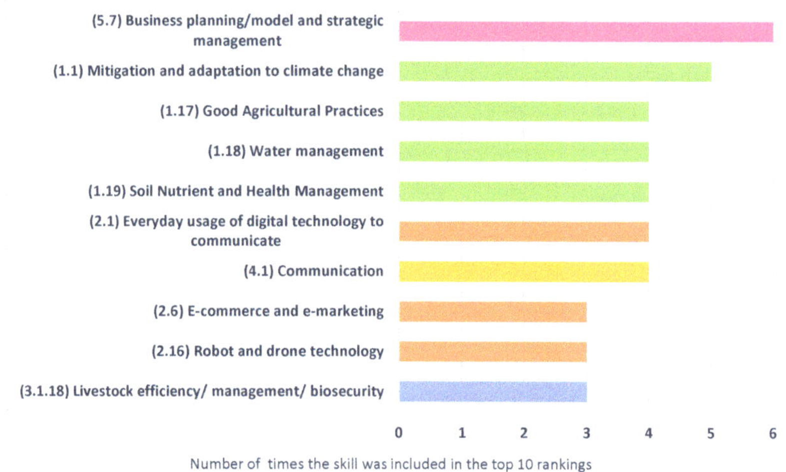

Figure 3. Most selected skills by farmers. In brackets are the skill numbers from Appendix A tables.

Following from *business planning/model and strategic management*, farmers deemed sustainability skills to include *mitigation and adaptation to climate change*, *good agricultural practices*, *water management*, and *soil nutrient and health management* very important. Previous studies suggest that agricultural professionals are not prepared to promote sustainable agriculture [8,11,37]. On water management, FG participants stressed the importance of water (quantity and quality) in agriculture: "It all starts with the water. If there is water, everything can be accomplished" (farmer, Greek FG). Planning and managing technologies and legislation were considered important aspects for training: "There are many advances in technology to do good water management, but we also have to go deep into planning and management, at the level of the European policies. Either we learn to manage properly in this climate change situation that increases conflicts, or we will do a disservice to the next generations" (farmer, Spanish/Portuguese FG). For soil nutrient and health management, the proper soil management and its relation to circularity and sustainability were seen as important topics for participants: "Since resources are becoming scarce, it is actually becoming much more important to operate more circularly and to look at the end how one can produce more sustainably" (farmer, French FG). Similar studies have identified areas where skill needs for a more sustainable agriculture are needed: precision technology; remote sensing to assess land capability; integrated pest management in plant protection;

agricultural reuse of organic residuals; drip irrigation and water-conserving technologies; renewable energy; and bioenergy and energy crops [38].

Two skills related to communication follow in the list: *everyday use of digital technology to communicate* (digital skill) and *communication* (soft skill). Communication skills were also highlighted by Bailey et al. [34], who found that young farmers' desire to enhance their ability to communicate with family members, other people within their farm or ranch, and to develop the skills to communicate to a broader audience, such as loan officers and even the general public. Other works have also found communication skills very important for farmers and other agricultural workers [9–11].

Two digital skills follow in the ranking: *e-commerce and e-marketing* and *robot and drone technology*, and with the same number of citations the list finishes with *livestock efficiency/management/biosecurity* (bioeconomy–technical skill). Regarding *E-commerce and e-marketing*, the need for training was stressed on this set of skills. As a farmer from the Italian FG said: "[. . .] it's not like you make a site and two minutes later you sell your products. With this, there's still some work to be done because logistics costs are not easy for farms. In Italy we always run into problems of logistics not so much with large-scale distribution but with private individual."; and from another farmer from the French FG: "The whole digitalization part is very important, we need it both for our crops, for farm management, and if we want to sell our products for direct sales. If the farmers who are selling directly are not professional enough in managing their site, then there's certainly something to work on that side of it".

3.2.2. Cooperatives

Cooperatives (values, legal framework and management) and *communication* are the most selected skills (Figure 4) among cooperative representatives.

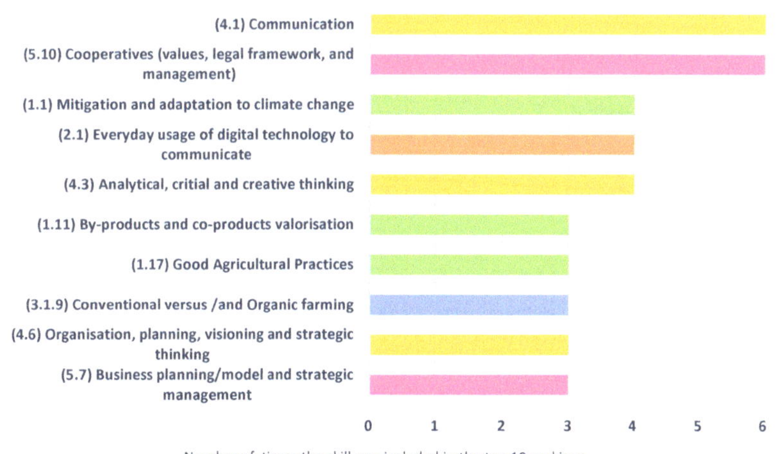

Figure 4. Most selected skills by cooperatives. In brackets skill numbers from Appendix A tables.

A cooperative participant in the Spanish FG remarked the importance of putting cooperatives in their current context: "[...] the objective of the farmer is to obtain a sustainable production and translate it into an appropriate income for his work and risk. This must be contextualized in terms of the problems of the sectoral organization, the positioning of farmers in the value chain and in relation to the context of climate change, use of natural resources, food security and globalization".

Regarding *communication*, as a participant in the Slovenian FG said: "Communication is essential. You need to know how to communicate with the members, to coordinate with them, also in the transaction itself, the purchase, such as for example why is it so, why not otherwise, why such price? If this works well, the cooperative has much less problems than if issues are handled without any discussions".

These findings are in concordance with the "Support for farmers' cooperatives" report [39], which indicated that general technical and entrepreneurial education and training of (future) employees, managers, and board members is necessary, as well as education and training on how cooperative identity translates into business activities. The report also suggested that all parties involved need to be aware of the specific characteristics of this form of collective entrepreneurship and to develop the capacity and the willingness to communicate with each other and jointly develop their businesses.

Skills for *mitigation and adaptation to climate change*; the *everyday use of digital technology to communicate*; and *analytical, critical and creative thinking* follow in the list. For the last, a cooperative participant from the Irish FG remarked on the lack of these skills in young alumni: "[...] I see the lack of these skills in people who have just graduated, at the moment that there is an inability to kind of decipher between facts and fiction".

Five skills follow in the list with the same number of mentions: *by-products and co-products valorization*; *good agricultural practices*; *organization, planning, visioning and strategic thinking*; *conventional vs./and organic farming*; and *business planning/model and strategic management.*

3.2.3. Agri-Food Companies

As observed for farmers as well as for agri-food company participants, *business planning/model and strategic management* (from the category business entrepreneurship skills) was the most selected skill followed by *communication* (soft skill); *ethics for food* (bioeconomy skill); *being resilient, adaptable and proactive* (soft skill); *organization, planning, visioning and strategic thinking* (soft skill); and *collaboration/cooperation across all sectors in the food chain* (business entrepreneurship skill) (Figure 5).

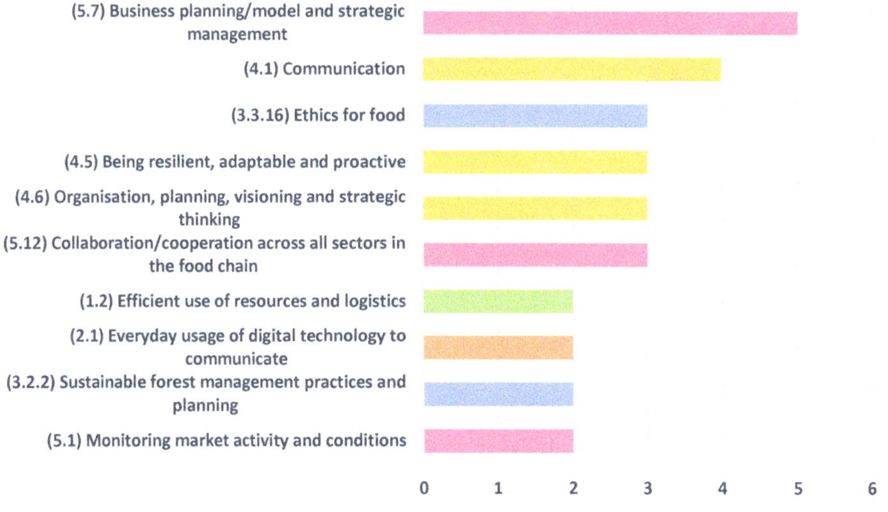

Figure 5. Most selected skills by agri-food companies. In brackets are the skill numbers from Appendix A tables.

Regarding the ability of *being resilient, adaptable, and proactive*, an agri-food company of the Slovenian FG said: "[...] specially in the sense that you are able to adapt to change rapidly and that you are proactive when tackling challenges". It was also viewed as a requirement for group leaders: "being resilient and adaptable and proactive are for sort of management of people, management of the teams etc. Having the ability to deal with day-to-day life and to keep themselves going strong, being adaptable and being able to see changes and make the right decisions on a day-to-day basis" (agri-food industry, Irish FG).

For *collaboration and cooperation across all sectors in the food chain*, an advisor from the German FG commented: "I find this cooperation very important because I often see that there are problems in communication and cooperation, simply because there is no interest or no time or no know-how". It was also viewed as an opportunity to grow businesses: "Open communication in the value chain and pursue win-win situations (advisor, Dutch FG).

Nazzaro et al. [22], found that companies have turned towards business models based on social and environmental protections. Indeed, consumers' purchasing behavior lead stakeholders to adopt sustainable, socially-oriented production models in which natural and environmental resources become a lever for competitive advantage. Consequently, competitiveness changes to address sustainability and citizen-consumer issues.

Akyazi et al. [15] identified current and near-future key skills and competencies emerging with Industry 4.0, demanded by different professional profiles, and generated a database of current and future professions, competencies, and skills. They used the ESCO database of knowledge, skills and competences [40] for current skills and competencies, and data from the European ICT Professional Role Profiles framework and from several sectorial and inter-sectorial European projects for future skill needs and competences. The database focuses future industry needs on digital and soft skills, and some similarities with skills from Figure 5 can be found: *advanced communication skills*, *use of digital communication tools*, and *adaptability and continuous learning*. The need for soft skills, and communication skills in particular, for food industry workers has often been reported in the literature. Jack et al. [41] found shortages in planning and organizing skills in Northern Ireland Food industry workers, especially among managers and supervisors. Flynn et al. [12], in a survey found that the most desired skills for food industry employers were soft skills and more specifically those related to communication. Mayor et al. [13], comparing Flynn results with a survey to food industry employees, confirmed the importance of soft skills also for food industry workers, and also found that marketing, consumer science, and financial skills were evaluated as low by both employers and employees, a situation that may be hindering entrepreneurship. Topliceanu et al. [42] also identified new skills requested by the food industry labor market, such as "people of character, able to work in teams, communicative and capable to cope with stressful situations". Lertpiromsuk et al. [43], assessing current levels of skills in small and medium enterprises in the Thai food industry, found that social skills (including communication, team working, negotiation, etc.) and personal skills (including leadership, flexibility, continuous learning, working under pressure, etc.) are deemed very important for Industry 4.0.

3.2.4. Forestry Sector

For the forestry sector (Forestry FG, Figure 6), overall bioeconomy–forestry and sustainability skills predominated with *sustainable forest management practices and planning* (the bioeconomy–forestry skill related to sustainability being the most selected skill followed by *everyday usage of digital technology to communicate* (digital skill) and *forest disease control and prevention* (bioeconomy–forestry skill); *mitigation and adaptation to climate change* (sustainability skill); *multifunctional forests and ecosystem services* (sustainability skill but related to forestry); *the prevention and management of natural disturbances* (bioeconomy–forestry skill); *communication* (soft skill); *national, EU, and international environmental policies; regulation, subsidy, and support programs* (sustainability skill); *water management* (sustainability skills); *the reforestation, afforestation, and restoration of forest ecosystems* (bioeconomy–forestry skill); and *new value chains/new business models* (business entrepreneurship skill).

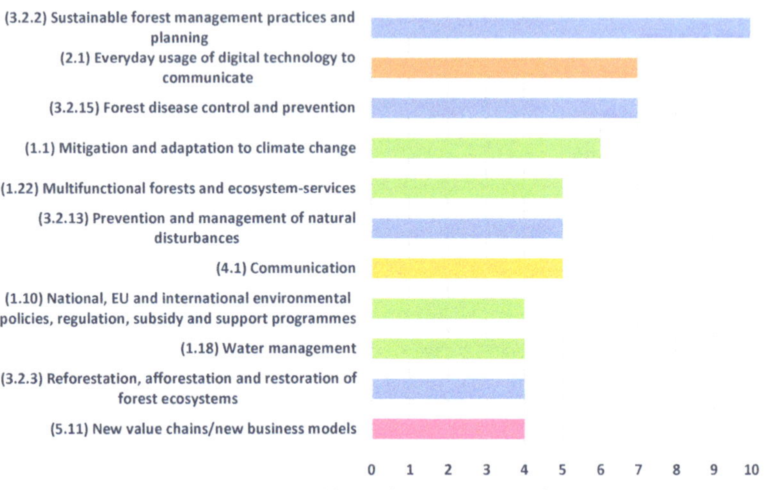

Figure 6. Most selected skills for the forestry FG. In brackets are the skill numbers from Appendix A tables.

As an education provider in the Forestry FG said, *forest disease control and prevention* will need training in the coming years: "Forest Disease Control and Prevention is going to be rising at least here in Finland. As for climate change, there's not that much pests or diseases yet, so we're going to need new skills on that". Additionally, the importance of skills for *prevention and management of natural disturbances* in the future were brought up: "Here in the open range regions of Austria we will have to deal much more with prevention and management of natural disturbances. First, climate change also damages forests and so people have to be more assertive in these strategies and in the interaction between the forest owners and the wood processing industry" (Advisor, Forestry FG). For *national, EU, and international environmental policies and regulation, subsidy, and support programs*, some quotations can be shown but not from the Forestry FG, although they are representative of all the FG sectors. Quotations stress the importance of these knowledge/skills because regulations affect the business strategy, they are helpful in risk prevention and important to obtain national and European funding. As a farmer of the Spanish/Portuguese FG said: "We are in a situation where administrative regulation has been put in place above science, so if we want to focus on making farms profitable, we have to combine the two aspects [...] regulation, subsidies and support programs. In the end this is what marks the path that we are going to follow as farmers, it absolutely conditions the agricultural approach of a farm".

Communication skills have been considered an important need in the forestry sector at least since mid of the past century [44]. Written and oral communication are the basics demanded by forestry employers [45–47] but also communication skills in general [48]. In a broader sense, Bullard [49] reported that forestry employers, recent graduates, and educators agree that entry-level foresters must be better prepared than they are now in terms of communicating relevance and building strong relationships with key forestry stakeholders throughout society.

Bullard et al. [48], in focus groups and surveys to forestry employers and alumni, found that areas of knowledge on emerging issues for society-ready forestry graduates included climate change; water availability and quality; and dealing with invasive plants, pathogens, and insects. However, the skill sets and abilities that involve dealing effectively with people (communication, conflict management, problem solving, etc.) were those that should be prioritized to strengthen the bachelor curricula. Moreover, assessing curricula content in forestry degree programs, Kelly and Brown [17] found that land-management skills should be the focus of curriculum improvement, while additionally, employers and students alike identified professional skills to be very important, especially behaving professionally and ethically and communicating effectively.

More recently, The Forestry Skills Forum [16], emphasized the need for enforcing the core silvicultural knowledge taught across all levels of forestry education, the importance of non-forestry skills, and the general lack of business and commercial acumen, digital skills, marketing/promotional skills, and those of communication and engagement.

3.2.5. Similarities and Differences among Sectors

Skills related to business planning, business management, and business modelling were very important for all FG participants, independent of the producing sector they pertained to. This is supported by the fact that *business planning/model and strategic management* was the most selected skill for farmers and agri-food companies and *cooperative skills* for the cooperatives sector and *sustainable forest management practices and planning* for the forestry sector. Essentially, strategic planning is about setting visions for an organization and realizing that visioning through goals, strategies, and actions; entailing on the one hand the ability to see the big picture and draw road maps, thereby planning, organizing, and executing tasks; and on the other hand, entailing the ability to communicate the vision, goals, tasks, and necessary steps to employees. Participants associated the basis of understanding business, the role of the company, its viability, and how value and profit are generated with this skill. Indeed, business planning and strategic management is very much related to the ability to adapt and respond to current and future technological, environmental, social, and economic challenges.

Communication skills appear to be also very important for all the sectors with *communication* and *everyday usage of digital technology to communicate* being the two skills among the most selected for all of them. The ability to use digital technologies to communicate and the ability to communicate overall were seen as fundamental skills in transferring information to others and in engaging with immediate stakeholders as a means of transferring information about complex concepts, such as sustainability, in an easily understandable way. Several FGs also discussed the broader concept of communication and the importance of engaging with civil society (particularly important for the forestry sector), connecting with consumers and other stakeholders not only for marketing and management purposes but also for sustainability purposes.

A shift towards business and soft skills is observed when advancing in the agri-food value chain from farmers to cooperatives and in the food industry where sustainability and digital skills predominate for farmers but are replaced by business and soft skills for cooperatives and the food industry. It seems that, from the FG participants' perspective, skills related to sustainability are critical for producing raw materials and less important when the raw produce is available in the food chain.

3.3. *Main Outcomes from the Policy Focus Group*

3.3.1. Focus Group Design and Implementation

With representatives from agriculture, food industry and forestry sectors, and education at European level, the FG on EU policy issues produced a set of recommendations on how to improve the current policy framework on skills and training in the agri-food and forestry sectors:

To set up the particular goals of this FG, the guiding topics were the following:

- from an EU perspective on skills needs, participants were given the opportunity to share their opinion on the skills lists developed by the FIELDS project partners linking with the work of the national FGs;
- reviewing participants' opinions on the current legislative framework: how EU policies are set in training and education and how they adapt to the sectoral needs,

particularly to the sectors specified. This matter had the intended outcome to have some recommendations on the next steps regarding EU policy.

3.3.2. Skill Lists Used in the National and Forestry Focus Groups

Policy FG participants found the skill lists exhaustive and appropriate for managing the expectations of the labor market. Nevertheless, since the "skills" are defined in a wide sense, it was recommended to expand the lists when used in the design of training and to be more specific in terms of knowledge, skills, and competences for each of the skills of the lists.

For the same purpose of training development and adaptation to the ESCO classification [40], participants recommended the organization of skills into a hierarchy in two ways: considering the level of education (basic, intermediate, higher education) and the level of application (general skills and specific skills that are not essential for all employees but needed in certain sectors or ESCO job profiles).

In accordance with national FGs, policy FG participants agreed on the importance of developing new skills and competences to better communicate with society and increase consumers' trust in the agri-food and forestry sectors, particularly in aspects, such as environmental and social sustainability and risk management.

3.3.3. Improvement of the EU Policy Landscape

The first aspect tackled by policy FG participants was the harmonization of national educational systems for job profiles and related skills. When it comes to training activities, there is often no equal recognition in different EU countries and there is a clear discrepancy between the national and international level training. Referring to the ESCO framework, which includes all skills [40], participants highlighted the challenge of harmonizing skill concepts between countries or regions, as it is often difficult to understand the names of job positions and the skills needed because of the language barrier and as the definition of a job position and related skills are different among EU countries. Further dialogue is needed to overcome these differences since skills and training might differ among national and international policies but needs do not.

Furthermore, policy FG participants stressed the lack of understanding of the current skill challenges faced by employers and employees in the agri-food and forestry sectors. Participants agreed that the "Social Dialogue" should be strengthened, fostering communication between employers and employees at both EU and member state level, and on relevant topics such as how to train the current workforce. A cross-sectorial stakeholder dialogue (policy, academic, corporate, etc., actors) was also identified as particularly important to develop skills addressing contemporary economic and sustainability issues [6].

Social partners must provide evidence-based practices for policy makers. Good examples in this field are the recognition of universities as capacity-building entities or projects that bring together a community of different organizations and experts around a specific topic. Another best practice, as suggested by participants, could be to ensure access to lifelong learning for the entire workforce, which is, in general, poorly trained in the farming sector [33]. In the end, the exchange of best practices is a powerful tool to improve EU policy.

Finally, the need to increase the attractiveness of the agri-food and forestry sectors was stressed as one of the main causes of the continuous decline of the labor force in the last decades [50].

3.3.4. Recommendations/Key Messages

The final recommendations/key messages of the policy FG are presented below:

- The skills gap should be explored and even forecasted to design the training of the future.
- Curricula at universities and training centers must be adapted to the sector needs, adjusting the homologation and recognition of skills and experience.

- European strategies on education and training for workers should take a holistic approach, addressing the complexity of food and forestry systems and ensuring the coherence of the skills and training provided at the EU level in the agri-food and forestry sectors [8,51].
- The agri-food sector is the largest in Europe; there is a need to establish bigger alliances to reshape the scenario in order to support farmers and the food industry.
- There must be an evidence-based approach to provide guidelines to policy makers in the field of education in the agri-food sector. These recommendations should be given by sectorial and educational representatives in collaboration with policy makers.
- The Social Dialogue should be reinforced to promote the interaction between employers and employees in order to set the basis for the needed training and skills.
- There is a need to increase the attractiveness of the agri-food and forestry sectors for the younger generation.

4. Conclusions

Skills related to business planning, business management. and business modelling were very important for all the FG participants, independently of the producing sector they pertained to. As such, *business planning/model and strategic management* was the most selected skill for farmers and agri-food companies, *cooperatives* for the cooperatives sector, and *sustainable forest management practices and planning* for the forestry sector.

Communication skills appear to be also very important for all sectors, *communication* and the *everyday usage of digital technology to communicate* being the two skills that were among the most selected.

A shift towards business and soft skills was observed when advancing in the agri-food value chain (farmers–cooperatives–food industry). Sustainability and digital skills predominate in the lists for farmers, but these skills are replaced by business and soft skills for cooperatives and the food industry.

Differences among educational systems at a national level were discussed in the EU-policy FG and the need for further harmonization between EU and national policies regarding education was agreed. Education and training curricula must be adapted to sectorial needs, addressing the complexity of the food and forestry systems, establishing bigger sector alliances, and reinforcing the social dialogue. The need to increase the attractiveness of the agri-food and forestry sectors for the younger generation was also remarked.

Undoubtedly, the study has limitations related to the use of FGs that do not allow for the generalization of results; however, further empirical studies could be performed focusing on specific stakeholder groups in order to validate the research results. The results from the national and policy FGs fed into the complimentary development of a quantitative European-wide survey on skills needs and will form the basis for the implementation of a future scenario analysis at both sectorial and national levels. For that purpose, the FGs were found to be successful activities for the identification of the most important skills in the topics of sustainability, bioeconomy, digitalization, soft skills, and business entrepreneurship skills.

Author Contributions: Conceptualization: R.B., P.B. and D.R.; methodology: L.M. and L.F.L.; formal analysis: L.M., L.F.L., C.F.K., A.R., C.T., B.G., C.A., M.L., K.S., M.B., C.K., A.M., V.M., G.F. and J.L.-M.; investigation: F.S., C.T., B.G., C.A., M.L., K.S., M.B., C.K., A.M., V.M., G.F. and J.L.-M.; writing—original draft preparation: L.M., L.F.L., C.F.K. and A.R.; writing—review and editing: F.S., R.B., P.B., D.R., C.T., B.G., C.A., M.L., K.S., M.B., C.K., A.M., V.M., G.F. and J.L.-M.; supervision: L.M. and L.F.L.; project administration: R.B., F.S. and P.B.; funding acquisition: R.B., P.B. and D.R. All authors have read and agreed to the published version of the manuscript.

Funding: This research was co-funded by the ERASMUS+ programme of the European Union, grant agreement number 612664-EPP-1-2019-1-IT-EPPKA2-SSA-B. This publication reflects only the authors' view, and the European Union is not responsible for any use that may be made of the information it contains.

Informed Consent Statement: Informed consent was obtained from all subjects involved in the study. The study was conducted in accordance with the Declaration of Helsinki, and the protocol was approved by the High Steering Committee of the FIELDS ERASMUS+ project (612664-EPP-1-2019-1-IT-EPPKA2-SSA-B).

Data Availability Statement: The data presented in this study are available on request from the corresponding author. The data are not publicly available due to privacy.

Acknowledgments: The authors would like to express their gratitude to the FIELDS project partners who contributed to the conduction and reporting of the focus groups.

Conflicts of Interest: The authors declare no conflict of interest. The funders had no role in the design of the study; in the collection, analyses, or interpretation of data; in the writing of the manuscript, or in the decision to publish the results.

Appendix A Skill Lists

Table A1. Sustainability skills.

Skill No.	Climate Change
1.1	Mitigation and adaptation to climate change incl. climate change competences (weather extremes; interdependency of climate systems and biospheres) and climate change mitigation (e.g., fostered CO2 sequestration) and adaptation (e.g., species composition) via sustainable forest management
	Sustainable management of resources
1.2	Efficient use of resources and logistics
1.3	Improved agri-food system productivity incl. the sustainable management of water, streams, and energy in the food industry
1.4	Active management of natural resources
1.5	Integrated pest management (incl. the sustainable use of pesticides)
1.6	Biodiversity (incl. the detection and support of biodiversity of plants and animals)
1.7	Sustainable metrics and certification incl. public and private schemes for certification (e.g., green labels) and Life Cycle Assessment, Life Cycle Analysis data, including PEF (Product Environmental Footprint)
	Sustainable Business and Governance Models and Environmental Policy
1.8	Environmental Management Systems
1.9	Corporate social responsibility associated with sustainability reporting/press releases
1.10	National, EU, and international environmental policies, regulation, subsidy, and support programs
	Circularity
1.11	By-products and co-products valorization incl. the treatment and reuse of reclaimed water; inorganic waste management practices; the agricultural valorization of organic fertilizers; the management of slurry in livestock farms; biodegradable and compostable materials (incl. packaging); the valorization of forestry residues and new industrial technologies in pulp and paper manufacturing; the use of by-products of timber harvesting (nutrients circulation vs. nutrients removal); circular economy and recycling in the pulp and paper industry; and the reuse, recycling, and valorization of raw materials, contact materials (packaging), by-products, and waste in the food industry
	Energy
1.12	Generation, storage, and the use of renewable energies incl. the next generation bio-refineries and bio-product mills and their outlets, residual forest wood products to produce energy and design, and the building and operation of renewable energy systems
1.13	The identification of renewable energy systems suitable for farm/business enterprises
1.14	The identification of raw materials and waste for energy production in farm/business enterprises
1.15	The identification of energy consumption and demand on farm/business enterprises

Table A1. *Cont.*

Skill No.	Climate Change
1.16	National and EU energy markets
	Specific skills for sustainable agriculture
1.17	Good Agricultural Practices incl. the global GAP and international standards of good practices in agriculture
1.18	Water management incl. water quality control and protection, water saving cultivation, tools, and models for saving water and selecting the proper crop pattern at the farm level, irrigation management and techniques, advisor services for irrigation water management to improve training, information and knowledge transfer, and the optimization of irrigation system design and management and associated energies
1.19	Soil Nutrient and Health Management incl. soil protection and improvement, the maintenance of permanent vegetal soil cover and minimum tillage, and techniques for carbon sequestration in the soil
	Specific skills for sustainable forestry
1.20	The impact of timber harvesting and other forest management practices in wildlife populations and habitats
1.21	The protective role of forests and their management in mountainous areas
1.22	Multifunctional forests and ecosystem-services
	Specific skills for sustainable food industry
1.23	Organic production requirements
1.24	The analysis of contaminants
1.25	Sustainable packaging

Table A2. Digitalization skills.

Skill No.	General Digital Skills
2.1	Everyday usage of digital technology to communicate incl, the use of computers, tablets or mobile phones; word processing; sending emails; browsing the internet safely; making video calls; and social media networks
2.2	Data handling and analysis
2.3	Data protection
2.4	Cloud technology
2.5	Smart connected devices incl.., general principles, categories, requirements, limitations, and vulnerabilities
	Digitalization for business
2.6	E-commerce and e-marketing
2.7	Digital entrepreneurship
2.8	Digital information and services (e.g., product prices and standards, payment services, advisory services)
	Digital tools to support production and production management
2.9	Digital supplier management systems
2.10	Digital product quality management systems
2.11	Warehouse management systems
2.12	Digital food traceability systems
2.13	Digital reversed logistics systems
2.14	Digital pest control systems

Table A2. *Cont.*

Skill No.	General Digital Skills
2.15	Decision support systems incl. control technology with decision support tools (DST) and the use of web SIG platforms (or ICT platforms) including sensors network models and tools for DSS within a feedback process
2.16	Robot and drone technology
	Specific skills for digital (smart) farming
2.17	Farm Management Information Systems (FMIS)
2.18	Precision animal health and productivity management systems (incl. feed intake management)
2.19	Field operations management systems (incl. soil, plant, seed, and yield management systems)
2.20	Digital irrigation control systems
2.21	Digital soil nutrient control systems
2.22	Weather data management systems/software
2.23	Climate control systems (incl. indoor and roofed farming (greenhouses and roofed fields)
2.24	Robot and drone technology in agriculture

Table A3. Bioeconomy skills.

	(a) Bioeconomy Skills—Agriculture
Skill No.	**Basic Skills in Production Operations and Production Management**
3.1.1	Planning and coordinating production
3.1.2	Performing farming operations
3.1.3	Equipment maintenance
3.1.4	Logistics and storage incl. storage techniques and requirements of different raw materials, the transportation of livestock (incl. droving), produce and supplies and warehouse management
3.1.5	Calculating, handling, and managing risk
3.1.6	Health and safety management and operations incl. to drive and operate agricultural machines safely
3.1.7	Product traceability
	Technologies, products, and production approaches
3.1.8	Urban, peri-urban, and rural area agriculture
3.1.9	Conventional versus/and organic farming incl. organic farming and hybrid farming (the combination of organic and conventional farming methods)
3.1.10	Controlled Environment Agriculture
3.1.11	Crop diversification and rotation
3.1.12	New plant breeding techniques
3.1.13	Agricultural biodiversity
3.1.14	Genetically Modified Crops
3.1.15	New industrial crops and bioproducts for the bioeconomy incl. bioproducts: biofuels, bioplastics, biochemicals, textiles, cosmetics and pharmaceuticals and new industrial crops: cameline, hemp, castor, guayule, etc.
3.1.16	Biofertilizers, compost, and bio-digestates
	Healthy farm
3.1.17	Animal care and animal welfare during transport and production

Table A3. *Cont.*

	(a) Bioeconomy Skills—Agriculture
3.1.18	Livestock efficiency/management/biosecurity
3.1.19	Crop protection (incl. the prevention of crop disorders and the use of plant protection products)
3.1.20	Integrated pest/disease management
3.1.21	Plant and animal breeding for resilience and robustness
	(b) Bioeconomy skills—Forestry and related industries
Skill No.	**Production operations, technologies, and production approaches**
3.2.1	Characteristics of forests, geographical differences, and ownership patterns
3.2.2	Sustainable forest management practices, and planning
3.2.3	The reforestation, afforestation, and restoration of forest ecosystems
3.2.4	Forest equipment/machinery and maintenance
3.2.5	Health and safety management and operations
3.2.6	Calculating, handling, and managing risk
3.2.7	Products of forestry incl. harvesting on the focus of high quality/high value logs (right shaping of logs), logs for construction, timber for the pulp and paper industry, and timber for energy supply (material use before energy use)
3.2.8	Process operations in the pulp, paper, timber, and cork industry
3.2.9	Safety and health in the pulp, paper, timber, and cork industry
3.2.10	Equipment/machinery and maintenance in the pulp, paper, timber, and cork industry
3.2.11	Automation in the pulp, paper timber and cork industry
3.2.12	New technologies in pulp, paper, timber, and cork manufacturing
	Healthy forest
3.2.13	The prevention and management of natural disturbances (e.g., floods, drought, and forest fires)
3.2.14	Seedling damage incl. that caused by e.g., deer, moose, and other mammals
3.2.15	Forest disease control and prevention
3.2.16	Water quality in forests
	(c) Bioeconomy skills—Food industry
Skill No.	**Skills for food quality and food safety**
3.3.1	Quality management, quality assurance, and quality control incl. sensory evaluation
3.3.2	Food safety management, food hygiene, and food safety control
	Skills for food production and manufacturing (industrial performance)
3.3.3	Cleaning and preparation
3.3.4	Production operations and management (incl. milk processing)
3.3.5	Health and safety management
3.3.6	Engineering maintenance
3.3.7	Preservation and packaging
3.3.8	Shop floor control and other control operations
3.3.9	Risk assessment and management
3.3.10	Continuous improvement
	Logistics and supply chain skills
3.3.11	Supply to production and supplier management
3.3.12	Transportation (modalities and planning) and logistics management

Table A3. *Cont.*

	(a) Bioeconomy Skills—Agriculture
3.3.13	Management of inventories incl. goods received, pick and pack, storage and storage systems (FIFO), and stock management
3.3.14	Traceability
	Other skills
3.3.15	Food security
3.3.16	Ethics for food
3.3.17	Emerging technologies
3.3.18	Food labelling/certifications
3.3.19	Food defense
3.3.20	Food fraud

Table A4. Soft skills.

Skill No.	**Fundamental Soft Skills**
4.1	Communication with others at work and in the daily life, languages, reporting and briefing, public speaking, and press releases
4.2	Problem solving
4.3	Analytical, critical, and creative thinking
	Self-management skills
4.4	Demonstrating positive attitudes and behaviors
4.5	Being resilient, adaptable, and proactive
4.6	Organization, planning, visioning, and strategic thinking
4.7	Equality skills interculturalism, gender, empowerment, harassment
4.8	Safety awareness
4.9	Reflecting on own performance
	Team working and interpersonal skills
4.10	Team building incl. conflict resolution, negotiation, flexibility
4.11	Teamwork character incl. responsibility, honesty, empathy
4.12	Conflict management
4.13	Change management
	Business soft skills
4.14	Providing leadership
4.15	Managing personnel incl. delegating, motivating, assessing
4.16	Networking
4.17	Innovative thinking

Table A4. *Cont.*

Skill No.	Fundamental Soft Skills
	Education skills
4.18	Digital tools to support learning and distance learning
4.19	Learning at work incl. learning by doing, learning from others (mentoring, shadowing, etc.) and teaching each other
4.20	Learning continuously (lifelong learning)
4.21	Training others incl. training skills, training tools, course design, assessment, etc.
4.22	STEM knowledge (Science, Technology, Engineering and Mathematics) to understand and cooperate through the whole food/bioproducts value chains

Table A5. Business entrepreneurship skills.

Skill No.	Marketing Skills
5.1	Monitoring market activity and conditions
5.2	Direct marketing in agriculture, food industry, and forestry
5.3	Sales and marketing
5.4	Local marketing associations
5.5	Selling skills building buy-in to an idea, a decision, an action, a product, or a service
5.6	Customers service
	Financial skills
5.7	Business planning/model and strategic management incl. scenario foresighting/forecasting; recognizing and realizing business opportunities; Key Performance Indicators (KPIs) management, knowledge management, and stakeholder management
5.8	The basics of financial issues incl. balance sheets analysis
5.9	Purchasing/renting incl. equipment, structures, seeds, fertilizers, herbicides, animal feed, and other supplies)
	Fair, collaborative, and competitive value chains
5.10	Cooperatives (values, legal framework, and management)
5.11	New value chains/new business models (incl. values-based supply chains and short food supply chains)
5.12	Collaboration/cooperation across all sectors in the food chain
	Skills for research, development, and innovation
5.13	Social expectations/consumer science and behavior
5.14	Interdisciplinary knowledge to assess the whole value chain
5.15	Funding opportunities
5.16	Product development incl. laboratory and desk research
5.17	Project management
5.18	Knowledge transfer in the bioeconomy chains
5.19	Innovation management and its deployment on-site
5.20	Scale-up issues per sector incl. technical difficulties, costs, and volume calculations
5.21	The protection of intellectual property rights

Table A5. *Cont.*

Skill No.	Marketing Skills
	Compliance with policy and legislation
5.22	Fiscal basis and regulations
5.23	Specific sector legislation incl. agricultural policy and legislation, food policy and legislation, and forest- and nature-related policies (EU and national legislations and marked-based systems covering natural resources and their management)
5.24	Food labelling/certifications
5.25	Farm environmental management plan

References

1. IPCC. Summary for Policymakers Climate Change 2022: Impacts, Adaptation and Vulnerability. In *Proceedings of the Climate Change 2022: Impacts, Adaptation, and Vulnerability*; Pörtner, H.-O., Roberts, D.C., Poloczanska, E.S., Mintenbeck, K., Tignor, M., Alegría, A., Craig, M., Langsdorf, S., Löschke, S., Möller, V., et al., Eds.; Cambridge University Press: Cambridge, UK; New York, NY, USA, 2022; *in press*.
2. FAO. *Climate Change and Food Systems: Global Assessments and Implications for Food Security and Trade*; Elbehri, A., Ed.; Food and Agriculture Organization of the United Nations (FAO): Rome, Italy, 2015; ISBN 9251086990.
3. European Commission. *The European Green Deal*; Communication No. COM/2019/640; European Commission: Brussels, Belgium, 2019.
4. European Commission. Farm to Fork Strategy for a Fair, Healthy and Environmentally-Friendly Food System. In Proceedings of the Communication from the Commission to the European Parliament, the Council, the European Economic and Social Committee and the Committee of the Regions 2020, Brussels, Belgium, 14 January 2020.
5. International Labour Organisation; European Centre for the Development of Vocational Training; European Training Foundation; Organisation for Economic Co-operation and Development. *Skill Needs Anticipation: Systems and Approaches: Analysis of Stakeholder Survey on Skill Needs Assessment and Anticipation 2017*; ILO: Geneva, Switzerland, 2017.
6. Aver, B.; Fošner, A.; Alfirević, N. Higher Education Challenges: Developing Skills to Address Contemporary Economic and Sustainability Issues. *Sustainability* **2021**, *13*, 12567. [CrossRef]
7. Dias, C.S.L.; Rodrigues, R.G.; Ferreira, J.J. Agricultural Entrepreneurship: Going Back to the Basics. *J. Rural Stud.* **2019**, *70*, 125–138. [CrossRef]
8. Sorensen, L.B.; Germundsson, L.B.; Hansen, S.R.; Rojas, C.; Kristensen, N.H. What Skills Do Agricultural Professionals Need in the Transition towards a Sustainable Agriculture? A Qualitative Literature Review. *Sustainability* **2021**, *13*, 13556. [CrossRef]
9. Erickson, B.; Fausti, S.; Clay, D.; Clay, S. Knowledge, Skills, and Abilities in the Precision Agriculture Workforce: An Industry Survey. *Nat. Sci. Educ.* **2018**, *47*, 1–11. [CrossRef]
10. Easterly, R.G., III; Warner, A.J.; Myers, B.E.; Lamm, A.J.; Telg, R.W. Skills Students Need in the Real World: Competencies Desired by Agricultural and Natural Resources Industry Leaders. *J. Agric. Educ.* **2017**, *58*, 225–239. [CrossRef]
11. Charatsari, C.; Lioutas, E.D. Is Current Agronomy Ready to Promote Sustainable Agriculture? Identifying Key Skills and Competencies Needed. *Int. J. Sustain. Dev. World Ecol.* **2019**, *26*, 232–241. [CrossRef]
12. Flynn, K.; Wahnström, E.; Popa, M.; Ruiz-Bejarano, B.; Quintas, M.A.C. Ideal Skills for European Food Scientists and Technologists: Identifying the Most Desired Knowledge, Skills and Competencies. *Innov. Food Sci. Emerg. Technol.* **2013**, *18*, 246–255. [CrossRef]
13. Mayor, L.; Flynn, K.; Dermesonluoglu, E.; Pittia, P.; Baderstedt, E.; Ruiz-Bejarano, B.; Geicu, M.; Quintas, M.A.C.; Lakner, Z.; Costa, R. Skill Development in Food Professionals: A European Study. *Eur. Food Res. Technol.* **2015**, *240*, 871–884. [CrossRef]
14. Handayani, M.N.; Ali, M.; Mukhidin, D.W. Industry Perceptions on the Need of Green Skills in Agribusiness Vocational Graduates. *J. Tech. Educ. Train.* **2020**, *12*, 24–33.
15. Akyazi, T.; Goti, A.; Oyarbide, A.; Alberdi, E.; Bayon, F. A Guide for the Food Industry to Meet the Future Skills Requirements Emerging with Industry 4.0. *Foods* **2020**, *9*, 492. [CrossRef]
16. Forestry Skills Forum. Forestry Workforce Research; 2021. Available online: https://www.lantra.co.uk/sites/default/files/2021-08/Forestry%20Workforce%20Research%20Final%20Report%2013.08.21.pdf (accessed on 31 July 2022).
17. Kelly, E.C.; Brown, G. Who Are We Educating and What Should They Know? An Assessment of Forestry Education in California. *J. For.* **2019**, *117*, 95–103. [CrossRef]
18. Blanc, S.; Lingua, F.; Bioglio, L.; Pensa, R.G.; Brun, F.; Mosso, A. Implementing Participatory Processes in Forestry Training Using Social Network Analysis Techniques. *Forests* **2018**, *9*, 463. [CrossRef]
19. Marrelli, A.F. Collecting Data through Focus Groups. *Perform. Improv.* **2008**, *47*, 39–45. [CrossRef]
20. Farrell, M.; Murtagh, A.; Weir, L.; Conway, S.F.; McDonagh, J.; Mahon, M. Irish Organics, Innovation and Farm Collaboration: A Pathway to Farm Viability and Generational Renewal. *Sustainability* **2021**, *14*, 93. [CrossRef]
21. Subramony, D.P.; Lindsay, N.; Middlebrook, R.H.; Fosse, C. Using Focus Group Interviews. *Perform. Improv.* **2002**, *41*, 38–45. [CrossRef]

22. Nazzaro, C.; Stanco, M.; Marotta, G. The Life Cycle of Corporate Social Responsibility in Agri-Food: Value Creation Models. *Sustainability* **2020**, *12*, 1287. [CrossRef]
23. Williams, A.; Katz, L. The Use of Focus Group Methodology in Education: Some Theoretical and Practical Considerations, 5 (3). *IEJLL Int. Electron. J. Leadersh. Learn.* **2001**, *5*. Available online: https://journals.library.ualberta.ca/iejll/index.php/iejll/article/view/496 (accessed on 31 July 2022).
24. Carvalho, C.; Almeida, A.C. The Adequacy of Accounting Education in the Development of Transversal Skills Needed to Meet Market Demands. *Sustainability* **2022**, *14*, 5755. [CrossRef]
25. Robertson, R.W. Local Economic Development and the Skills Gap: Observations on the Case of Tampa, Florida. *High. Educ. Ski. Work. Based Learn.* **2018**, *8*, 451–468. [CrossRef]
26. Halliday, M.; Mill, D.; Johnson, J.; Lee, K. Let's Talk Virtual! Online Focus Group Facilitation for the Modern Researcher. *Res. Soc. Adm. Pharm.* **2021**, *17*, 2145–2150. [CrossRef]
27. Ramalho, A.; Goodburn, B.; Lindner, L.F.; Mayor, L.; Knöbl, K.F.; Trienekens, J.; Rossi, D.; Sanna, F.; Berruto, R.; Busato, P. Skill Needs for Sustainable Agri-Food and Forestry Sectors (II): Insights of a European Survey. *Sustainability* **2022**, *submitted*.
28. Bröring, S.; Vanacker, A. Designing Business Models for the Bioeconomy: What Are the Major Challenges? *EFB Bioeconomy J.* **2022**, *2*, 100032. [CrossRef]
29. Commission, E.; Media, D.-G. *For the I.S. and Survey of Schools: ICT in Education: Benchmarking Access, Use and Attitudes to Technology in Europe's Schools*; Publications Office: Luxembourg, 2013.
30. Bikse, V.; Grinevica, L.; Rivza, B.; Rivza, P. Consequences and Challenges of the Fourth Industrial Revolution and the Impact on the Development of Employability Skills. *Sustainability* **2022**, *14*, 6970. [CrossRef]
31. European Commission, Executive Agency for Small and Medium-sized Enterprises. *Blueprint for Sectoral Cooperation on Skills: Towards an EU Strategy Addressing the Skills Needs of the Steel Sector: European Vision on Steel-Related Skills of Today and Tomorrow*; Publications Office: Luxembourg, 2019.
32. McElwee, G. A Taxonomy of Entrepreneurial Farmers. *Int. J. Entrep. Small Bus.* **2008**, *6*, 465–478. [CrossRef]
33. Augère-Granier, M.-L. *Agricultural Education and Lifelong Training in the EU*; European Parliamentary Research Service: Brussels, Belgium, 2017; Available online: https://www.europarl.europa.eu/RegData/etudes/BRIE/2017/608788/EPRS_BRI(2017)608788_EN.pdf (accessed on 2 June 2020).
34. Bailey, N.E.; Arnold, S.K.; Igo, C.G. Educating the Future of Agriculture: A Focus Group Analysis of the Programming Needs and Preferences of Montana Young and Beginning Farmers and Ranchers. *J. Agric. Educ.* **2014**, *55*, 167–183. [CrossRef]
35. Phelan, C.; Sharpley, R. Exploring Entrepreneurial Skills and Competencies in Farm Tourism. *Local Econ. J. Local Econ. Policy Unit* **2012**, *27*, 103–118. [CrossRef]
36. Couzy, C.; Dockes, A.C. Are Farmers Businesspeople? Highlighting Transformations in the Profession of Farmers in France. *Int. J. Entrep. Small Bus.* **2008**, *6*, 407–420. [CrossRef]
37. George, D.A.; Clewett, J.F.; Wright, A.; Birch, C.; Allen, W. Improving Farmer Knowledge and Skills to Better Manage Climate Variability and Climate Change. *J. Int. Agric. Ext. Educ.* **2007**, *14*, 5–19. [CrossRef]
38. Silva, L.L.; Baptista, F.; Cruz, V.F. Analysis of Skills Needs for Agricultural Workers for a "Sustainable Agriculture." Silva: 2017. Available online: http://www.sagriproject.eu/wp-content/uploads/2018/11/D2.1_Analysis-of-skills-needs-for-agricultural-workers_EN_VFinal.pdf (accessed on 31 July 2022).
39. Bijman, J.; Iliopoulos, C.; Poppe, K.J.; Gijselinckx, C.; Hagedorn, K.; Hanisch, M.; Hendrikse, G.W.J.; Kühl, R.; Ollila, P.; Pyykkönen, P. *Support for Farmers' Cooperatives*; Wageningen UR: Wageningen, The Netherlands, 2012.
40. European Commission (EC). Directorate-General for Employment, Social Affairs and Inclusion. In *ESCO, European Classification of Skills/Competences, Qualifications and Occupations: The First Public Release: A Europe 2020 Initiative*; European Union: Brussels, Belgium, 2014.
41. Jack, C.; Anderson, D.; Connolly, N. Innovation and Skills: Implications for the Agri-Food Sector. *Educ. Train.* **2014**, *56*, 271–286. [CrossRef]
42. Topliceanu, L.; Bibire, L.; Nistor, D. Professional Competences of the Personnel Working on Quality Control and Food Safety in the Food Industry. *Procedia Soc. Behav. Sci.* **2015**, *180*, 1030–1037. [CrossRef]
43. Lertpiromsuk, S.; Ueasangkomsate, P.; Sudharatna, Y. Skills and Human Resource Management for Industry 4.0 of Small and Medium Enterprises. In *Proceedings of Sixth International Congress on Information and Communication Technology*; Springer: Berlin/Heidelberg, Germany, 2022; pp. 613–621.
44. Barrett, J.W. The Role of Humanities and Other Liberal Courses in the Professional Forestry Curriculum. *J. For.* **1953**, *51*, 574–578.
45. Sample, V.A.; Ringgold, P.C.; Block, N.E.; Giltmier, J.W. Forestry Education: Adapting to the Changing Demands on Professionals. *J. For.* **1999**, *97*, 4–10. [CrossRef]
46. Brown, T.L.; Lassoie, J.P. Entry-Level Competency and Skill Requirements of Foresters: What Do Employers Want? *J. For.* **1998**, *96*, 8–14.
47. Straka, T.J.; Marsinko, A.P.; Childers, C.J. Individual Characteristics Affecting Participation in Urban and Community Forestry Programs in South Carolina, US. *Arboric. Urban For.* **2005**, *31*, 131–137. [CrossRef]
48. Bullard, S.H.; Stephens Williams, P.; Coble, T.; Coble, D.W.; Darville, R.; Rogers, L. Producing "Society-Ready" Foresters: A Research-Based Process to Revise the Bachelor of Science in Forestry Curriculum at Stephen F. Austin State University. *J. For.* **2014**, *112*, 354–360. [CrossRef]

49. Bullard, S.H. Forestry Curricula for the 21st Century—Maintaining Rigor, Communicating Relevance, Building Relationships. *J. For.* **2015**, *113*, 552–556. [CrossRef]
50. Schuh, B.; Maucorps, A.; Munch, A.; Brkanovic, S.; Dwyer, J.; Vigani, M.; Khafagy, A.; Coto Sauras, M.; Deschelette, P.; Lòpez, A. Research for AGRI Committee–The EU Farming Employment: Current Challenges and Future Prospects, European Parliament, Policy Department for Structural and Cohesion Policies, Brussels. 2019. Available online: https://www.europarl.europa.eu/thinktank/en/document/IPOL_STU(2019)629209 (accessed on 31 July 2022).
51. Francis, C.A.; Jensen, E.S.; Lieblein, G.; Breland, T.A. Agroecologist Education for Sustainable Development of Farming and Food Systems. *Agron. J.* **2017**, *109*, 23–32. [CrossRef]

Article

The Influence of Governmental Agricultural R&D Expenditure on Farmers' Income—Disparities between EU Member States

Mirela Stoian, Raluca Andreea Ion, Vlad Constantin Turcea *, Ionut Catalin Nica and Catalin Gheorghe Zemeleaga

The Department of Agrifood and Environmental Economics, Faculty of Agrifood and Environmental Economics, The Bucharest University of Economic Studies, 010961 Bucharest, Romania

* Correspondence: turceavlad14@stud.ase.ro; Tel.: +40-72-295-9353

Abstract: This article investigates how governmental agricultural R&D expenditure affect economic prosperity and sustainable development, attempting to verify the hypothesis that agricultural research and development expenditures are among the key factors influencing the farmers' income, as one of the sustainable development indicators. Statistical data were retrieved from European international databases for the period of 2004–2020 and were analyzed using the regression model. The results of the study indicate positive effects for most of the EU member states. The countries where the results validate the hypothesis are Austria, Belgium, Bulgaria, the Czech Republic, Germany, Estonia, Finland, France, Greece, Croatia, Ireland, Latvia, Poland, Slovakia, Slovenia, and the United Kingdom, as a former member state of the EU. Further, the model confirms that a significant portion of farmers' income growth is explained by the governmental R&D expenditure. These findings may change the methods and directions regarding the agricultural R&D expenditure, underpinning the macroeconomic policy and agriculture in rural areas along the pathway to achieving the sustainable development goals.

Keywords: sustainable agriculture; governmental agricultural R&D expenditure; farmers' income; sustainable development goals

Citation: Stoian, M.; Ion, R.A.; Turcea, V.C.; Nica, I.C.; Zemeleaga, C.G. The Influence of Governmental Agricultural R&D Expenditure on Farmers' Income—Disparities between EU Member States. *Sustainability* **2022**, *14*, 10596. https://doi.org/10.3390/su141710596

Academic Editors: Mariarosaria Lombardi, Erica Varese and Vera Amicarelli

Received: 5 August 2022
Accepted: 18 August 2022
Published: 25 August 2022

Publisher's Note: MDPI stays neutral with regard to jurisdictional claims in published maps and institutional affiliations.

1. Introduction

The sustainable development of agriculture is essential for the economic prosperity of the European Union's rural communities. Statistical data show that rural areas represented 83% of the total EU area, and that agricultural land, forest, and natural areas represented 80% of the total EU area, in 2018, as stated by Eurostat. Representative percentages explain the role of the rural communities in the economy, and discussions about macroeconomic policy may focus on a sectors' prioritization, when speaking about funding, investment, and public expenditure and their implementation effectiveness. When investments contribute to the economic prosperity of the rural communities, the macroeconomic policy is considered to be effective. Economic prosperity, in turn, may be described by numerous indicators; among these, the farmers' income can illustrate the economy's current status. A deeper analysis, which explores this indicator and its drivers, is needed in order to visualize the sustainable development of rural areas. Our study starts from the assumption that research and development (R&D) investments in agriculture are among the key factors influencing the levels of farmers' income.

Sustainable agriculture represents the equilibrium point of several aspects, including social, economic, and environmental, in both rural and urban agri-based contexts. For the agriculture to be designated as "sustainable", it is mandatory to have a versatile position, to offer an easy scalability, and to be continuously adaptable [1].

Agricultural industry is also becoming more data-centric, and novel technologies are offering advantages to worldwide farmers. Several break-through sustainable agricultural practices have been highlighted in the research literature, and arguments have been offered for consideration [2].

Many researchers have emphasized the importance of R&D expenditure in agriculture [3–8], arguing a direct and causal relationship between R&D investment in agriculture and farmers' income. However, most of these studies are based on a theoretical and conceptual framework, and only a few of them have conducted an empirical analysis for the EU member states, divided into old and new, with a more or less important agricultural sector. Our study explores the relationship between farmers' income and governmental agricultural R&D expenditure, using concatenated statistical methods for time series evaluation. Statistical data for the period of 2004–2020 were retrieved from the Eurostat database. Because agriculture plays different roles in European Union member states' economies, we have considered it necessary to analyze the situations separately according to two criteria—the share of agriculture in GDP, and the time of country's accession to the EU. Thus, the countries with large shares of agriculture in the GDP, above the average of the European Union (1.63%), are Bulgaria, Cyprus, the Czech Republic, Spain, Estonia, Finland, France, Greece, Croatia, Hungary, Lithuania, Latvia, Poland, Portugal, Romania, the Slovak Republic, and Slovenia, as seen in Figure 1. The date refers to 2021. The second criterion, the time of accession of the country to the EU, divided them into old and new member states, considering 2004 as the threshold.

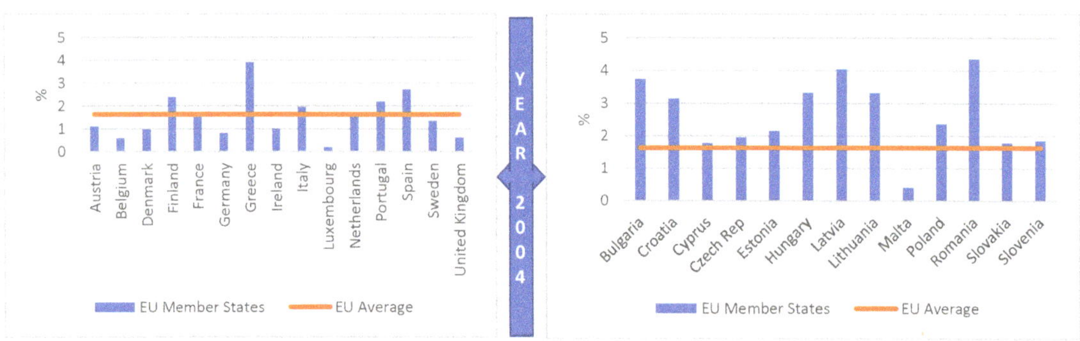

Figure 1. Year 2021 share of agriculture, forestry, and fishing value added in the GDP of EU28 member states (**left**: old member states; **right**: new member states). Source: edited by the authors from The World Bank [9].

The contribution of this paper to the literature is two-fold. First, two essential indicators of the Second Sustainable Development Goal (SDG 2), Zero Hunger, have been studied together—governmental agricultural R&D expenditure and farmers' income. Furthermore, several types of farmers' income indexes are used to measure sustainable farmers' income, considering that the results can faithfully reflect how agricultural R&D expenditure affect farmers' income. Secondly, the paper investigates farmers' income, within the context that ensuring a fair standard of living for farmers and contributing to the stability of their incomes are essential objectives for the Common Agricultural Policy of the European Union [10].

The paper is structured with five parts; following the introduction, the literature review describes the state of research in the fields of sustainable agriculture, sustainable development goals and income targets, agricultural investments, and research and development expenditure, in part two. The dataset regarding the R&D expenditure in agriculture and the farmers' income are analyzed, using the regression model, in part three, and the results are then discussed in part four. Finally, conclusions are drawn in part five.

2. Literature Review

The topic of sustainable agriculture is even more current as the recent trends in food prices and the unstable political, social, and economic conditions raise concerns about agricultural market equilibrium, food security, and farmers' income stability. Researchers [11] argue that agricultural R&D is a significant determinant of agricultural production and productivity and as a result, food prices and poverty.

Sustainable development of agriculture represents one decisive goal for the near future, and most states have adopted this approach in policy definition. The term, in essence, has diverse meanings, depending on the context, and also includes varying explanations and practices, such as farming methods and ecological stability. The concept is also known for aggregating economic aspects with resource conservation, maintenance, and improvement, concentrating on both the environmental and the ecological aspects. Depending on the reporting context, the sustainable agriculture focus can also vary, from yield improvement, crop diversity, and income prosperity to environmental stresses.

Gherardelli [12] claimed that one of the major challenges for governments is to ensure sufficient food for the population, taking into account the global emergencies of population and income growth, changes in diets, and decreasing availability of natural resources. These challenges call for increasing agricultural production, but in the context of a more economically, socially, and environmentally sustainable agriculture. Promoting a significant expansion of agricultural R&D and its funding could address this challenge.

Furthermore, trade-offs between agricultural productivity and sustainability have started to be studied. FAO reports [13] show that a major challenge for agriculture is to acknowledge and explore the potential trade-offs and contradictions between sustainability, with its environmental and social dimensions, and productivity, as its economic dimension.

The Second Sustainable Development Goal (SDG 2), Zero Hunger, is part of the UN's 17 Goals aimed at transforming the current world. They represent a ready-for-action guideline for governments all around the world, no matter the economical profile, that ensure sustainable three-layered growth (social, economic, and environmental). SDG 2 or the Zero Hunger ambition, is a comprehensive strategy with tangible goals and targets that tackle hunger elimination, food and nutrition security accomplishment, and agricultural sustainable development [14].

Particularly, our research will draw attention to the importance of R&D investment in promoting a sustainable agricultural development in countries with various agri-profiles; specifically, the concerned SDG 2 targets consist of targets 2.3—agricultural incomes and 2.A—increased investment in agricultural research. The first one, agricultural incomes, is relevant to be studied because supporting farmers' incomes and stabilizing them remain essential objectives of the Common Agricultural Policy of the European Union, as stated by the European Commission [10]. As declared by the UN [14], SDG target 2.3 aims to double the agricultural productivity and revenues of small-scale food producers, family farmers, etc., including secure and equal access to resources and inputs, knowledge, and financial services. The second goal, SDG target 2.A, aims to increase investment in rural infrastructure, agricultural research and extension services, technology development, etc. [14]. The linkages between the two targets are explored in the current research, assuming that investments in agricultural R&D are one of the key drivers of farmers' income growth, as suggested by FAO [15].

Investments in SDG 2, specifically through continuous worldwide focus on parents receiving more socio-economic attention in order to provide the needed food for their children, as well as on smallholder farmer empowerment and agricultural sustainable development, gender equity in farming and socio-farming, institutional de-formalization, have to be overseen as a holistic SDG approach that definitely boost collateral targets and objectives [16].

Research literature highlights how to intervene regarding the global food security topic; however, effective actions, strategies, and assessment methods remain challenging. Authors summarize some aspects that could improve general food security achievement through targeted agricultural interventions in food security, measurable actions that address food security, and improved systematic methodological reviews as methods for agricultural interventions [17].

Scientific papers also suggest leveraging know-how into concrete action plans towards reducing SDGs disparities through research, industrial, political, and consumer collaboration; there is also an acute need of implementation [18]. Another important aspect besides implementation is the threat-constrain identification in applying existing knowledge, as the incapacity to deliver could further increase the SDG gaps.

Evidence from the research literature also points towards an existing causal relationship between agricultural output and domestic agricultural investment [19], whereas the current paper emphasizes the governmental agricultural R&D investment as a possible defining pillar of sustainable rural economic growth.

Other studies have reviewed the foreign non-governmental investment in agriculture and several short-run and long-run effects have been noted, with the prompt recommendation of improving the absorption capacity and administrative fluency [20].

Some studies have discussed the controversial effects of the agricultural technology investment on farmers' income. On the one hand, researchers found that the agricultural R&D had a positive effect on the growth of farmers' income [21]. The authors of [22] explored the possible ways to double the farmers' income and concluded that this objective can be achieved if the stakeholders follow a comprehensive and targeted approach regarding income opportunities, including investment in agricultural R&D and infrastructure. It has been found [23] that, when the agricultural modernization starts, per capita income increases. On the other hand, some studies show that the agricultural investments have limited positive effect on farmers' income [24].

Income and, as such, the economic growth, enter into discussions about sustainability. The authors of [25] claimed that economic growth is sustainable only if it is compatible with environmental quality. This argument is controversial: some researchers argue that the economic growth is deteriorating the environment [26], while others [27] state that the link between income growth and environmental degradation is insignificant.

Governmental agricultural research plays an essential role in the development of modern agricultural scientific breakthroughs, high quality economical results, and productivity increases [28]. Generally, according to economic theory [29], R&D expenditure is considered to be essential for economic growth, development, and sustainability. The agricultural R&D expenditure have positive effects on famers' functional distribution and scale distribution of income, while agricultural technology promotion expense has negative effects. In comparison, the first relationship mentioned has a stronger link than the second [7]. The authors of [30], in their analysis of the impact of R&D on productivity, claimed that R&D positively drives productivity.

The basic economic theory provides few ideas about R&D expenditure. Thus, researchers tried to fill in this gap with findings regarding either the relationship between agricultural output and R&D expenditure or the R&D as a source of efficiency [31,32], or the long-run relationship between productivity, as a dependent variable, and R&D expenditure, as an independent variable [33].

Evidence from the research literature also proved that R&D capital investment is mandatory to determine productivity amplification in the agri-food industry; while in the study of [34], it is also stated that the R&D expenditure should improve at a faster pace than the output's, as a direct result of the transferability drawback in the technological sector of agriculture across countries or businesses, and also due to the fact that conservative research levels should be prevented from plunging.

There is a large body of empirical studies on the research and development expenditure in agriculture. When interrogating the Web of Science (WoS) database using the following parameters: "agricultural R&D" and "agricultural income", 295 results were reported, most studies being written in the following domains: environmental science (21%), economics (15%), environmental studies (12%), etc. The researchers' interest in the way the agricultural R&D expenditure impacted the farmers' income increased after 2010, when, on average, 10 articles were published in the journals indexed in WoS, reaching a peak of 48 in 2021. Using the VoS Viewer Software to see the linkages between agricultural R&D and income and other related topics, five clusters were identified. They comprise themes such as yield, management, performance, efficiency, quality, systems, soil, sustainability, emissions, economic growth, agricultural productivity, food security, trade, consumption, income, poverty, policy, and strategies, demonstrating the interest of researchers in these topics in connection with agricultural R&D investments.

Regarding the relationship between agricultural R&D expenditure and income, the authors of [35] argued that agricultural R&D has made important contributions to reducing poverty in South Asia in the post Green Revolution period. The authors of [3] found that public agricultural R&D has a positive impact on rural household income. It was found in [7] that the agricultural R&D expenditure has positive effects on famers' functional distribution and the scale distribution of income, while agricultural technology promotion expenditure has negative effects. In comparison, the impact of the former is larger than that of the latter. Some authors [36] explored the R&D expenditure for new technology in livestock farming and argued that investments in research and development lead to more efficient and sustainable resource management for developing countries.

Innovation-led economic growth has proved to gain popularity among governments, a process that is also known as smart growth [37], which is growth that is also applicable in the agricultural sector, therefore highlighting the importance of R&D as a main driver of innovation.

Income in agricultural holdings has been studied in the context of the sustainable development of agriculture [38], the authors exploring the inequalities among farms and demonstrating that the process of the concentration of land and capital led to an increase in farmers' income disparities. Farmers' income has been explored in relation to rural tourism [39], and it was found that rural tourism positively and significantly affects sustainable farmers' revenues, although, among different types of farmers' income, the magnitude varies.

Based on the analysis of previous studies, the question is raised regarding how governmental agricultural R&D expenditure affect farmers' revenues, expecting a direct and intensive relationship between them. Thus, a hypothesis is put forward as follows:

Hypothesis 1 (H1). *Governmental agricultural R&D expenditure influences farmers' income, and the extent varies between countries with different agricultural profiles.*

The research objective is to investigate the relationship between governmental agricultural R&D expenditure and farmers' revenues, with the final purpose of better managing the paths and directions of governmental agricultural R&D investments towards achieving the sustainable development goals, including farmers' income growth.

3. Materials and Methods

Statistical data regarding the government support of agricultural research and development in the EU member states are presented in Figure 2. Significant structural differences between countries can be noticed across multiple states.

The government support of agricultural research and development, representing the independent variable (R&D) in the current paper's tested model, is part of the European Union Sustainable Development goals, targeting SDG 2, Zero Hunger. It is referring to the governmental allocation of the budget for each member state for research and development activities in the agricultural sector, signaling how prioritized the public funding of research and development is for each state.

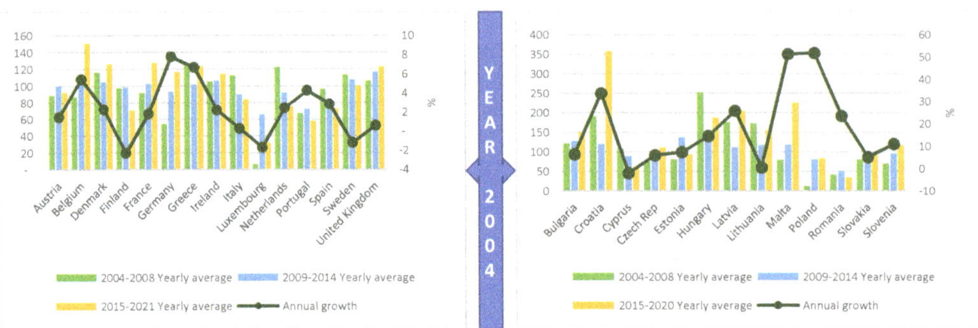

Figure 2. Annual average for government support of agricultural research and development (**left** axis) (Variable R&D in the model) and the average annual growth (**right** axis) at the EU28 member state level, in percentages (**left**: old member states; **right**: new member states). Source: edited by the authors from Eurostat [40].

The visual representation of the dependent variable tested in the econometric model, Governmental Support to Agricultural R&D (variable R&D), is illustrated in Figure 2, which concatenates the yearly average for the mentioned interval on the left hand axis and the year over year average growth on the right hand axis. Both representations are illustrated in percentages. It aims to describe the most representative financing period for the research side in agriculture and signals in which member state this expenditure is gaining popularity. In the case of yearly growth, there are a few states that raise attention, as their values surpass 20% of annual growth, such as Poland (52%), Croatia (34%), Latvia (26%), and Romania (24%), and the remaining countries account for less than 10% of the yearly increase. Meanwhile, Cyprus, Finland, Luxembourg, and Sweden recorded negative values.

As the variable is described through indexes, the three-colored chart would best describe which period recognized R&D as being essential. What can be easily observed is that either period 2004–2008 or period 2015–2020 are more visible for most of the states. As the country comparison could not be performed for the annual average, due to the reporting method of the dataset, and as each year's value for a given member state is reported for the year 2010, the average annual increase was compared.

The most representative sectors of the economy where R&D has been financed are clearly indicated in Figure 3. Western European countries set the trends, but also the orientation of the investment profile. The largest investment in R&D has been recorded in the German R&D industry. In the agricultural sector, the countries with the largest investments are Germany (USD 1883 million), Spain (USD 1510 million), and the United Kingdom (USD 1016 million), for the old EU member states, and for new member states, the largest investments in agricultural R&D are recorded in Poland (USD 261 million), the Czech Republic (USD 181 million), and Hungary (USD 162 million). For the rest of the sectors for both old and new countries, only the largest investment will be presented: environment R&D in Germany (USD 1908 million) and Poland (USD 318 million); energy R&D in Germany (USD 2989 million) and the Czech Republic (USD 157 million); defense R&D in France (USD 5297 million) and Poland (USD 335 million); education R&D in Germany (USD 947 million) and Poland (USD 302); and industry R&D in Germany (USD 8397 million) and the Czech Republic (528 million).

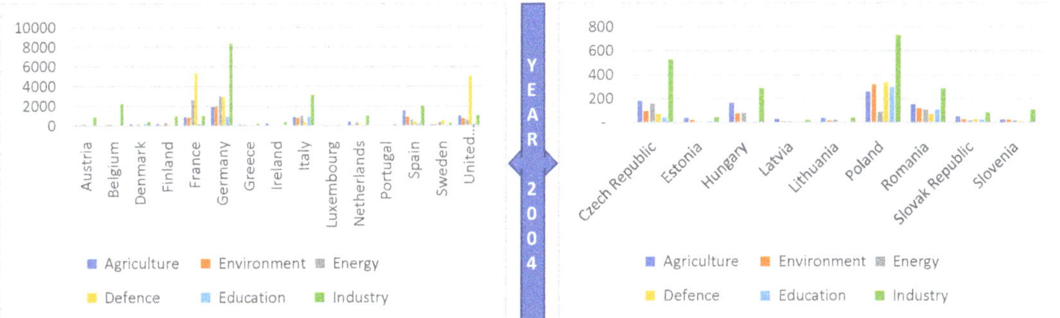

Figure 3. Cumulative R&D governmental expenditure by economic sector for the 2005–2020 period (in million USD, 2015 constant) (**left**: old member states; **right**: new member states). Source: edited by the authors based on available data [41].

As top R&D investments have already been discussed in Figure 3, Figure 4 presents the R&D expenditure share by sector for each individual member state. Industrial R&D was expected to represent a significant part of total governmental R&D for the developed countries. The largest agricultural R&D shares were recorded in Ireland (33.5%), Spain (27%), and The Netherlands (20.4%), for old member states, while for new member states, Latvia has recorded the largest agricultural R&D share with 38.2%, followed by Lithuania (29.7%), and Estonia (28.2%).

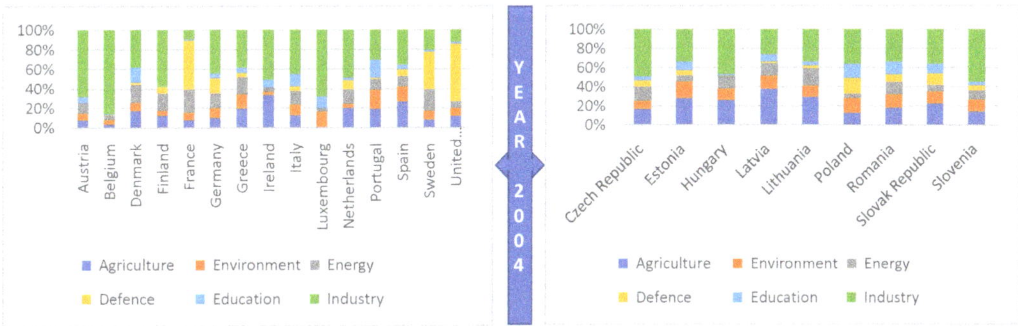

Figure 4. Share of cumulative R&D governmental expenditure by economic sector in percentages (**left**: old member states; **right**: new member states). Source: edited by the authors based on available data [41].

The dependent variable of the model tested in this particular research, entitled real income of factors in agriculture per annual work unit (INC), is also part of the European Union Sustainable Development Goals indicator set. It aims to monitor the SDG progress towards reaching the Zero Hunger ambition. The indicator has an additional scope as part of the Common Agricultural Policy objectives, to be more precise. The described indicator is an aggregation of income factors generated by agricultural activities, such as remunerated factors of production—capital, wages, and land, either owned, borrowed or rented, according to Eurostat, the issuing entity—and it represents a description of all factors of production (inputs, depreciation, taxes, and subsidies).

A visual representation of the independent variable tested in the econometric model, real income of factors in agriculture (variable INC), is illustrated in Figure 5, which concatenates the yearly average for the mentioned interval on the left hand axis and the year over year average growth on the right hand axis. Both representations are reflected in percentages, aiming to describe the most representative financing period for the agricultural income; therefore, the annual average country comparison could not be performed due to the reporting method of the dataset, as each year's value for a given member state is reported for the year 2010 against the average annual growth, which can be compared.

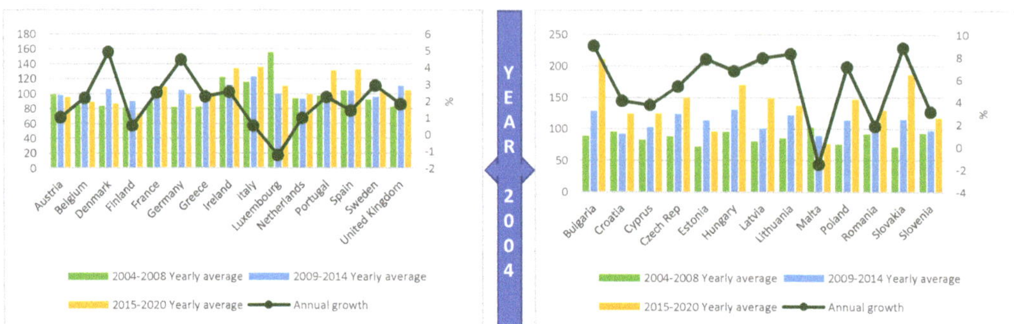

Figure 5. Annual average of real income of factors in agriculture per annual work unit (Variable INC in the model) (**left** axis) and average annual growth (**right** axis) at the EU28 member state level, in percentages (**left**: old member states; **right**: new member states). Source: edited by the authors from Eurostat [42].

Significant yearly growths, have been recorded for the following states: Bulgaria and Slovakia (9%), Estonia, Latvia, and Lithuania (8%), and Hungary and Poland (7%); while on the opposite side, besides Luxembourg and Malta's negative growth, Austria, Finland, Italy, The Netherlands, and Spain all recorded just 1% annual growth. For the 2004–2008 period, Luxembourg and Ireland recorded the largest annual average for the agricultural income, with Slovakia the lowest average, while for the period 2009–2014, Hungary reported the highest value, and for the period 2015–2020, Bulgaria recorded the largest growth.

The EU28 territory is explored in this particular research, as the EU countries are known to possess either developed or developing status, and it has already been demonstrated how important R&D investment is for sustainable development. The potential for agricultural-rural socio-economic development lies in the governmental investment profile, especially expenses that concentrate on bringing novelty.

As referring to the research methodology, the scientific literature indicates similar statistical approach when farmers' incomes are assessed; the R-squared concept, together with linear regression, have been used to determine the influence level of several socio-economic variables regarding the agricultural cooperative income [43]. Other research papers applied similar econometric approaches to draw the interdependence between agricultural product purchasing power and investment in the agricultural and processing sector, along with other independent variables, to demonstrate the importance that agriculture plays in sustainable economic development [44].

In order to be able to compute the statistical data as extracted from Eurostat, multiple transformation procedures were required, especially for the independent variable (R&D), as originally it has been described in absolute values (as seen in Figure 6). Therefore, a duplication was mandatory for the indexing methodology of INC, determining annual figures to be reported up to the year 2010.

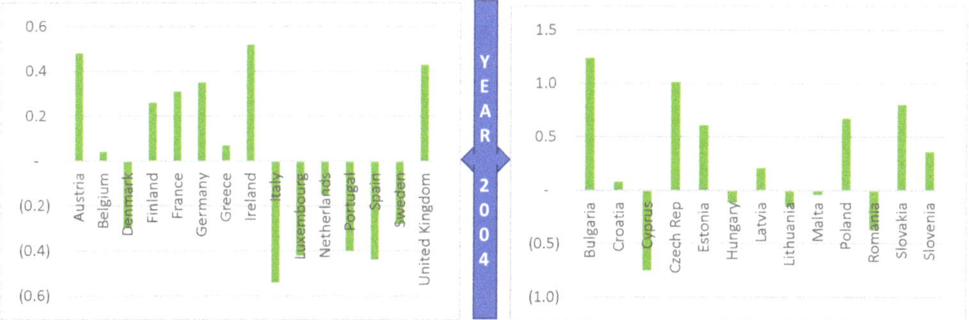

Figure 6. Coefficient of the independent variable (governmental agricultural R&D expenditure) in the regression equation (**left**: old member states; **right**: new member states). Source: edited by the authors.

4. Results and Discussions

The regression model has been performed individually, across all the EU28 member states, for the dependent variable, farmers' income, and the independent variable, agricultural R&D expenditure. Using statistical software to perform the analysis for each country, the resulting econometric R-squared values are graphically represented in Figure 7 for each individual country, together with the independent variable coefficient, as seen in Figure 6.

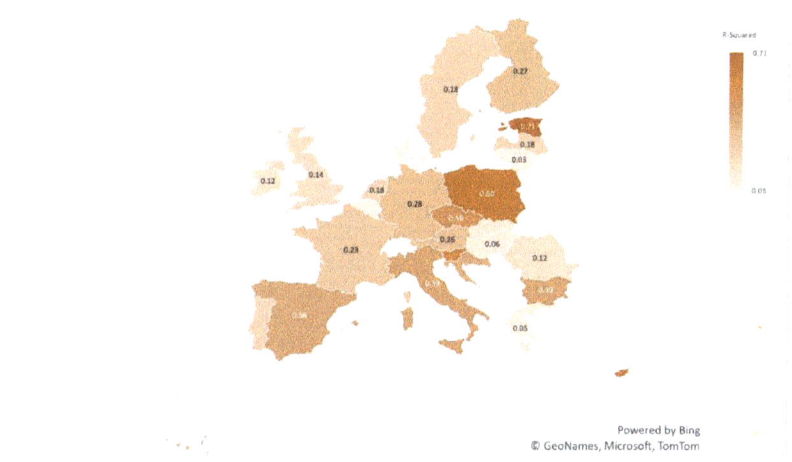

Figure 7. R-squared values for the EU28 member states of the agricultural governmental R&D (R&D) expenditure's capacity for explaining the real income of factors in agriculture per annual work unit. Source: edited by the authors.

Top performing R-squared parameters in the present model have been registered in the countries of Estonia (0.71), Cyprus (0.63), Poland (0.60), Slovenia (0.54), and the Czech Republic (0.46), while the lowest performing parameters have been recorded in Belgium and Latvia (0.03), Denmark (0.04), Greece (0.05), Hungary (0.06), and Slovakia (0.07). Thus, for the first group of countries, the levels of governmental agricultural R&D expenditure strongly influence the levels of farmers' income, and any change in their allocation may significantly change the farmers' well-being and rural economies' prosperity. For the second group of states, the levels of governmental agricultural R&D investments slightly influenced the levels of farmers' income. Each member state value reflects a specific national profile, including the agricultural profile; therefore, drawing a one-sided direction line would be a matter of partial subjectivity. Moreover, other variables of the economic, natural, technological, social, or political kind, which are not presented in the current analysis, clearly impact the farmers' incomes.

$$\text{INC} = \text{R\&D} \times \text{Coeff} + \text{C} \quad (1)$$

Equation (1) R&D capability of explaining INC (using the coefficients stated in Figure 8, if the hypothesis is valid). Source: edited by the authors.

Member State	R2	X coeff
Austria	0.26	0.48
Belgium	0.03	0.04
Denmark	0.04	(0.30)
Finland	0.27	0.26
France	0.23	0.31
Germany	0.28	0.35
Greece	0.05	0.07
Ireland	0.12	0.52
Italy	0.39	(0.54)
Luxembourg	0.26	(0.42)
Netherlands	0.18	(0.15)
Portugal	0.17	(0.40)
Spain	0.36	(0.44)
Sweden	0.18	(0.28)
United Kingdom	0.14	0.43

YEAR 2004

Member State	R2	X coeff
Bulgaria	0.39	1.24
Croatia	0.37	0.08
Cyprus	0.63	(0.75)
Czech Rep	0.46	1.01
Estonia	0.71	0.61
Hungary	0.06	(0.11)
Latvia	0.18	0.21
Lithuania	0.03	(0.15)
Malta	0.09	(0.04)
Poland	0.60	0.67
Romania	0.12	(0.37)
Slovakia	0.07	0.80
Slovenia	0.54	0.36

Figure 8. Directions and intensities of the relationships between governmental agricultural R&D expenditure and farmers' income for each EU member states and the UK (**left**: old member states; **right**: new member states). Source: edited by the authors.

The values of R^2 and of the coefficient of governmental R&D expenditure in agriculture in relation to farmers' income is summarized in Figure 8. Four clusters have been identified:

(i) Countries where governmental R&D expenditure in agriculture are among the key factors of farmers' income growth and positively impact them: Estonia, Poland, and Slovenia;
(ii) Countries where governmental R&D expenditure in agriculture are among the key factors of farmers' income and negatively impact them: Cyprus;
(iii) Countries where governmental R&D expenditure in agriculture are not considered among the key factors of farmers' income growth and positively impact them: Bulgaria, the Czech Republic, Finland, France, Greece, Croatia, Latvia, and Slovakia;
(iv) Countries where R&D governmental expenditure in agriculture are not considered among the key factors of farmers' income growth and negatively impact them: Spain, Hungary, Lithuania, Portugal, and Romania.

In order to assess how a 1% increase in INC is determined by coefficient % increase in R&D and properly use Equation (1) for a member state, it is also important to take into consideration the C (constant) value that varies from −21.53 to 177.1, with an average value of 91.75.

The threshold value of R^2 from which R&D governmental expenditure in agriculture is considered to be among the key factors of farmers' income is 0.5. The countries that registered values of the share of agriculture in GDP lower than the average of the European Union, 1.63%, as seen in Figure 1, may be excluded from the results' analysis: Austria,

Belgium, Germany, Denmark, UK, Ireland, Italy, Luxembourg, Malta, The Netherlands, and Sweden.

For some member states, as seen in Figure 6, as the governmental R&D expenditure in agriculture increases, the resulted coefficients indicate a positive relationship. The positively impacted farmers' income due to R&D expenditure growth are registered in the following countries: Bulgaria, the Czech Republic, Finland, France, Greece, Croatia, Latvia, Poland, Slovakia, and Slovenia. Similar results have been found by other authors [38], who demonstrated that the budgetary support for agriculture reduced the polarization and the inequalities of farmers' income. As stated in FAO reports [15], evidence from many countries shows that governmental agricultural R&D, education, and access to information for farmers lead to income growth. These results are consistent with those of other studies [21,22], showing that the agricultural R&D expenditure has a positive effect on the growth of farmers' income. Generalizing, investments in rural areas, not only in agriculture, have a positive impact on per capita income, as argued in the research literature [23].

For other member states, as the governmental R&D expenditure in agriculture increases, the resulted coefficients indicate a negative relationship (Figure 6). The negatively impacted farmers' income due to R&D expenditure increase are recorded in the following countries: Cyprus, Spain, Portugal, Romania, Lithuania, and Hungary. Since only for Cyprus is the relationship between the variable strong ($R^2 = 0.63$), while for the rest of the countries, the relationships are medium to weak, we may argue that these results do not change the assumption established at the beginning of the research. Moreover, although controversial, these results are consistent with those found in the literature [24], showing that the agricultural investments have a limited positive effect on farmers' income.

In order to find possible explanations for which in six countries with large shares of agriculture in the GDP and for which the governmental agricultural R&D expenditure negatively impacts the farmers' income (Figures 1 and 8), the government support for agricultural research and development in each EU member state in absolute values are taken into account (Figure 9). A total of 22 EU member states have allocated less than EUR 100 million per year for the governmental agricultural R&D expenditure. In all six countries, except for Spain, the levels of government support for agricultural research and development are very low, less than EUR 50 million. Future research should investigate the reasons why, in the case of Spain, the farmers' income is negatively impacted by the agricultural R&D, although in this member state, the government support is significant, at EUR 488 million.

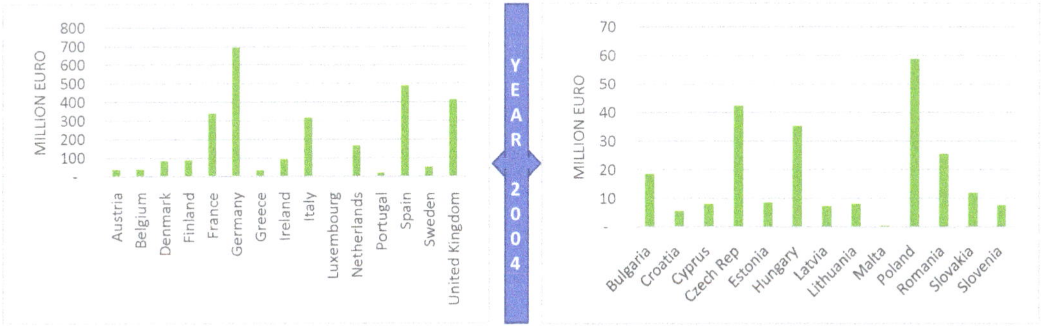

Figure 9. Government support for agricultural research and development by EU member state in absolute figures in millions of euros (annual average of 2004–2020 period) (**left**: old member states; **right**: new member states). Source: edited by the authors from Eurostat [40].

Forthcoming strategic directions may take into consideration improving the competitiveness of rural areas and creating new income and employment opportunities for farmers and their families, as these remain major aims for the future of the European Union [45]. The results of this study have broader implications that go beyond the economic prosperity of farmers and rural areas towards those involving social issues. Income growth affects the farmer's willingness to remain in agriculture and to continue to produce food, while a decrease in income would create negative pressures on social welfare, and migration from rural to urban conditions [46]. As a result of this situation, production amounts will decrease, food prices will increase, food security will be jeopardized, and pressure on governmental financing will increase. Strategies and macroeconomic policies should consider not only the economic results, but also the social and ecological consequences, and the trade-offs and contradictions between sustainability, with its environmental and social dimensions, and productivity, as its economic dimension, should be acknowledged and explored.

5. Conclusions

This study presents the impact of the governmental agricultural research and development expenditure on farmers' income, in the sustainable development context. Statistical data were analyzed, using the regression model, for the time period of 2004–2020 for each member state of the European Union.

The results of the model are diverse, emphasizing the diversity of the EU's economy itself. For some countries, a significant portion of farmers' income growth is explained by governmental R&D expenditure in agriculture, e.g., Estonia, Poland, and Slovenia; for others, the farmers' incomes are partially explained by R&D investments in agriculture, e.g., Austria, Belgium, Bulgaria, the Czech Republic, Germany, Finland, France, Greece, Croatia, Ireland, Latvia, and Slovakia. Controversial results have been found for countries where the governmental R&D expenditure in agriculture negatively impacts the farmers' income. However, in the case of these countries, the farmers' income is weakly impacted by R&D expenditure, as the values of R^2 show. Thus, the hypothesis (H1): Governmental agricultural R&D expenditure influences farmers' income, and the extent varies between countries with different agricultural profiles, has been validated.

Two main categories of results were obtained—countries where farmers' income are influenced by the governmental agricultural R&D expenditure, and countries where the influence is weaker. For the first category of states, the governmental expenditure should continue to focus on ensuring sustainable income for farmers. The implications go beyond improving farmers' revenues, generating on- and off-farm employment, and contributing to strengthening the economic prosperity of the European Union rural communities. For the second group of countries, where the influence of agricultural R&D investments is weaker, the governmental expenditure may be directed towards strengthening rural development, promoting food quality, meeting safety standards and food security, and fostering animal welfare.

Considering this variety of results, their implications are also diverse. The findings should change the methods and directions for using the governmental agricultural R&D expenditure; for example, in countries where the R&D investments in agriculture are among the key factors of farmers' income growth, the governmental expenditure should be carefully underpinned by economic analysis. As such, the macroeconomic policy in rural areas and agriculture become effective in its pathway to achieving the sustainable development indicators.

Bearing in mind the controversial results of the research, in conclusion, in order to achieve the Sustainable Development Goals, Goal 2, Zero Hunger, and to ensure a fair standard of living for farmers and the stability of their incomes, as declared by the objectives of the EU's Common Agricultural Policy, a specific role should be attributed differently to agricultural policy instruments, in general, and governmental agricultural R&D expenditure, in particular, in each member state of the European Union.

The study has its limitations, since it takes into consideration only one of the farmers' income drivers, the governmental agricultural R&D expenditure. Future research and in-depth analysis are required, which should include other factors influencing the agricultural income, such as weather conditions, market prices, factor productivity, production costs, supply chain fluency, economic and social crises, etc.

Author Contributions: Conceptualization, M.S., V.C.T. and R.A.I.; methodology, V.C.T.; software, V.C.T.; validation, V.C.T., M.S. and R.A.I.; formal analysis, I.C.N. and C.G.Z.; investigation, C.G.Z.; resources, V.C.T. and I.C.N..; data curation, V.C.T.; writing—original draft preparation, V.C.T.; writing—review and editing, R.A.I.; visualization, I.C.N.; supervision, M.S.; project administration, M.S.; funding acquisition, M.S. All authors have read and agreed to the published version of the manuscript.

Funding: This research received no external funding.

Institutional Review Board Statement: Not applicable.

Informed Consent Statement: Not applicable.

Data Availability Statement: The data presented in this study are available on request from the corresponding author.

Conflicts of Interest: The authors declare no conflict of interest.

References

1. Smith, G.; Nandwani, D.; Kankarla, V. Facilitating resilient rural-to-urban sustainable agriculture and rural communities. *Int. J. Sustain. Dev. World Ecol.* **2016**, *24*, 485–501. [CrossRef]
2. Khan, N.; Ray, R.; Sargani, G.; Ihtisham, M.; Khayyam, M.; Ismail, S. Current Progress and Future Prospects of Agriculture Technology: Gateway to Sustainable Agriculture. *Sustainability* **2021**, *13*, 4883. [CrossRef]
3. Zhan, J.; Ma, Y.; Hu, W.; Chen, C.; Lu, Q. Enhancing rural income through public agricultural R&D: Spatial spillover and infrastructure thresholds. *Rev. Dev. Econ.* **2022**, *26*, 1083–1107. [CrossRef]
4. Nin-Pratt, A. Agricultural R&D investment intensity: A misleading conventional measure and a new intensity index. *Agric. Econ.* **2021**, *52*, 317–328. [CrossRef]
5. Trusova, N.V.; Kohut, I.A.; Osypenko, S.A.; Radchenko, N.G.; Rubtsova, N.N. Implementation of the results of fiscal decentralization of Ukraine and the countries of the European union. *J. Adv. Res. Law Econ.* **2019**, *10*, 1649.
6. Lee, J.; Koh, M.; Jeong, G. Analysis of the impact of agricultural R&D investment on food security. *Appl. Econ. Lett.* **2017**, *24*, 49–53. [CrossRef]
7. Wu, D.L.; Liu, F.C. Research on the Effects of Agricultural Technology Investment on Farmers' Income Structure. In *2015 International Conference on Social Science, Education Management and Sports Education*; Atlantis Press: Amsterdam, The Netherlands, 2015; pp. 1335–1338.
8. Alston, J.M.; Pardey, P.G.; James, J.S.; Andersen, M.A. The Economics of Agricultural R&D. *Annu. Rev. Resour. Econ.* **2009**, *1*, 537–566. [CrossRef]
9. Anon. Agriculture, Forestry, and Fishing, Value Added (% of GDP)—The World Bank Database. 2022. Available online: https://data.worldbank.org/indicator/NV.AGR.TOTL.ZS?name_desc=true (accessed on 2 August 2022).
10. European Commission. *Communication from the Commission to the Council and the European Parliament on the Mid-Term Review of the Common Agricultural Policy*; COM(2002) 394 Final; European Commission: Brussels, Belgium, 2002. Available online: https://eur-lex.europa.eu/legal-content/EN/TXT/PDF/?uri=CELEX:52002DC0394&from=EN (accessed on 28 April 2020).
11. Pardey, P.G.; Alston, J.M.; Chan-Kang, C. Public agricultural R&D over the past half century: An emerging new world order. *Agric. Econ.* **2013**, *44*, 103–113. [CrossRef]
12. Gherardelli, A. *Increasing Agricultural Productivity and Production in a Socially, Economically and Environmentally Sustainable Manner*; Committee on World Food Security; FAO: Rome, Italy, 2017. Available online: https://policycommons.net/artifacts/2026869/increasing-agricultural-productivity-and-production-in-a-socially-economically-and-environmentally-sustainable-manner/2779312/ (accessed on 28 July 2022).
13. Food and Agriculture Organization of the United Nations. *A Summary to Assess Synergies and Trade-offs among the Twenty Interconnected Sustainable Food and Agriculture (SFA) Actions*; FAO: Rome, Italy, 2020. Available online: https://www.fao.org/3/ca9923en/ca9923en.pdf (accessed on 28 July 2022).
14. United Nations. The 17 sustainable development goals, UN general assembly. In *Transforming Our World: The 2030 Agenda for Sustainable Development*; UN Resolution A/RES/70/1; United Nations: New York, NY, USA, 2015.
15. Food and Agriculture Organization of the United Nations. *Transforming Food and Agriculture to Achieve the SDGs*; FAO: Rome, Italy, 2018. Available online: https://www.fao.org/3/I9900EN/i9900en.pdf (accessed on 28 July 2022).
16. Ramanujam, N.; Richardson, S.B. Ending Child Malnutrition Under SDG 2: The moral imperative for global soli-darity and local action. *Soc. Altern.* **2018**, *37*, 18–24.

17. Bizikova, L.; Jungcurt, S.; McDougal, K.; Tyler, S. How can agricultural interventions enhance contribution to food security and SDG 2.1? *Glob. Food Secur.* **2020**, *26*, 100450. [CrossRef]
18. Mensi, A.; Udenigwe, C.C. Emerging and practical food innovations for achieving the Sustainable Development Goals (SDG) target 2.2. *Trends Food Sci. Technol.* **2021**, *111*, 783–789. [CrossRef]
19. Al-Mulali, U.; Fereidouni, H.G.; Bin Mohammed, M.A.H.; Lee, J.Y.M. Agriculture investment, output growth, and CO2emissions relationship. *Energy Sources Part B Econ. Plan. Policy* **2016**, *11*, 665–671. [CrossRef]
20. Djokoto, J.G. Foreign direct investment into agriculture: Does it crowd-out domestic investment? *Agrekon* **2021**, *60*, 176–191. [CrossRef]
21. Liu, Y.; Xiu, C. Rural Finance Development, Agricultural Science and Technology Advance and the Growth of Farmers' Income. *J. Agrotech. Econ.* **2013**, *9*, 92–100.
22. Kumar, S.; Chahal, V.P. Doubling farmers' income. *Indian Farming* **2018**, *68*, 95–96.
23. Yang, D.T.; Zhu, X. Modernization of agriculture and long-term growth. *J. Monet. Econ.* **2013**, *60*, 367–382. [CrossRef]
24. Tang, G. Empirical Analysis on the Relationship between Rural Public Goods Supply and Farmers' Income Growth. *J. Hunan Agricult. Univ.* **2009**, *10*, 19–24.
25. Arrow, K.; Bolin, B.; Costanza, R.; Dasgupta, P.; Folke, C.; Holling, C.; Jansson, B.-O.; Levin, S.; Mäler, K.-G.; Perrings, C.; et al. Economic growth, carrying capacity, and the environment. *Ecol. Econ.* **1995**, *15*, 91–95. [CrossRef]
26. Apergis, N.; Ozturk, I. Testing Environmental Kuznets Curve hypothesis in Asian countries. *Ecol. Indic.* **2015**, *52*, 16–22. [CrossRef]
27. Sadorsky, P. Renewable energy consumption and income in emerging economies. *Energy Policy* **2009**, *37*, 4021–4028. [CrossRef]
28. Heisey, P.W.; Fuglie, K.O. Public agricultural R&D in high-income countries: Old and new roles in a new funding environment. *Glob. Food Secur.* **2018**, *17*, 92–102. [CrossRef]
29. Du, J.; Dabuo, F.T.; Madzikanda, B.; Boamah, K.B. The Influence of R&D in Mining on Sustainable Development in China. *Sustainability* **2021**, *13*, 5289. [CrossRef]
30. Balcombe, K.; Rapsomanikis, G. An Analysis of the Impact of Research and Development on Productivity Using Bayesian Model Averaging with a Reversible Jump Algorithm. *Am. J. Agric. Econ.* **2010**, *92*, 985–998. [CrossRef]
31. Alston, J.M.; Norton, G.W.; Pardey, P.G. *Science Under Uncertainty: Principles and Practice for Agricultural Research Evaluation and Priority Setting*; Cornell University Press: Ithaca, NY, USA, 1995.
32. Esposti, R.; Pierani, P. *Building the Knowledge Stock: Lags, Depreciation and Uncertainty in Agricultural R&D*; Working paper No. 145; Departmento di Economia, Universita Politecnica delle Marche: Ancona, Italy, 2001.
33. Salim, R.A.; Islam, N. Exploring the impact of R&D and climate change on agricultural productivity growth: The case of Western Australia*. *Aust. J. Agric. Resour. Econ.* **2010**, *54*, 561–582. [CrossRef]
34. Fuglie, K. R&D Capital, R&D Spillovers, and Productivity Growth in World Agriculture. *Appl. Econ. Perspect. Policy* **2017**, *40*, 421–444. [CrossRef]
35. Hazell, P.B. Chapter 68 An Assessment of the Impact of Agricultural Research in South Asia Since the Green Revolution. *Handb. Agric. Econ.* **2010**, *4*, 3469–3530. [CrossRef]
36. Spada, A.; Fiore, M.; Monarca, U.; Faccilongo, N. R&D Expenditure for New Technology in Livestock Farming: Impact on GHG Reduction in Developing Countries. *Sustainability* **2019**, *11*, 7129. [CrossRef]
37. Mazzucato, M. Mission-oriented innovation policies: Challenges and opportunities. *Ind. Corp. Chang.* **2018**, *27*, 803–815. [CrossRef]
38. Kata, R.; Wosiek, M. Inequality of Income in Agricultural Holdings in Poland in the Context of Sustainable Agricultural Development. *Sustainability* **2020**, *12*, 4963. [CrossRef]
39. He, Y.; Wang, J.; Gao, X.; Wang, Y.; Choi, B.R. Rural Tourism: Does It Matter for Sustainable Farmers' Income? *Sustainability* **2021**, *13*, 10440. [CrossRef]
40. Anon. Government Support to Agricultural Research and Development—Products Datasets—Eurostat. 2022. Available online: https://ec.europa.eu/eurostat/web/products-datasets/-/sdg_02_30 (accessed on 17 May 2022).
41. Anon. OECD Statistics. 2022. Available online: https://stats.oecd.org/# (accessed on 17 May 2022).
42. Anon. Economic Accounts for Agriculture—Agricultural Income (Indicators A, B, C)—Products Datasets—Eurostat. 2022. Available online: https://ec.europa.eu/eurostat/web/products-datasets/-/aact_eaa06 (accessed on 17 May 2022).
43. Sibuea, M.B.; Sibuea, F.A. Contribution of village cooperation unit in improving farmers incomes. *Iop Conf. Ser. Earth Environ. Sci.* **2018**, *122*, 012021. [CrossRef]
44. Dokić, A.; Užar, D.; Petković, G.; Stojković, D. Impact of commercial and investment activities in agriculture on local development. *Econ. Agric.* **2020**, *67*, 569–584. [CrossRef]
45. European Commission. *Green Deal*; European Commission: Brussels, Belgium, 2021. Available online: https://ec.europa.eu/info/strategy/priorities-2019-2024/european-green-deal_ro (accessed on 29 July 2022).
46. Doğan, H.P.; Aydoğdu, M.H.; Sevinç, M.R.; Cançelik, M. Farmers' Willingness to Pay for Services to Ensure Sustainable Agricultural Income in the GAP-Harran Plain, Şanlıurfa, Turkey. *Agriculture* **2020**, *10*, 152. [CrossRef]

MDPI

Article

Forecasting the Optimal Sustainable Development of the Romanian Ecological Agriculture

Ana Ursu [1] and Ionut Laurentiu Petre [1,2,*]

[1] The Agricultural Economics Office, Research Institute for Agriculture Economy and Rural Development, 010961 Bucharest, Romania

[2] The Department of Agrifood and Environmental Economics, The Bucharest University of Economic Studies, 010961 Bucharest, Romania

* Correspondence: laurentiu.petre@eam.ase.ro

Abstract: Organic farming is an important objective of the European Commission, translated into the European Green Pact through the Farm to Fork Strategy and the Biodiversity Strategy, with EU member countries having to find solutions to meet the target of at least 25% of agricultural land being used for organic cultivation by 2030. The aim for Romania can be achieved by modelling the distribution of crops in terms of cultivated areas and production yields obtained in organic and conventional systems according to the population size. Applying quantitative and qualitative analysis of EUROSTAT data for the above-mentioned indicators, the geomean function, linear programming, and the simplex method were used, depending on the set objectives. To demonstrate that organic farming can be sustainable and in line with the three pillars of sustainability, economic, social and environmental, we related the agricultural area to the population of Romania to highlight the average annual growth rate for the 2020–2030 tine horizon. The results showed an increase in agricultural area per capita of 0.708 ha (4.91%), compared to 0.69 ha as the average for the period 2012–2020, which correlated with organic production yields 32% lower than conventional agriculture. Through modelling, the reduction in organic farm yield was found to be less than or equal to the increase in area per capita, thus reaching the proposed target. The results of this study have long-term implications for supporting the transition to organic farming in the sense that the study argues that reaching the target of 25% of agricultural land that can enter organic farming is in line with the sustainability trilogy. The approach used can be followed and replicated according to national agricultural policies.

Keywords: modelling organic crops; organic area; strategies; common agricultural policy

Citation: Ursu, A.; Petre, I.L. Forecasting the Optimal Sustainable Development of the Romanian Ecological Agriculture. *Sustainability* **2022**, *14*, 14192. https://doi.org/10.3390/su142114192

Academic Editors: Mariarosaria Lombardi, Erica Varese and Vera Amicarelli

Received: 3 October 2022
Accepted: 26 October 2022
Published: 31 October 2022

1. Introduction

Agriculture is an important sector for Romania, with an average utilised agricultural area (UAA) of 13.6 million hectares [1]. Agriculture contributes 3.8% toward Romanian GDP (in 2020). Agriculture is an activity that competes for land, so any policy change that affects one land use has the potential to induce changes in the other [2]. Sustainable land use involves considering the range of social, economic and environmental goods and services provided in a given region [3]. Sustainable land use also involves careful consideration of the long-term attributes of resilience and robustness that maintain the underlying ecosystem processes. Population density and GDP are useful indicators in relation to the two dimensions of human activity, the social and economic aspects, which are connected to land use characteristics. The presence of a larger population density requires a higher intensity of land use. On the other hand, increasing economic production requires more intervention on the land [4]. To carry out this study and in line with the set target of increasing arable land in organic farming by 25% by 2030, we analysed the input indicators established for analysis and modelling.

1.1. Conventional Agriculture

The utilised agricultural area (UAA) decreased by 4.98% in 2020 (13,048.80 thousand ha) compared to 2012 (13,733.14 thousand ha), with Romania ranking 6th in the EU-27, after France, Spain, Germany, Poland and Italy (Table 1). Within the structure of land use categories, the largest share, 64%, is occupied by arable land, with a decrease recorded in 2020 (8482.86 thousand ha) of 3.6% compared to 2012 (8797.65 thousand ha). Romania ranks 5th in the EU-27 in this indicator, after France, Spain, Germany and Poland.

Table 1. The breakdown of the Romanian utilized agricultural area (UAA).

	2012	2013	2014	2015	2016	2017	2018	2019	2020	Avg. (%)
Utilised agricultural area (1000 ha)	13,733	13,905	13,830	13,858	13,521	13,378	13,414	13,826	13,049	100.0
Arable land (%)	64.1	62.9	63.5	63.3	63.5	63.9	64.8	64.8	65.0	64.0
Permanent grassland (%)	32.7	33.9	33.5	33.6	33.4	33.0	32.0	32.2	31.9	32.9
Permanent crops (%)	2.4	2.4	2.2	2.3	2.4	2.4	2.5	2.3	2.3	2.3
Kitchen gardens (%)	0.8	0.8	0.8	0.8	0.7	0.7	0.7	0.7	0.8	0.8

Source: Calculations based on EUROSTAT data series, years 2012–2022 https://ec.europa.eu/eurostat/databrowser/product/page/TAG00025__custom_3351494 (accessed on 14 September 2022) Utilised agricultural area by categories [TAG00025__custom_3351494].

Although Romania is among the top EU countries in terms of cultivated areas, it has lower production yields per ha. The extremely severe 2020 and 2022 droughts have worsened these deficits [5]. The relative economic performance of organic and conventional agriculture is determined by the ratio of production costs to production value. Both organic and conventional farmers are vulnerable to fluctuations in input and output prices. The future of material prices is uncertain. However, changes in commodity prices may have a greater impact on conventional farmers [6].

1.2. Organic Farming

Organic farming has been present in Romania since 2000 (17,388 ha) and the land area in production was relatively constant until 2007 (131,529 ha). Since 2007 and until now both the area and the number of organic operators has increased, at a variable pace. In the National Rural Development Plan (NRDP) 2007–2013, Romania did not benefit from compensatory payments, because no measure was implemented in the programme [7]. In the National Rural Development Plan (NRDP) 2014–2020, organic farming benefited from Measure 11- Organic farming, with support being directed towards conversion, methods and maintenance of organic farming practices.

Data presented by "[8]" on organic farming in the EU reveal that at the end of 2020, there were 14.9 million ha (9.2% of total production) of organic land in the European Union managed by more than 349 thousand producers (Table 2, col 2 and col 12). The countries with the largest organic agricultural areas are France (2.5 million ha), Spain (2.4 million ha), Italy (2.1 million ha), and Germany (1.7 million ha). Romania has an organic agricultural area of over 469 thousand ha (3.5%) managed by 9647 producers. The organic areas, for the countries mentioned, are composed of grassland (minimum 26% (Bulgaria) and maximum 89% (Ireland)), arable crops (minimum 21% (Spain) and maximum 74% (Poland)), permanent crops (0% (Ireland) and maximum 39% (Malta)) (Table 2).

From a policy perspective, the Farm to Fork strategy target of having "at least 25% of EU farmland in organic farming by 2030" is seen as a challenge, with many stakeholders questioning whether this ambition can be achieved. According to the data presented in Table 2, the lowest proportions of organic land in the EU are found in Romania (3.5%), Bulgaria (2.3%), Ireland (1.7%) and Malta (0.6%).

Table 2. Organic land use in Europe, 2020.

Country	Organic Land Area (1000 ha)	Percentage of Organic Agricultural Land (%)	Organic Land Use								Producers (nr)	Processors (nr)
			Grassland (ha)	%	Arable Crops (ha)	%	Permanent Crops (ha)	%	Other (ha)	%		
Austria	680	26.5	392,168	58	275,403	41	12,301	2	No d	0	24,480	22,689
Italy	2095	16.0	583,781	28	1,016,287	49	495,296	24	No d	0	71,590	20,087
Spain	2438	10.0	1,273,392	52	502,074	21	662,425	27	No d	0	44,493	5561
Germany	1702	10.2	880,000	54	735,727	49	25,132	2	No d	0	53,255	19,311
EU-27	14,900	9.2	6,290,847	42	6,750,565	46	1,680,385	11	84,525	1	349,551	778,416
France	2549	8.8	879,244	34	1,445,221	57	193,731	8	30,481	0	41,632	16,651
Hungary	301	6.0	180,961	60	105,562	35	14,907	5	No d	0	5128	521
Polonia	509	3.5	85,741	17	375,939	74	47,615	9	No d	0	18,598	668
Romania	469	3.5	155,038	33	291,628	62	22,221	5	No d	0	9647	201
Bulgaria	116	2.3	30,154	26	61,269	53	24,849	21	No d	0	5942	249
Ireland	75	1.7	66,488	89	8103	11	75	0	No d	0	1777	180
Malta	0.1	0.6	No d	0	41	61	26	39	No d	0	25	8

Source: Austrian Federal Ministry of Agriculture Forestry—Environment and Water Management—Eurostat. Research Institute of Organic Agriculture (FiBL) and Agricultural Market Information Company (AM). Data compiled by Fibl based on Eurostat and national data sources. https://www.organicseurope.bio/about-us/organic-in-europe/ (accessed on 14 September 2022).

1.3. The Population of Romania

Romania has an area of 238,369 km^2 and a population recorded in 2020 of 19,281,118 inhabitants, representing approximately 4.3% of the EU-27 population [9].

Rural areas have substantial sources of development, representing 87% of the national territory, and the rural population in 2020 was 8.9 million, approximately 46.4% of the Romanian population.

The rural population decreased (measured in number of inhabitants) by 4.3% due to negative changes in the main demographic indicators: population ageing, declining birth rate and migration of the labour force, especially young people, from villages to cities and especially abroad. It is predicted that some countries in Europe will lose more than 15% of their population by 2050 due to international migration, Romania being one of these countries [10,11].

1.4. Policies and Strategies

The 2030 Agenda includes global objectives to guide the actions of international communities until 2030, and is relevant for both developed and developing countries. The transition to sustainable food production and agriculture will require major improvements in resource efficiency, environmental protection and system resilience [12].

Increasing the share of organic agriculture in the EU is part of the Action Plan for the development of organic production, with the objective of having "at least 25% of the EU's agricultural land in organic agriculture and a significant increase in organic aquaculture by 2030", contained within the From Farm to Consumer Strategy [13,14].

Organic agriculture is one of the many approaches and paradigms found to fulfil the objectives of sustainable development of agriculture [15,16].

According to IFOAM EU, reaching 25% of organic agricultural land area in the EU by 2030 is achievable if the CAP provides the necessary remuneration for the benefits of ecological conversion and maintenance through existing rural development policies or innovative instruments such as ecological schemes [17].

1.5. The Purpose and Hypothesis of the Research

In view of the above, the aim of this paper is to determine how organic farming can be developed sustainably, i.e., to determine the size of the areas that can be converted to organic farming so that, on the one hand, the share specified by EU strategies is achieved and, on the other hand, low yields do not affect food security and thus sustainable development objectives.

Based on the information, data and literature, the research hypothesis can be concretized. We believe that in Romania, the ecological agricultural system can be developed, given that the loss of yield can be balanced or cushioned by the fact that the population is decreasing, so there is a possibility that the agricultural area per capita can increase.

In relation to the share of organic farming in the total agricultural area and its expansion, we believe that large areas of grassland and meadows can be converted, as they contribute essentially to this objective.

2. Literature Review

Studies reveal a multitude of approaches regarding sustainability and ecological agriculture. The dynamics of organic agriculture certification in Romania was studied, starting from the hypothesis that the slow pace of certifications is due to some subjective barriers that can be eliminated if incentive measures are applied to support certification [18].

Regarding the EU Action Plan for organic agriculture, axis 1, stimulating and ensuring consumer confidence in the context of the sustainability and competitiveness of organic farms [19], proposes the implementation of ecological marketing strategies that would stimulate both consumption and production, thus contributing to sustainability and business development.

Another research study [20] addressed the issue of the limiting factors on the development of the organic food sector. The study used the qualitative analysis method with semi-structured interviews applied to 10 large and medium-sized companies active in the ecological sector. The limiting factors indicated by the managers refer not only to the legislation, the lack of constant supply of organic raw materials and increased competition on the domestic and international markets, but also to the instability of the financial situation, regarding financial liquidity, costs, capital and credits [20].

Rasche and Steinhauser [21], investigated how an increase agricultural area would affect yield differences between conventional and organic systems. Through the accounting tool FABLE, they evaluated the changes in consumption of available calories per person-year/day and the extent of cultivated lands, pastures and areas where natural processes predominate, until the year 2050. It was concluded that by increasing the ecological surface, there will be a caloric deficit of 7–80 kcal/person/day, corresponding to a surface of 1000–5000 km^2 of land cultivated. It was also estimated that the deficit would disappear without any changes to the system by 2045 due to demographic and technological development, and that would be no need for additional cultivated land at all if crop productivity were to increase.

Eneizen [22], used exploratory qualitative analysis combined with empirical research results to determine the main obstacles that must be solved for the expansion of ecological agriculture. The findings of the study, based upon interviews with organic farmers, suggest that obstacles to adoption of organic farming are: the absence of an organization to certify organic products, high cost of certification, lack of financing sources, low yield, high price, lack of specific markets for organic food, the low awareness of farmers, unsuccessful agricultural reforms, and lack of coordination between interested parties and institutional changes. The authors recommend that organic farming be carried out by qualified farmers using modern organic farming techniques that can contribute to increasing production yields and cost efficiency. The authors also recommend improving communication between the interested parties of organic agriculture, from farms to markets, including any relevant intergovernmental departments, to develop organic agriculture.

To answer the question of what the contribution of ecological agriculture to the sustainable development of agriculture is, Kilker [23] refers to the trilogy of sustainability, socio-economic and environmental development, which would help producers and exporters to improve their incomes and living conditions, especially in poorer countries. From an economic and social point of view, organic farming reduces the risk of production failure, stabilises profits and improves the quality of life of small farmers' families, while from an ecological point of view, it improves soil fertility and preserves biodiversity, leading to ecosystem stability, reduced susceptibility to drought and pest attack. These benefits appear if production methods adapted to local conditions are applied, synthetic chemical pesticides and fertilizers are avoided, and crop diversity is maximized [24].

In another case study, organic farming is seen as a multi-functional business through which sustainable profits can be obtained, creating economic opportunities for people which can help society develop in a sustainable manner. The research was based on visits to organic farms and organic markets, as well as interviews with farmers. This was a model for the local community and for wider communities, thus contributing to the fulfilment of some among the objectives of sustainable development [25].

Sher [26] investigated the barriers to adopting green entrepreneurial agriculture to obtain economic growth through the minimal use of resources. Of the 34 barriers identified, 20 were considered critical barriers. Based on factor analysis, the 20 barriers were grouped into six major categories: training and development, entrepreneurial orientation, market orientation, customer orientation, innovation orientation, and barriers related to the provision of ecological support. The dominant barrier was training and development, as well as the marginal role of the government in carrying out such efforts.

For Romania, organic farming can become a technological alternative to conventional agriculture, as land conversion is within the reach of managers, and this opportunity is further enhanced by the high level of land fragmentation and the high number of small farms in agriculture [27,28].

3. Materials and Methods

The focus of this study was to determine the areas cultivated in an ecological system for each crop in Romania in order to reach the threshold imposed by the European Union regulations, regarding the share of organic agriculture in the total agricultural area of 25%. It is desired that development of ecological agriculture results in as little damage as possible in terms of yield and productivity; thus, a sustainable expansion of this farming system is desired.

For this purpose, data taken from European databases (Eurostat) on areas, production, and crop yields in Romania, both for organic and conventional agriculture, were analysed quantitatively and qualitatively in order to determine yield differences.

For the expansion of organic farming to be sustainable, the agricultural area per capita, especially its dynamics, was determined and forecasted for the year 2030, when each Member State must contribute to 25% of the agricultural area being farmed organically. This will compare the potential increase in agricultural area per capita (given the demographic decline in Romania) with the reduction in yields on the organic area (the 25%), so that the reduction in productivity is less than or equal to the increase in area per capita.

The forecast agricultural area per capita will be determined by relating the agricultural area to the population forecast by FAOSTAT, which is forecast using the average annual rate method, this indicator having the following formula [29]:

$$\overline{R} = \left(\overline{I} - 1\right) \times 100$$

and

$$\overline{I} = \sqrt[n-1]{\prod I_{t/t-1}}$$

where: $\overline{R}$—average rate; $\overline{I}$—average index; I—individual levels of chain-based indices.

Linear programming and the Simplex method were used to determine an optimum yield (tending towards the minimum point), with certain conditions that satisfy both the requirements of European Union regulations and the soil and crop structure specific to Romania.

Programming problems involve the efficient use or allocation of limited resources to achieve desired goals. These problems are characterized by many solutions that satisfy the basic conditions of each problem. Choosing a specific solution as the best solution to a problem depends on the goals or overall objectives contained in the problem statement. The solution that satisfies both the problem conditions and the given objective is called the optimal solution [30].

Linear programming is an important cornerstone of optimization theory. Many real-world problems can be formulated with linear mathematical models. The simplex algorithm is the most used tool for solving linear programming [31].

Maximum efficiency means minimizing effort and maximizing output, and the concept of optimal is defined as a program that minimizes or maximizes an objective function while satisfying all techno-economic constraints.

Assuming that each component of the line vector "c" measures the efficiency of one unit of the output of an activity, then the linear function can be introduced [32]:

$$f_{(X)} = c_1 \times X_1 + c_2 \times X_2 + c_3 \times X_3 + \ldots + c_n \times X_n$$

Summarizing, we obtain the following linear programming equations:

$$\begin{Bmatrix} optimum\left[f_{(X)}\right] & (A) \\ \sum_{j=1}^{n} a_{ij} \times x_j \leq b_i & (B) \\ \sum_{j=1}^{n} a_{kj} \times x_j \leq b_k & (C) \\ x_j \geq 0 & (D) \\ j = 1, n & \end{Bmatrix}$$

Relations A–D together constitute the general model of a linear programming problem, each having a specific role: Relation (A) is called the efficiency objective function of the problem, relation B represents resource constraints, and relation C refers to techno-economic constraints.

Constructing the model of the linear programming problem led to the following system of equations. The objective function was minimising yield losses, i.e., losses in organic production compared to conventional production:

$$f_{(X)_{(min)}} = \frac{\sum_{i=1}^{n} \Delta\%\overline{Q} \times X_i}{\sum_{i=1}^{n} X_i}$$

Xi—The variables taken into account (areas of organic crops cultivated in Romania);

$\Delta\%\overline{Q}$—Relative yield differences for each organic crop compared to the same crop in a conventional system.

For the objective function, it was desired that the weighted average of the yield differences be as small as possible, so each (relative) yield difference between organic and conventional farming for each variable (crop) was multiplied by the area cultivated relative to the total area cultivated organically.

This objective function was conditioned by a series of equations in order to make the expansion of areas sustainable and to be able to determine as correctly as possible the extent of organic crops. Together, the following equations form the system of conditions for the linear programming problem.

$$\begin{Bmatrix} \frac{\sum_{i=1}^{n} X_i}{UUA} = 0.25 \\ \frac{\sum_{i=1}^{n} \Delta\%\overline{Q} \times X_i}{\sum_{i=1}^{n} X_i} \times 0.25 \leq 4.91\% \\ X_i \geq X_{i2020} \\ \frac{X_i'}{UAA} \leq 0.25 \\ ('for\ X_i\ with\ \Delta\%\overline{Q} > 0) \end{Bmatrix}$$

The first condition in the previous system of equations refers to the main target of the European Union strategy, i.e., that the share of organic crops should reach 25%, so that the sum of the organic areas to be established for the year 2030, in relation to the utilised agricultural area (the projected one) should reach 25%.

The second condition is the one that provides a sustainable direction for this expansion of organic areas, i.e., the relative yield gap between organic and conventional agriculture for the 25% of the agricultural area to be less than or equal to 4.91%, which is the potential degree by which the agricultural area per inhabitant will increase by 2030, given that Romania's population is decreasing faster than the agricultural area.

The third equation requires that the organic areas should start from the year 2020, i.e., the last year for which data have been recorded in European statistics, and the last equation requires that the areas of organic crops with positive differences in rankings should not exceed 25%, i.e., the average increase in areas in order to avoid situations in which the development of organic farming is based on 2–3 crops, which currently have very low proportions.

Therefore, the research stages to be presented will start with the determination and forecast of the dynamics of the agricultural area per inhabitant, so that on the basis of the expected increase in the indicator, it will be possible to determine the percentage that Romania can assume in terms of productivity losses on the 25% of the organic areas. Subsequently, the data on yield loss for each crop will be entered into the linear programming model and these conditions related to the proportion of area and the correlation of losses with agricultural area per capita will be introduced in order to determine the exact size of the ecological area for each crop analysed.

4. Results

To identify the areas that should be extended for each organic crop in Romania to reach the threshold of 25% of the agricultural area, we started with a quantitative analysis of statistical data on both organic and conventional agriculture.

It can be assumed that organic farming is in its infancy, even if there are data as early as 2012, or perhaps organic farming existed in practice before this period, but this statement is based on the proportion of organic areas in the total utilised agricultural area, as shown in Table 3.

Table 3. Dynamics of total area under organic farming in Romania, hectares.

Time	2012	2013	2014	2015	2016	2017	2018	2019	2020
UAA Eco	103,093	138,125	190,430	175,571	149,613	149,106	171,594	211,487	275,965
%	0.75%	0.99%	1.38%	1.27%	1.11%	1.11%	1.28%	1.53%	2.11%

Source: processing based on Eurostat data.

The area cultivated organically in Romania increased from 103 thousand hectares in 2012 to approximately 276 thousand hectares in 2020, which represents an increase of 168%. We also observed an average annual growth of 13.1% during the period analysed. However, it can be seen that the expansion of the organic land area has not been constant and strictly increasing; there is a decrease in the middle of the period, with 2016 and 2017 recording slightly smaller areas. These years coincided with the interval between the two programming periods of the Common Agricultural Policy. The subsidies and funds for agriculture were lower in this period. The standard deviation was approximately 50 thousand hectares, a variation of ±28%.

Table 3 shows the proportion of organic area in the total agricultural area in Romania, which increased from 0.75% to 2.11%. However, as mentioned above, this proportion is low compared to other EU countries, so the development of organic farming up to 25% of the agricultural area will be a challenge.

In order for this development to be sustainable, without economic (drastic reduction in yields), social (transition to food insecurity) and environmental (high resource consumption) implications, it is hoped that there is a possible situation in which the difference in yield and decrease in productivity for that 25% of the agricultural area is covered by the increase in agricultural area per capita, given the demographic decline in Romania.

This will determine the agricultural area per inhabitant by 2030, the deadline for meeting the EU biodiversity strategy target.

From Figure 1 it can be seen that utilised agricultural area and population are both decreasing, but by analysing the trend equation of the two indicators, we found that population is decreasing faster than the utilised agricultural area. Over the period 2012–2020, the utilised agricultural area decreased from 13.73 million hectares to approximately 13.05 million hectares, representing a decrease of 4.95% and an average annual rate of change of −0.64%.

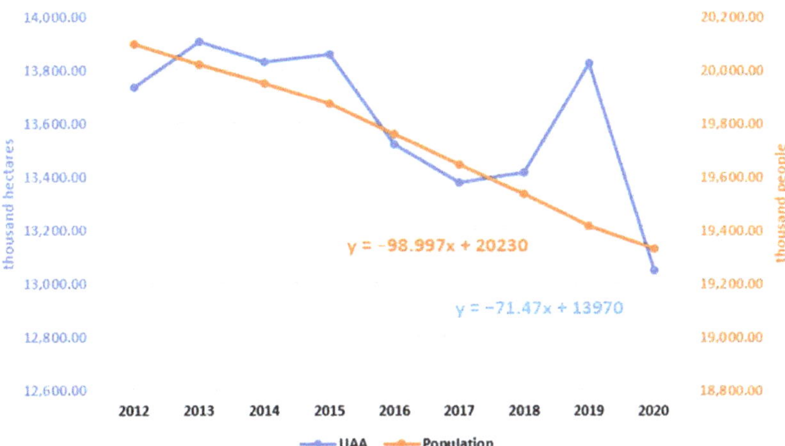

Figure 1. Dynamics of agricultural land use and population in Romania. Source: Eurostat data.

Based on the average annual rate of change in utilised agricultural area, as well as FAOSTAT population forecasts, which estimate that the population will reach 18.3 million in 2030, it was possible to determine and forecast the agricultural area per capita and the dynamics of this indicator.

In regard to the period 2012–2020, for which precise data have been recorded for both utilised agricultural area and population, there were no increases as perhaps expected, given the steady decrease in population, because the utilised agricultural area has also fluctuated with both negative and positive variations. The utilised agricultural area per capita ranged from 0.68 hectares per capita to 0.71 hectares per capita, with an average of 0.69 hectares per capita over the period and a standard deviation of 0.01 hectares per capita from this average, giving a variation of ±1.6%. (Figure 2).

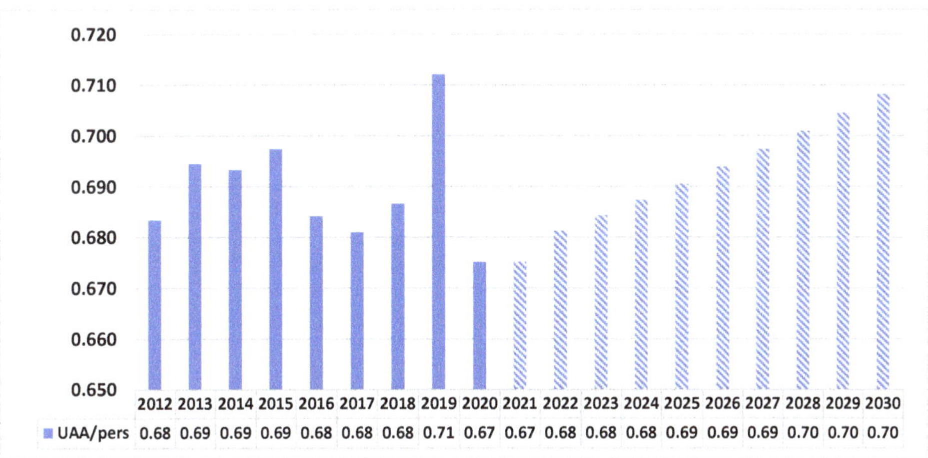

	2012	2013	2014	2015	2016	2017	2018	2019	2020	2021	2022	2023	2024	2025	2026	2027	2028	2029	2030
UAA/pers	0.68	0.69	0.69	0.69	0.68	0.68	0.68	0.71	0.67	0.67	0.68	0.68	0.68	0.69	0.69	0.69	0.70	0.70	0.70

Figure 2. Determining and forecasting the agricultural area used per capita (ha/capita).

Forecasting this indicator on the basis of the utilised agricultural area determined on the basis of the average annual rate of change and on the basis of the population according to the FAO forecast, it is estimated that the agricultural area per capita will follow an increasing trend until 2030, reaching a level of 0.71 hectares per capita.

In order to determine the degree of sustainability in terms of yield reduction for organic farming, the dynamics of the agricultural area per capita was determined, i.e., the relative difference between the target year (2030) and the last year with exact data, i.e., 2020, the agricultural area per capita will be expected to increase by 4.91%, which allows for a slight decrease in agricultural productivity given the characteristics of the organic farming system.

Next, the areas and yields for all crops recorded in the Eurostat databases, both for organic and conventional farming, were researched in order to finally determine the yield differences, which are essential in the second part of the work on minimizing the decrease in agricultural productivity in organic versus conventional farming, depending on the areas of the crops studied. The determination of the yields for the two cropping systems in agriculture and their levels can be seen in Tables A1 and A2.

Table 4 presents the percent yield differences for each crop in Romania grown organically, according to the Eurostat data, compared to conventional yields, for the period 2012–2020, where data were available. Analysing the average percent differences across years, there are organic crops for which yields are higher than in the conventional system. These crops include berries (excluding strawberries), whose yield in organic system was 158% higher; a second crop is hops, but the average was determined over a short period of time, so there is a larger margin of error. The average yield of organic hops is 65% higher. Oats and spring cereal mixtures had yields in organic system higher by 9.26%. While all these crops were higher yielding in organic systems, their area share was not very high. The situation is different for grain maize and corn-cob-mix, which is only 2% higher yielding in organic systems, but the area cultivated is about 19% of the total organic area, being the largest single crop.

However, for the most part, organic crop yields are lower than conventional yields; among the closest but still lower yields are sunflower (−4.34%), oilseed rape (−10.9%) and common wheat (−12.46%). At the other end of the scale, there are crops whose organic yields are much lower, more than half, especially for fruit and vegetables, where organic yields are more than 70% lower than conventional ones.

On average, for all the organic crops analysed, there was a yield gap of 32% against organic compared to conventional farming. However, it should be noted that this is a simple arithmetic average; without considering the share of cultivated areas, by taking this average weighted by the areas of the main crops, the yield in organic farming was about 10% lower than in conventional farms, so, as is natural, farmers turned to crops with potential to risk as little as possible and eliminate losses. However, given that Romania must expand its organic farming area, farmers will no longer be able to focus on certain crops, and expansion will most likely widen the gap between weighted yields.

At the same time, in addition to determining the yield differences, which will represent the coefficients of each variable of the linear programming function, the weight of each crop will also be used (Figure 3), given that, until now, there are areas cultivated in an organic system, these will have to be extended from now on, so the values of the variables will have to be higher than or at least equal to those at present.

As mentioned above, the area under organic maize has a significant share, but this crop is in first place, with a share of 19% of the total area under organic cultivation, followed by wheat and spelt, with 15.7%, then sunflower with 8.1%, followed by plants harvested green from arable land with 6.5%, barley with 3.4%, rapeseed with 3%, and then hops with only 0.002%.

Therefore, having created this context with which we can realize and determine the areas that should be cultivated in 2030 in order to reach the European Union target, we constructed a linear programming model that led to the following system of equations.

Table 4. Determination of yield differences for each organic crop in Romania compared to the yield in the conventional system (%).

Relative Differences [Eco–Conv (%)]	2012	2013	2014	2015	2016	2017	2018	2019	2020	Average
Apples	–45.13	–46.05	–40.42	–70.59	–40.89	–14.55	–57.50	–27.49	–70.02	–45.85
Apricots	x	–74.93	–98.99	–81.26	–87.80	–82.33	–98.92	–94.53	–96.94	–89.46
Aromatic, medicinal and culinary plants	x	–30.62	5.84	–12.01	74.84	2.47	–25.37	–5.72	–20.74	–1.41
Barley	–31.95	–35.72	–3.38	–15.39	–8.72	–16.67	–48.21	–41.78	–15.82	–24.18
Berries (excluding strawberries)	x	–19.17	515.72	41.55	692.70	–14.13	–25.32	40.91	30.93	157.90
Brassicas	x	–23.51	–64.53	–94.13	–71.68	–98.83	–50.71	–41.59	–75.86	–65.10
Cereals (excluding rice) for the production of grain (including seed)	x	–34.47	2.38	21.49	–1.21	–16.78	–38.38	–31.78	–30.84	–16.20
Cereals for the production of grain (including seed)	x	–33.94	3.69	22.79	–2.02	–16.43	–38.09	–31.28	–29.74	–15.63
Cherries	x	–74.83	–85.40	–84.80	–70.20	–40.81	–86.32	–85.11	–92.74	–77.53
Common wheat and spelt	x		13.31	6.64	–0.30	–17.80	–31.66	–27.66	–29.71	–12.46
Dry pulses and protein crops for the production of grain (including seed and mixtures of cereals and pulses)	x	–20.09	40.10	15.10	–22.74	–0.51	–1.25	1.59	–3.14	1.13
Durum wheat	x	x	–39.29	4.38	–21.84	–43.04	–49.78	41.48	–21.10	–18.45
Fibre crops	x	x	–68.99	–98.17	47.20	29.50	9.83	–11.88	–91.41	–26.27
Fresh pulses	x	–17.97	–87.51	–58.99	–47.98	–82.12	–71.71	58.55	49.57	–32.27
Fresh vegetables (including melons)	x	x	–90.58	–68.31	–74.27	–84.41	–84.42	–72.52	–66.71	–77.32
Fresh vegetables (including melons) and strawberries	x	x	–90.44	–68.16	–74.11	–83.74	–84.24	–72.65	–66.19	–77.08
Fruits, berries and nuts (excluding citrus fruits, grapes and strawberries)	x	x	–51.88	–81.73	–62.61	–63.95	–82.40	–70.72	–83.18	–70.92
Grain maize and corn–cob–mix	18.70	–30.93	12.22	66.02	18.56	0.81	–22.52	–21.54	–23.51	1.98
Grapes	–16.83	–30.19	–40.35	9.24	–59.03	–23.43	51.24	–17.14	–37.46	–18.22
Hops	x	x	43.37	4.55	147.62	x	x	x	x	65.18
Industrial crops	x	–28.73	1.91	x	x	x	x	x	x	–13.41
Leafy and stalked vegetables (excluding brassicas)	x	–41.90	–57.22	–92.11	–64.08	–63.91	–91.33	–6.83	–68.06	–60.68
Linseed (flax)	x	x	0.14	–12.56	33.68	4.89	–7.03	–24.83	2.28	–0.49
Nuts	x	–89.54	–95.61	–98.14	–96.30	–99.09	–97.36	–98.55	–98.31	–96.61
Oats and spring cereal mixtures (mixed grain other than maslin)	71.03	–26.85	5.42	100.94	19.45	–30.34	–26.71	–0.09	–29.57	9.26
Oilseeds	x	–27.85	3.07	14.57	3.45	–23.22	–26.64	–29.16	–5.37	–11.39
Other oilseed crops n.e.c.	x	x	58.17	–12.06	–2.77	–59.48	–69.35	15.33	70.71	0.08
Other pome fruits n.e.c.	x	x	x	–95.41	–93.36	–92.16	–95.54	–90.43	–88.51	–92.57
Other root crops n.e.c.	x	–67.49	–67.52	–80.41	–84.17	–74.34	–71.70	–70.69	–95.43	–76.47
Peaches	x	–18.47	–57.29	–26.65	–70.05	–61.92	–86.18	–61.64	–83.64	–58.23
Pears	x	–74.68	–86.62	–89.76	–77.50	–76.96	–94.58	–83.94	–82.29	–83.29
Permanent crops for human consumption	x	x	–41.20	–63.83	–54.51	–53.49	–60.12	–56.48	–73.79	–57.63
Plants harvested green from arable land	x	x	–38.79	–27.58	–4.81	–19.20	–34.98	–16.45	–27.94	–24.25
Plums	x	–58.25	–59.80	–85.58	–80.16	–74.34	–88.92	–87.16	–84.46	–77.33
Pome fruits	x	x	–47.77	–77.23	–58.52	–37.55	–62.76	–34.96	–72.03	–55.83
Potatoes (including seed potatoes)	x	–43.32	–32.97	–45.67	–37.42	–57.79	–71.61	–57.48	–37.21	–47.93

Table 4. *Cont.*

Relative Differences [Eco–Conv (%)]	2012	2013	2014	2015	2016	2017	2018	2019	2020	Average
Rape and turnip rape seeds	x	x	−8.46	1.93	−10.83	−18.78	−16.87	−3.20	−20.06	−10.90
Rice	−21.33	−29.35	66.10	18.64	−25.96	0.26	−18.50	−7.46	19.56	0.22
Root crops	x	−43.58	−19.42	−10.53	2.28	−43.54	−42.84	−11.58	21.34	−18.48
Root, tuber and bulb vegetables	x	−25.61	−83.24	−32.29	−54.34	−80.74	−64.43	−49.79	−51.51	−55.24
Rye and winter cereal mixtures (maslin)	42.60	−20.74	−19.21	−3.51	−3.72	−46.30	−54.44	−24.18	−1.13	−14.51
Soya	x	x	0.99	2.58	−6.52	−17.45	−30.30	−32.02	−13.92	−13.80
Stone fruits	x	x	−64.14	−84.58	−77.91	−70.82	−89.15	−87.28	−85.93	−79.97
Strawberries	x	x	−44.93	−54.36	−59.16	50.52	−68.65	−61.91	−11.88	−35.77
Sugar beet (excluding seed)	x	−44.77	−8.87	−27.34	−36.77	−65.58	−63.12	−54.86	17.68	−35.45
Sunflower seed	x	x	7.36	24.18	20.39	−23.55	−28.18	−32.83	2.27	−4.34
Tobacco	x	−33.27	x	x	x	x	x	x	x	−33.27
Vegetables cultivated for fruit (including melons)	x	−39.05	−50.95	−68.37	−59.33	−69.00	−77.23	−74.05	−68.76	−63.34
Wheat and spelt	−2.47	−30.79	12.41	5.85	−1.89	−18.35	−32.20	−27.54	−29.80	−13.87

Source: authors' calculations, x—no data available.

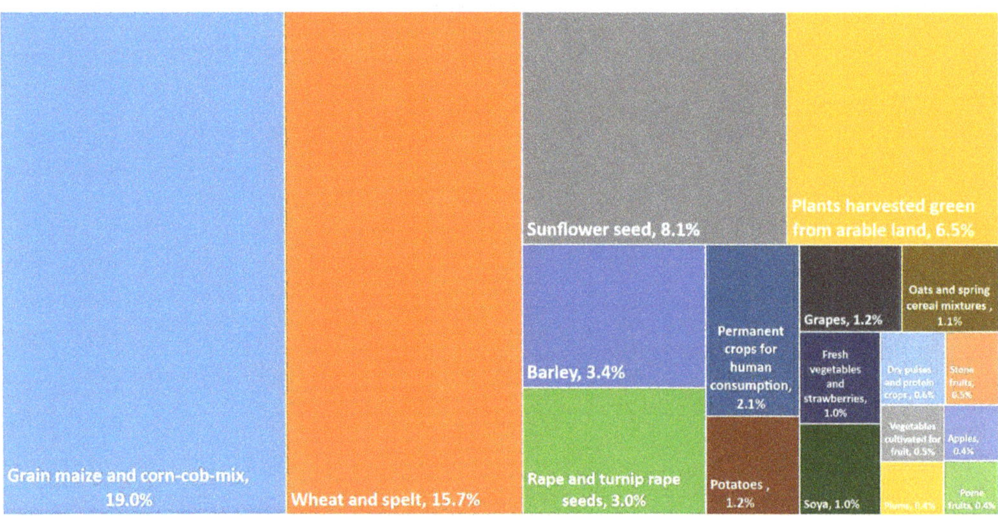

Figure 3. Share of main organic crops in Romania.

The variables considered and the main coefficients of the variables or equations are presented in Table A3.

Table A3 shows the 40 variables, i.e., the 40 crops grown organically in Romania, for which data were available, with the related coefficients, i.e., the relative difference in yield between organic and conventional farming, and the share of each crop in the total agricultural area used. All these crops will be included in the Simplex method, and the change in each area for each crop will fulfil both the conditions presented above and the objective function.

Following the application of the simplex method, which led to the optimal solution that fulfils both the objective function (Table 5), where productivity do not decrease very much, and the set of conditions imposed, the areas for the 40 organic crops were identified for which Romania would reach the share of 25% of the utilised agricultural area.

Table 5. Solution of the objective function as a function of each variable.

Crt. No.	Crop	Variable	Values X_i (ha)	Proportion in Organic Area	Proportion in UAA
1	Apples	X1	7994.49	0.247%	0.06%
2	Apricots	X2	300.80	0.009%	0.00%
3	Aromatic, medicinal and culinary plants	X3	738.00	0.023%	0.01%
4	Barley	X4	363,833.67	11.225%	2.81%
5	Berries (excluding strawberries)	X5	1660.00	0.051%	0.01%
6	Brassicas	X6	14.45	0.000%	0.00%
7	Cherries	X7	315.30	0.010%	0.00%
8	Dry pulses and protein crops for the production of grain (including seed and mixtures of cereals and pulses)	X8	4173.00	0.129%	0.03%
9	Durum wheat	X9	1732.00	0.053%	0.01%
10	Fresh pulses	X10	24.27	0.001%	0.00%
11	Fresh vegetables (including melons) and strawberries	X11	1330.45	0.041%	0.01%
12	Grain maize and corn-cob-mix	X12	614,446.52	18.956%	4.74%
13	Grapes	X13	1867.00	0.058%	0.01%
14	Hops	X14	0.00	0.000%	0.00%
15	Leafy and stalked vegetables (excluding brassicas)	X15	7.10	0.000%	0.00%
16	Linseed (flax)	X16	1252.00	0.039%	0.01%
17	Nuts	X17	29,288.03	0.904%	0.23%
18	Oats and spring cereal mixtures (mixed grain other than maslin)	X18	1118.00	0.034%	0.01%
19	Other oilseed crops n.e.c.	X19	478.00	0.015%	0.00%
20	Other pome fruits n.e.c.	X20	299.34	0.009%	0.00%
21	Other root crops n.e.c.	X21	1.00	0.000%	0.00%
22	Peaches	X22	33.96	0.001%	0.00%
23	Pears	X23	160.58	0.005%	0.00%
24	Permanent crops for human consumption	X24	254,849.78	7.862%	1.97%
25	Plants harvested green from arable land	X25	1014,334.26	31.293%	7.82%
26	Plums	X26	17,385.66	0.536%	0.13%
27	Pome fruits	X27	13,371.47	0.413%	0.10%
28	Potatoes (including seed potatoes)	X28	92.26	0.003%	0.00%
29	Rape and turnip rape seeds	X29	98,396.75	3.036%	0.76%
30	Rice	X30	1424.00	0.044%	0.01%
31	Root, tuber and bulb vegetables	X31	135.36	0.004%	0.00%
32	Rye and winter cereal mixtures (maslin)	X32	127.00	0.004%	0.00%
33	Soya	X33	14,536.00	0.448%	0.11%
34	Stone fruits	X34	25,398.29	0.784%	0.20%
35	Strawberries	X35	20.27	0.001%	0.00%
36	Sugar beet (excluding seed)	X36	72.09	0.002%	0.00%

Table 5. *Cont.*

Crt. No.	Crop	Variable	Values X_i (ha)	Proportion in Organic Area	Proportion in UAA
37	Sunflower seed	X37	262,274.88	8.091%	2.02%
38	Tobacco	X38	0.00	0.000%	0.00%
39	Vegetables cultivated for fruit (including melons)	X39	360.68	0.011%	0.00%
40	Wheat and spelt	X40	507,576.02	15.659%	3.91%
41	TOTAL		3241,422.74	100%	25%

Source: authors' calculations.

As can be seen from Table 5, the total organic area would need to be 3.241 million hectares to reach the 25% share. Of this total, the largest and most extensive crop share should be plants harvested green from arable land with 1.014 million hectares, which represents 31.29% of the organic area and 7.8% of the total utilised agricultural area.

Grain maize and corn-cob-mix is the second most important crop, with an area of 614.4 thousand hectares, i.e., a share of 4.7% of the utilised agricultural area. Wheat and spelt is in third place with an organic area of 507.5 thousand hectares, representing 3.9% of the utilised agricultural area.

According to the optimal solution, i.e., according to the values of the 40 variables, this results in an objective function value of −19.64%, i.e., the smallest difference in yield/productivity between organic and conventional farming according to the organic crop structure shown in Table 5 and Figure 4. With this organic crop structure, the first condition is met, i.e., the organic area is 25% of the utilised agricultural area; the second condition is also met at the limit, but it can be seen that this yield decrease of 19.64% applied to 25% of the agricultural area results in a fixed total decrease in agricultural production of 4.91%, which does not exceed the increase of agricultural area per capita, so we can consider that this extension of agricultural area can be considered as sustainable. The other conditions have also been met.

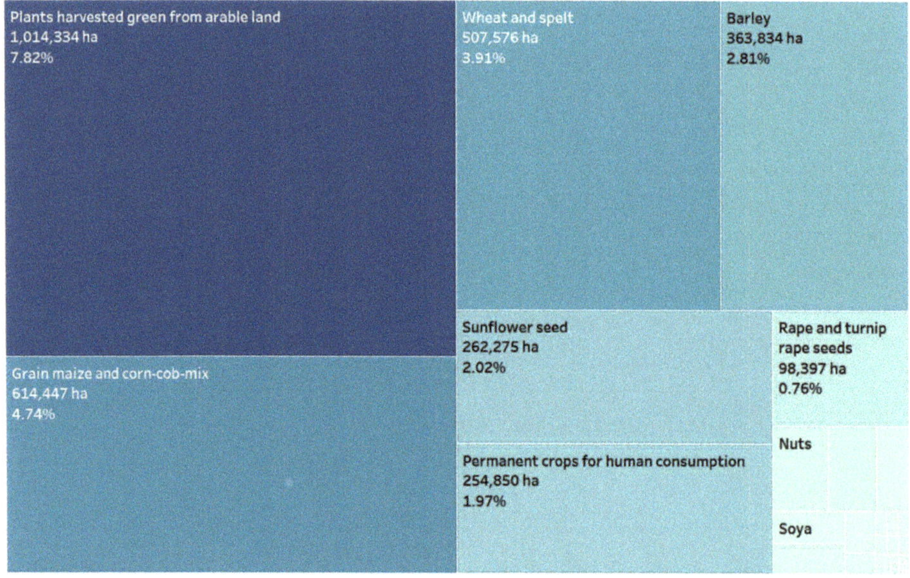

Figure 4. Structure of ecological surfaces according to the optimal solution.

5. Discussion

The study was based on the premise that the share of agricultural land cultivated organically should reach 25% of the country's agricultural land by 2030, and the aim of the study was to determine the amount of each crop that should be cultivated organically in order to reach this share with certain sustainability restrictions.

Given that the average yield of organic crops is just over 30% lower, this would negatively affect food security and sustainable development goals. Thus, in determining the size of the areas cultivated for each crop, it was decided to minimise productivity losses so that, applying these losses to the 25% of the area, the total loss of production would be less than or equal to the gain in agricultural area per inhabitant determined by the decrease in Romania's population.

After identifying the optimal solution, when the crop structure and the size of the crop areas resulting in the smallest loss of productivity, which is about 20%, are applied to the 25% area, then a total loss of 5% of agricultural production results, which is recovered by increasing the agricultural area per capita.

Furthermore, following the identification of the optimal solution, it is observed that the hypothesis becomes true, namely that the main crop to be cultivated in the highest proportion will be the one related to green plants, thus leading us to the idea of cultivating and converting the cultivated areas to grassland and meadows.

The economic impact, not just the technical one, must also be discussed. Given the loss of yield and the high cost of inputs in organic farming, the final cost per unit of product will be higher, and this will be reflected in the market price. As the price of organic products is higher than conventional products, this means a higher value for organic products per unit of product, but this price aspect has not been taken into account, as this is the only technical condition of the European Union's area ratio strategies. If the price of products, the value of production and the price difference between the two systems had been taken into account, the crop structure would probably have looked different, with uncertainty as to whether the main condition of area assurance would still be met. Unfortunately, however, this analysis could not be carried out due to the limitations of the static data, as there are no price data for organic products.

6. Conclusions

In this paper, the aim was to determine the ecological areas that should be cultivated in Romania by 2030, so that their share meets the targets of the European Union strategy for biodiversity, namely 25%. Although this share represents the EU average of organic crop area in the total agricultural land, this paper assumes that Romania should ensure this share in a sustainable way.

Given that organic crop yields are lower than conventional ones, in order to achieve sustainable growth, the agricultural area per capita in Romania was determined, so even if the dynamics of agricultural area and population are decreasing, the rate of population decline is faster, so the indicator of agricultural area per capita is expected to increase in the period 2020–2030, with a forecasted increase of up to 5%. This would therefore be considered as the upper limit of yield losses in organic farming for the 25% of the area.

In order to determine as accurately and optimally as possible the areas of organic crops to be sown, the average yields per hectare for the intersection of crops recorded in Romania between the two systems were analysed, thus determining the relative differences in productivity between the organic and conventional systems for each crop and the average relative difference, which was 32% against organic farming.

In addition, we mention the fulfilment of the proportion of 25% of the total agricultural area as organic cultivation, as well as the difference in yield for the section of 25% of the agricultural area as less than or equal to the growth rate of agricultural area per capita, so that sustainable growth of organic agriculture will be present in Romania.

It was concluded that the organic area would have to increase by 11.7 times, i.e., to reach a size of 3.24 million ha, to ensure the proportion recommended in the EU strategy.

Solving the linear programming problem led to the determination of the size of the areas to be cultivated organically for each crop in order to fulfil the objective function of minimising yield loss. This gives the smallest yield loss in organic farming compared to conventional farming, 16.94%. At the same time, this yield decrease, applied to 25% of the agricultural area, leads to a loss of up to 5% of production, which is sustainably covered by the increase in agricultural area per capita due to the decrease in population.

Author Contributions: Conceptualization, A.U. and I.L.P.; methodology, I.L.P.; software, I.L.P.; validation, A.U. and I.L.P.; formal analysis, A.U.; investigation, A.U.; resources, A.U. and I.L.P.; data curation, I.L.P.; writing—original draft preparation, A.U. and I.L.P.; writing—review and editing, A.U. and I.L.P.; visualization, A.U. and I.L.P.; supervision, A.U. and I.L.P.; project administration, A.U.; funding acquisition, A.U. All authors have read and agreed to the published version of the manuscript.

Funding: This study was funded by the ADER 23.1.1 project "Technical-economic substantiation of production costs and estimates of the valorisation prices of the main crop and livestock products obtained in conventional and organic farming", by the Ministry of Agriculture and Rural Development (MADR), phase 4, 2022.

Institutional Review Board Statement: Not applicable.

Informed Consent Statement: Not applicable.

Data Availability Statement: Data is contained within the article.

Conflicts of Interest: The authors declare no conflict of interest.

Appendix A

Table A1. Conventional crop yield (t/ha).

Conventional Yield	2012	2013	2014	2015	2016	2017	2018	2019	2020
Apples	8.195	8.344	8.951	8.330	8.228	6.107	11.768	9.342	10.269
Apricots	11.128	9.511	14.191	11.340	13.395	15.469	17.477	14.422	13.187
Aromatic, medicinal and culinary plants	0.741	0.936	1.319	1.318	1.280	1.279	1.213	1.114	1.159
Barley	2.325	3.111	3.319	3.461	3.773	4.186	4.417	4.188	2.582
Berries (excluding strawberries)	1.105	1.382	1.149	1.239	1.155	1.436	1.915	2.414	2.365
Brassicas	19.739	18.604	23.178	21.386	21.371	22.640	23.331	20.543	27.218
Cereals (excluding rice) for the production of grain (including seed)	2.352	3.853	4.055	3.532	3.963	5.224	5.998	5.458	3.398
Cereals for the production of grain (including seed)	2.357	3.854	4.054	3.534	3.964	5.224	5.997	5.458	3.399
Cherries	9.946	10.927	12.360	11.468	11.535	9.218	12.333	12.061	11.971
Citrus fruits	0.000	0.000	0.000	0.000	0.000	0.000	0.000	0.000	0.000
Common wheat and spelt	2.653	3.469	3.587	3.781	3.945	4.890	4.796	4.754	2.967
Cotton seed	0.000	0.000	0.000	0.000	0.000	0.000	0.000	0.000	0.000
Cultivated mushrooms	465.500	439.500	488.000	1096.000	726.000	758.500	172.333	173.375	716.000
Dry pulses and protein crops for the production of grain (including seed and mixtures of cereals and pulses)	1.139	1.377	1.427	1.394	1.646	2.510	1.426	2.023	1.124

Table A1. *Cont.*

Conventional Yield	2012	2013	2014	2015	2016	2017	2018	2019	2020
Durum wheat	2.476	3.000	4.889	2.845	3.511	4.305	3.687	2.885	2.501
Fibre crops	0.667	0.467	5.921	4.300	5.435	1.544	1.870	2.210	2.488
Fresh pulses	3.251	4.160	4.090	4.173	4.008	4.104	4.339	3.949	5.209
Fresh vegetables (including melons)	14.523	16.336	17.659	16.546	16.091	17.835	18.632	16.632	20.379
Fresh vegetables (including melons) and strawberries	14.420	16.244	17.525	16.410	15.946	17.617	18.389	16.413	19.998
Fruits from subtropical and tropical climate zones	0.000	0.000	0.000	0.000	0.000	0.000	0.000	0.000	0.000
Fruits, berries and nuts (excluding citrus fruits, grapes and strawberries)	7.602	8.418	8.827	8.428	8.593	7.176	12.726	10.556	11.233
Grain maize and corn-cob-mix	2.180	4.488	4.770	3.459	4.158	5.956	7.637	6.500	3.974
Grapes	4.205	5.586	4.465	4.514	4.209	6.065	6.601	5.523	5.633
Hops	0.739	0.708	1.125	0.957	0.808	0.522	0.846	0.880	0.840
Industrial crops	1.320	2.072	2.307	0.000	0.000	0.000	0.000	0.000	0.000
Leafy and stalked vegetables (excluding brassicas)	9.442	11.064	12.208	12.674	13.522	14.395	15.372	10.197	8.945
Linseed (flax)	1.203	1.332	1.567	1.670	1.646	1.676	1.538	2.013	1.199
Nectarines	6.750	5.909	7.417	7.333	6.083	5.000	5.571	5.250	6.143
Nuts	18.025	18.110	14.711	15.191	13.910	19.034	23.827	19.857	19.004
Oats and spring cereal mixtures (mixed grain other than maslin)	1.743	2.051	2.124	1.999	2.239	2.460	2.376	2.243	1.941
Oilseeds	1.322	2.079	2.312	1.964	2.207	2.823	2.835	2.662	1.923
Olives	0.000	0.000	0.000	0.000	0.000	0.000	0.000	0.000	0.000
Other fresh vegetables n.e.c.	0.000	0.000	0.000	0.000	0.000	0.000	0.000	0.000	0.000
Other oilseed crops n.e.c.	0.667	0.896	0.920	0.870	1.144	1.234	0.841	0.816	0.680
Other permanent crops for human consumption n.e.c.	0.000	0.000	0.000	0.000	0.000	0.000	0.000	0.000	0.000
Other pome fruits n.e.c.	14.603	13.855	17.736	17.000	16.817	21.209	25.630	22.890	22.339
Other root crops n.e.c.	22.502	28.457	29.766	30.625	25.268	28.250	28.272	23.879	21.878
Other stone fruits n.e.c	0.000	0.000	0.000	0.000	0.000	0.000	0.000	0.000	0.000
Peaches	8.426	9.321	13.893	12.355	13.357	11.204	13.262	9.994	9.549
Pears	13.267	16.379	16.910	14.643	15.810	14.756	18.439	14.987	15.094
Permanent crops for human consumption	5.722	6.876	6.416	6.240	6.148	6.558	9.319	7.712	8.165
Plants harvested green from arable land	5.906	6.713	6.995	6.423	5.847	6.371	7.029	6.201	5.553
Plums	6.031	7.367	7.278	7.397	7.713	6.515	12.594	10.562	11.310
Pome fruits	8.686	8.960	9.610	8.819	8.808	6.906	12.425	9.946	10.802

Table A1. *Cont.*

Conventional Yield	2012	2013	2014	2015	2016	2017	2018	2019	2020
Potatoes (including seed potatoes)	10.752	15.846	17.365	13.769	14.442	18.186	17.442	15.086	15.849
Rape and turnip rape seeds	1.496	2.408	2.604	2.499	2.835	2.798	2.546	2.264	2.150
Rice	4.450	4.528	3.509	4.428	4.569	4.689	5.195	5.320	4.112
Root crops	12.968	18.913	21.516	17.606	17.989	21.834	20.495	18.305	19.181
Root, tuber and bulb vegetables	10.156	12.314	13.301	11.552	10.767	11.873	12.012	11.365	12.884
Rye and winter cereal mixtures (maslin)	2.104	2.217	2.395	2.533	2.479	2.936	2.791	2.797	2.532
Soya	1.384	2.344	2.687	2.164	2.190	2.521	2.908	2.783	2.018
Stone fruits	6.584	7.804	8.107	7.978	8.314	7.070	12.705	10.772	11.370
Strawberries	6.771	9.784	9.079	8.398	8.434	8.305	7.976	6.855	6.979
Sugar beet (excluding seed)	26.366	36.575	44.711	39.129	40.618	41.649	38.035	40.350	33.684
Sunflower seed	1.310	1.993	2.187	1.765	1.955	2.917	3.041	2.783	1.858
Tobacco	1.063	1.447	1.640	1.440	1.785	1.525	1.370	1.344	1.307
Vegetables cultivated for fruit (including melons)	15.519	18.196	18.769	18.049	17.560	20.083	21.105	18.784	23.987
Wheat and spelt	2.652	3.468	3.590	3.780	3.944	4.888	4.793	4.749	2.966

Source: authors' calculations.

Table A2. Organic crop yield (t/ha).

Organic Yield	2012	2013	2014	2015	2016	2017	2018	2019	2020
Apples	4.497	4.502	5.333	2.450	4.863	5.219	5.002	6.774	3.079
Apricots	x	2.385	0.143	2.125	1.635	2.733	0.188	0.788	0.403
Aromatic, medicinal and culinary plants	x	0.650	1.396	1.160	2.237	1.311	0.906	1.051	0.919
Barley	1.582	2.000	3.207	2.928	3.444	3.488	2.288	2.438	2.173
Berries (excluding strawberries)	x	1.117	7.074	1.754	9.157	1.233	1.430	3.401	3.096
Brassicas	x	14.231	8.222	1.256	6.051	0.265	11.500	12.000	6.571
Cereals (excluding rice) for the production of grain (including seed)	x	2.525	4.152	4.291	3.915	4.348	3.696	3.723	2.350
Cereals for the production of grain (including seed)	x	2.546	4.204	4.339	3.884	4.366	3.713	3.750	2.388
Cherries	x	2.750	1.805	1.743	3.438	5.456	1.688	1.796	0.869
Citrus fruits	x	x	x	x	x	x	x	x	x
Common wheat and spelt	x	x	4.065	4.032	3.933	4.019	3.278	3.439	2.085
Cotton seed	x	x	1.500	x	x	x	x	x	x
Cultivated mushrooms	x	x	x	x	x	x	x	x	x

Table A2. *Cont.*

Organic Yield	2012	2013	2014	2015	2016	2017	2018	2019	2020
Dry pulses and protein crops for the production of grain (including seed and mixtures of cereals and pulses)	x	1.100	1.999	1.605	1.272	2.497	1.408	2.055	1.089
Durum wheat	x	x	2.968	2.970	2.744	2.453	1.852	4.081	1.973
Fibre crops	x	x	1.836	0.079	8.000	2.000	2.054	1.947	0.214
Fresh pulses	x	3.413	0.511	1.711	2.085	0.734	1.228	6.261	7.792
Fresh vegetables (including melons)	x	x	1.664	5.244	4.141	2.780	2.903	4.570	6.785
Fresh vegetables (including melons) and strawberries	x	x	1.675	5.225	4.128	2.865	2.898	4.488	6.761
Fruits from subtropical and tropical climate zones	x	x	x	x	x	x	x	x	x
Fruits from temperate climate zones	x	3.959	4.219	1.646	2.781	3.147	2.855	3.838	2.280
Fruits, berries and nuts (excluding citrus fruits, grapes and strawberries)	x	x	4.247	1.540	3.213	2.587	2.239	3.091	1.889
Grain maize and corn-cob-mix	2.587	3.100	5.352	5.743	4.930	6.004	5.917	5.100	3.040
Grapes	3.497	3.900	2.664	4.931	1.724	4.644	9.982	4.577	3.523
Hops	x		1.613	1.000	2.000				
Industrial crops	x	1.477	2.351	2.203	2.283	2.140	2.054	1.875	1.806
Leafy and stalked vegetables (excluding brassicas)	x	6.429	5.222	1.000	4.857	5.195	1.333	9.500	2.857
Linseed (flax)	x		1.569	1.460	2.200	1.758	1.430	1.513	1.226
Nectarines	x	x	x	x	x	x	x	x	x
Nuts	x	1.895	0.646	0.282	0.514	0.174	0.630	0.288	0.321
Oats and spring cereal mixtures (mixed grain other than maslin)	2.980	1.500	2.239	4.016	2.674	1.714	1.742	2.241	1.367
Oilseeds	x	1.500	2.383	2.251	2.283	2.167	2.080	1.886	1.820
Olives	x	x	x	x	x	x	x	x	x
Other cereals (including triticale and sorghum)	3.199	1.557	2.484	1.575	0.978	2.383	1.814	1.183	1.562
Other fresh vegetables n.e.c.	x	x	0.910	21.610	3.262	3.695	1.937	1.283	5.778
Other industrial crops including energy crops n.e.c.	x	x	0.000	3.800	0.000	1.346	1.503	0.000	0.000
Other oilseed crops n.e.c.	x	x	1.455	0.765	1.112	0.500	0.258	0.941	1.161
Other permanent crops for human consumption n.e.c.	x	x	0.000	0.838	0.000	0.353	0.862	0.040	0.064
Other pome fruits n.e.c.	x	x	0.000	0.781	1.117	1.663	1.144	2.190	2.567

Table A2. *Cont.*

Organic Yield	2012	2013	2014	2015	2016	2017	2018	2019	2020
Other root crops n.e.c.		9.250	9.667	6.000	4.000	7.250	8.000	7.000	1.000
Other stone fruits n.e.c	x	x	3.409	2.000	8.607	5.988	1.522	1.158	1.774
Peaches	x	7.600	5.933	9.063	4.000	4.267	1.833	3.833	1.563
Pears	x	4.147	2.263	1.500	3.557	3.400	1.000	2.406	2.672
Permanent crops for human consumption	x	x	3.773	2.257	2.797	3.050	3.717	3.356	2.140
Plants harvested green from arable land	x	x	4.282	4.652	5.566	5.148	4.570	5.181	4.001
Plums	x	3.076	2.926	1.067	1.530	1.672	1.396	1.356	1.758
Pome fruits	x		5.019	2.008	3.654	4.312	4.627	6.469	3.021
Potatoes (including seed potatoes)	x	8.982	11.640	7.481	9.038	7.675	4.952	6.415	9.952
Rape and turnip rape seeds	x	x	2.384	2.548	2.528	2.273	2.117	2.191	1.719
Rice	3.501	3.199	5.829	5.254	3.383	4.701	4.234	4.923	4.916
Root crops	x	10.670	17.338	15.753	18.400	12.328	11.716	16.185	23.275
Root, tuber and bulb vegetables	x	9.160	2.229	7.821	4.917	2.287	4.273	5.706	6.248
Rye and winter cereal mixtures (maslin)	3.000	1.757	1.935	2.444	2.387	1.577	1.272	2.121	2.504
Soya	x	x	2.713	2.220	2.047	2.081	2.027	1.892	1.737
Stone fruits	x	x	2.907	1.230	1.837	2.063	1.378	1.370	1.600
Strawberries	x	x	5.000	3.833	3.444	12.500	2.500	2.611	6.150
Sugar beet (excluding seed)	x	20.200	40.743	28.431	25.684	14.336	14.026	18.216	39.638
Sunflower seed	x	x	2.348	2.192	2.353	2.230	2.184	1.869	1.900
Tobacco	x	0.966	x	x	x	x	x	x	x
Vegetables cultivated for fruit (including melons)	x	11.090	9.206	5.709	7.142	6.225	4.805	4.874	7.494
Wheat and spelt	2.586	2.400	4.035	4.001	3.869	3.991	3.250	3.441	2.082

Source: authors' calculations, x—no data available.

Table A3. Definition of variables and initial coefficients.

Crt. No.	Organic Farming (Area)	Variable	Relative Difference in Yield ($\Delta\%\overline{Q}$)	Share of Ecological Area in UAA
1	Apples	X1	–45.85	0.408%
2	Apricots	X2	–89.46	0.016%
3	Aromatic, medicinal and culinary plants	X3	–1.41	0.023%
4	Barley	X4	–24.18	3.417%
5	Berries (excluding strawberries)	X5	157.90	0.004%
6	Brassicas	X6	–65.10	0.205%

Table A3. *Cont.*

Crt. No.	Organic Farming (Area)	Variable	Relative Difference in Yield ($\Delta\%\overline{Q}$)	Share of Ecological Area in UAA
7	Cherries	X7	−77.53	0.043%
8	Dry pulses and protein crops for the production of grain (including seed and mixtures of cereals and pulses)	X8	1.13	0.621%
9	Durum wheat	X9	−18.45	0.042%
10	Fresh pulses	X10	−32.27	0.051%
11	Fresh vegetables (including melons) and strawberries	X11	−77.08	0.972%
12	Grain maize and corn–cob–mix	X12	1.98	18.956%
13	Grapes	X13	−18.22	1.162%
14	Hops	X14	65.18	0.002%
15	Leafy and stalked vegetables (excluding brassicas)	X15	−60.68	0.003%
16	Linseed (flax)	X16	−0.49	0.017%
17	Nuts	X17	−96.61	0.015%
18	Oats and spring cereal mixtures (mixed grain other than maslin)	X18	9.26	1.112%
19	Other oilseed crops n.e.c.	X19	0.08	0.032%
20	Other pome fruits n.e.c.	X20	−92.57	0.009%
21	Other root crops n.e.c.	X21	−76.47	0.084%
22	Peaches	X22	−58.23	0.012%
23	Pears	X23	−83.29	0.024%
24	Permanent crops for human consumption	X24	−57.63	2.099%
25	Plants harvested green from arable land	X25	−24.25	6.484%
26	Plums	X26	−77.33	0.446%
27	Pome fruits	X27	−55.83	0.401%
28	Potatoes (including seed potatoes)	X28	−47.93	1.211%
29	Rape and turnip rape seeds	X29	−10.90	3.036%
30	Rice	X30	0.22	0.066%
31	Root, tuber and bulb vegetables	X31	−55.24	0.222%
32	Rye and winter cereal mixtures (maslin)	X32	−14.51	0.074%
33	Soya	X33	−13.80	0.964%
34	Stone fruits	X34	−79.97	0.517%
35	Strawberries	X35	−35.77	0.019%
36	Sugar beet (excluding seed)	X36	−35.45	0.185%
37	Sunflower seed	X37	−4.34	8.091%
38	Tobacco	X38	−33.27	0.006%
39	Vegetables cultivated for fruit (including melons)	X39	−63.34	0.472%
40	Wheat and spelt	X40	−13.87	15.659%

References

1. EUROSTAT. Utilised Agricultural Area by Categories [TAG00025_custom_3351494]. 2022. Available online: https://ec.europa.eu/ontextu/databrowser/product/page/TAG00025__custom_3351494 (accessed on 14 September 2022).
2. Plummer, L.M. Assessing benefit transfer for the valuation of ecosystem services. *J. Front. Ecol. Environ.* **2009**, *7*, 38–45. Available online: https://esajournals.onlinelibrary.wiley.com/doi/full/10.1890/080091 (accessed on 14 September 2022). [CrossRef]
3. Pérez-Soba, M.; Petit, S.; Jones, L.; Bertrand, N.; Briquel, V.; Omodei-Zorini, L.; Contini, C.; Helming, K.; Farrington, J.H.; Mossello, M.T.; et al. Land use functions—A multifunctionality approach to assess the impact of land use changes on land use sustainability. *J. Sustain. Impact Assess. Land Use Chang.* **2014**, *6*, 375–404. Available online: https://www.researchgate.net/publication/37790528_Land_use_functions_-_a_multifunctionality_approach_to_assess_the_impact_of_land_use_changes_on_land_use_sustainability (accessed on 14 September 2022). [CrossRef]
4. EU-LUPA. European Land Use Pattern. The ESPON Programme and Projects. 2013. Available online: www.espon.eu (accessed on 14 September 2022).
5. EUROSTAT. Crop Production in EU Standard Humidity. 2022. Available online: https://appsso.eurostat.ec.europa.eu/nui/submitModifiedQuery.do (accessed on 13 September 2022).
6. Cacek, T.; Langner, L.L. The economic implications of organic farming. *Am. J. Altern. Agric.* **1986**, *1*, 25–29. Available online: https://www.eap.mcgill.ca/MagRack/AJAA/AJAA_2.htm (accessed on 14 September 2022). [CrossRef]
7. Teonce, I.; Petcu, V. *Agricultura Ecologică–Fapte și Cifre Cheie și Perspective în Ontextual PAC Post 2021 [Organic Farming–Key Facts and Figures and Perspectives in the post 2021 CAP context]*; Editura Club România: București, Romania, 2019; pp. 674–699. ISBN 978-606-94561-4-9.
8. Research Institute of Organic Agriculture (FiBL) and Agricultural Market Information Company (AM). Data Compiled by Fibl Based on Eurostat and National Data Sources. Available online: https://www.organicseurope.bio/about-us/organic-in-europe/ (accessed on 14 September 2022).
9. EUROSTAT. Population Change–Demographic Balance and Crude Rates at National Level [Demo_Gind]. 2022. Available online: https://migrationresearch.com/item/population-change-demographic-balance-and-crude-rates-at-national-level/216612 (accessed on 14 September 2022).
10. Policy and Strategy Directorate. *Rural Areas in Decline, Challenges, Actions and Perspectives for Territorial Governance, Rural Decline in ROMANIA*; Romanian Government: Bucharest, Romania, 2020; p. 5. Available online: https://www.mdlpa.ro\T1\guilsinglrightarticole\T1\guilsinglrightattachments (accessed on 14 September 2022).
11. Infographic Material. International Development Agenda for the Next 15 Years, Special Report–International Development Days for EurActiv Development Goals. 2022. Available online: https://www.euractiv.ro/we-develop (accessed on 14 September 2022).
12. FAO. *Sustainable Agriculture, FAO and the 2030 Agenda for Sustainable Development*; Food of Agriculture Organization of the United Nations (FAO): Rome, Italy, 2016; Available online: https://www.fao.org/sustainable-development-goals/overview/fao-and-the-2030-agenda-for-sustainable-development/sustainable-agriculture/en/ (accessed on 15 September 2022).
13. Farm to Fork Strategy. Planul de Actiune Pentru Dezvoltarea Productiei Ecologice [Action plan for the Development of Organic Production]. 2020. Available online: https://www.consilium.europa.eu/ro/policies/from-farm-to-fork/ (accessed on 9 September 2022).
14. European Commission–Action Plan for Organic Production in the EU. Agriculture and Rural Development. 2021. Available online: https://agriculture.ec.europa.eu/farming/organic-farming/organic-action-plan_en (accessed on 9 September 2022).
15. Mahadeo, S.D.; Tanaji, N.G. Organic Farming for Sustainable Agricultural Development, Kurukshetra. *J. Rural. Dev.* **2019**, *6*, 52. Available online: https://www.researchgate.net/publication/335960856_Organic_Farming_for_Sustainable_Agricultural_Development (accessed on 15 September 2022).
16. Regulation (EU) 2018/848 of the European Parliament and of the Council on Organic Production and Labelling of Organic Products and Repealing Council Regulation (EC) No. 834/2007. Available online: https://www.madr.ro/agricultura-ecologica/legislatie-comunitara-ae.html (accessed on 28 September 2022).
17. Cuoco, E. The Farm to Fork Strategy and EU Organic Action Plan; Comments Offon the Farm to Fork Strategy and EU Organic Action Plan, Director of IFOAM Organics Europe. 2022. Available online: https://www.organic-cities.eu/2022/02/09/the-farm-to-fork-strategy-and-eu-organic-action-plan/) (accessed on 15 September 2022).
18. Jităreanu, A.F.; Mihăilă, M.; Robu, A.-D.; Lipșa, F.-D.; Costuleanu, C.L. Dynamic of Ecological Agriculture Certification in Romania Facing the EU Organic Action Plan. *Sustainability* **2022**, *14*, 1105. [CrossRef]
19. Aceleanu, M.I. Sustainability and Competitiveness of Romanian Farms through Organic Agriculture. *Sustainability* **2016**, *8*, 245. [CrossRef]
20. Górska-Warsewicz, H.; Żakowska-Biemans, S.; Stangierska, D.; Świątkowska, M.; Bobola, A.; Szlachciuk, J.; Czeczotko, M.; Krajewski, K.; Świstak, E. Factors Limiting the Development of the Organic Food Sector—Perspective of Processors, Distributors, and Retailers. *Agriculture* **2021**, *11*, 882. [CrossRef]
21. Rasche, L.; Steinhauser, J. How will an increase in organic agricultural area affect land use in Germany? *Org. Agric.* **2022**, 1–18. [CrossRef]
22. Eneizan, B.M. Critical Obstacles to Adopt the Organic Farming in Jordan: From Marketing perspective. *Eur. J. Bus. Manag.* **2017**, *9*, 2017. Available online: https://www.researchgate.net/publication/332951113_Critical_Obstacles_to_Adopt_the_Organic_Farming_in_Jordan_From_Marketing_Perspective (accessed on 16 September 2022).

23. Kilker, L. How Organic Agriculture Contributes to Sustainable Development. *J. Agric. Rural. Dev. Trop. Subtrop.* **2007**, *6*, 89. Available online: https://www.researchgate.net/publication/237529824_11_How_Organic_Agriculture_Contributes_to_Sustainable_Development (accessed on 16 September 2022).
24. Iancu, T.; Tudor, V.C.; Dumitru, E.A.; Sterie, C.M.; Micu, M.M.; Smedescu, D.; Marcuta, L.; Tonea, E.; Stoicea, P.; Vintu, C.; et al. A Scientometric Analysis of Climate Change Adaptation Studies. *Sustainability* **2022**, *14*, 2945. [CrossRef]
25. Khoso, K.K.; Buarod, T.; Kaewpijit, J. Sustainable Impact of Organic Farming: A Social Constructivist Perspecitve. *Humanit. Arts Soc. Sci. Stud.* **2021**, *21*, 150–161. Available online: https://www.researchgate.net/publication/350447599_SUSTAINABLE_IMPACT_OF_ORGANIC_FARMING_A_SOCIAL_CONSTRUCTIVIST_PERSPECTIVE (accessed on 14 September 2022).
26. Sher, A.; Mazhar, S.; Zulfiqar, F.; Wang, D. Green Entrepreneurial Framing: Adream or Reality? *J. Clean. Prod.* **2020**, *220*, 1131–1142. Available online: https://www.researchgate.net/publication/331289481_Green_Entrepreneurial_Farming_A_Dream_or_Reality (accessed on 14 September 2022). [CrossRef]
27. Andrei, J.V.; Popescu, C.; Ion, R.A.; Dobre, I. From conventional to organic in Romanian agriculture–Impact assessment of a land use changing paradigm. *Land Use Policy* **2015**, *46*, 258–266. Available online: https://www.sciencedirect.com/science/article/pii/S0264837715000599 (accessed on 14 September 2022). [CrossRef]
28. Sterie, C.M.; Chetroiu, R. Analysis of economic indicators of agricultural holdings specialized in raising dairy cows in Romania. Case studies. *Sci. Pap. Manag. Econ. Eng. Agric. Rural. Dev.* **2021**, *21*, 4.
29. Anghelache, C.; Manole, A. Seriile dinamice/cronologice (de timp) [Dynamic/time series]. *Rom. Stat. Rev.* **2012**, *10*, 68–77. Available online: https://www.revistadestatistica.ro/wp-content/uploads/2014/02/RRS_10_2012_A5_ro.pdf (accessed on 19 September 2022).
30. Gass, S.I. *Linear Programming: Methods and Applications*; Courier Corporation: North Chelmsford, MA, USA, 2003.
31. Nabli, H. An overview on the simplex algorithm. *Appl. Math. Com-Putation* **2009**, *210*, 479–489. [CrossRef]
32. Tiganescu, E.; Mitrut, D. Bazele Cercetarii Operationale, Catedra de Cibernetica Economica [Fundamentals of Operations Research, Department of Economic Cybernetics]. Available online: http://www.asecib.ase.ro/Mitrut%20Dorin/Curs/bazeCO/html/23PL.htm (accessed on 19 September 2022).

Article

Regional Patterns of Pesticide Consumption Determinants in the European Union

Raluca Georgiana Robu *, Ana-Maria Holobiuc, Alina Petronela Alexoaei, Valentin Cojanu and Dumitru Miron

Department of International Business and Economics, Bucharest University of Economic Studies, 010374 Bucharest, Romania
* Correspondence: raluca.robu@rei.ase.ro

Abstract: This article contributes to the discussion about the socioeconomic factors that reinforce pesticide dependence in the European Union and hinder the transition to more sustainable agricultural practices in light of the European Union's Green Deal objective of reducing the use of pesticides by 50% by 2030. The analysis has a two-pronged purpose: (1) to identify the determinants of pesticide consumption in the European Union by conducting a set of four seemingly unrelated regressions and (2) to emphasize the existence of regional patterns across EU countries formed by the factors that significantly impact pesticide consumption based on a cluster analysis. Per capita GDP, selling prices, population, and real income positively influence pesticide use, whereas subsidies and organic agricultural area negatively influence them. Pesticide use is most affected by GDP per capita and least affected by subsidies. Cluster analysis highlights regional differences reflected in three clusters: (1) the most recent EU member states, (2) the European countries with large population levels, and (3) the countries with the highest GDP per capita. Our findings may contribute to the EU's capacity to generate policy changes at the member state level and can be built into recommendations to address the persistent overuse of pesticides.

Keywords: Green Deal; Farm-to-Fork; pesticides; sustainable crop production; sustainable production policies

Citation: Robu, R.G.; Holobiuc, A.-M.; Alexoaei, A.P.; Cojanu, V.; Miron, D. Regional Patterns of Pesticide Consumption Determinants in the European Union. *Sustainability* **2023**, *15*, 2070. https://doi.org/10.3390/su15032070

Academic Editors: Mariarosaria Lombardi, Erica Varese and Vera Amicarelli

Received: 29 November 2022
Revised: 17 January 2023
Accepted: 18 January 2023
Published: 21 January 2023

1. Introduction

The term 'sustainability' in agriculture describes the need to meet the food needs of the growing human population while ensuring minimal impact on the environment and people as well as profitability for producers [1]. Most researchers agree that sustainability in the agricultural field should, by definition, address the environmental, economic, and social issues associated with its practice [2]. Food systems refer to the entire range of activities from production to consumption [3] that contain the 'environment, people, inputs, processes, infrastructures, institutions' and the 'socioeconomic and environmental outcomes' [4]. They are widely spread across multiple economic territories in various geographic regions, a fact that leaves them exposed to various risks [5]. The growing demand for food, which mainly comes from the growing population, simultaneously exerts pressures to keep food prices affordable to all people which is opposed by pressure to keep the businesses of agri-food producers profitable and address climate change issues. These pressures impact food systems at an unprecedented level and emphasize the need for sustainability. Furthermore, prices for agri-food inputs and outputs have increased significantly in recent years because of the COVID-19 pandemic and the war in Ukraine; these events have caused severe shortages in the supply chain [6,7].

Contrary forces are at work when attempting to ensure the sustainability of agriculture. Several agricultural practices that are employed to ensure that food is affordable and available to the growing population, while also generating enough revenue for farmers to maintain agricultural production, have a detrimental effect on both human health and the environment. According to the European Environment Agency, most of the existing operations of the EU agri-food systems have a direct impact on the deterioration of the environment and climate in Europe. These activities also have a negative impact on biodiversity and climate change and cause pollution [4]. Various pesticides on the market have been banned because of their extremely severe effects on both human health and the environment; other available pesticides are fake. Pesticides that are used legally may also cause diseases of different types and intensity levels [8]. Estimates indicate that there are 168,000 deaths worldwide and one to two million cases of pesticide poisoning each year [8].

Fertilizers and pesticides are the two types of chemicals that are widely used in current agricultural systems. The former increase soil fertility, allowing crops to produce higher yields, while the latter protect crops from diseases and pests [9]. Therefore, the use of pesticides in agriculture, along with other measures, has positive effects on the level of crop production. Production per unit of input is often referred to as intensity in agriculture [10]. Higher intensity can be obtained mainly through mechanization [11], improved seed productivity [12], reduced crop cycle, increased fertilizer consumption [13,14], and reduced losses due to the use of pesticides [15].

The Common Agricultural Policy (CAP) had a major contribution to agriculture intensification [16,17], providing incentives to farmers to increase productivity. The drawback, which was later acknowledged, is that agriculture intensification further degrades the soil, decreasing the concentration of organic resources [18,19], thus increasing the need for additional compensation. Current food support policies do not meet the requirements of a modern food system and have not kept up with the rapid structural changes that affect food systems or the difficulties caused by these changes [20]. Market price support, production payments, and the unrestricted use of inputs are among the most distorting government interventions and the most environmentally damaging agricultural support programs. They create incentives for increased input consumption, for the allocation of land to subsidized crops, and for the introduction of additional land for agricultural production. In the absence of adequate limitations, payments based on variable inputs could encourage the overuse of pesticides, resulting in significant damage to freshwater ecosystems and biodiversity [20].

Against the backdrop of growing concerns around the impact of pesticide use on human health and the environment, many sectoral policies address the current status of agriculture and biodiversity in Europe, encourage a systemic approach to the sustainability of the agriculture and food sector, and outline the primary objectives centred on sustainability. By 2030, the Green Deal, Farm-to-Fork, and biodiversity programs hope to reduce the loss of nutrients from both mineral and organic fertilizers by at least 50%, while ensuring that soil fertility does not deteriorate in the process. This ambitious target, along with other EU objectives, such as reaching 25% organic agriculture, is expected to result in a reduction in pesticide use. The use of organic farming is generally minimal or non-existent [21]. However, many aspects are likely to be contentious, given the dispute over the production impacts of all new regulations. A macrolevel assessment based on the Green Deal objectives (that is, 50% reduction in overall pesticide use and risks, 50% reduction in more hazardous pesticides, 20% reduction in fertiliser use, 10% of agricultural land under organic production), concluded that the Farm-to-Fork and biodiversity strategies will result in a 'decrease of the volumes produced per crop in the entire EU on average ranging from 10 to 20%' and prices for some crops (for example, wine, olives and hops) will increase. As a result, exports of EU crops will fall, while imports will increase (the volume of imports of products can double) [22].

Throughout the EU, agri-food systems differ significantly and their progress in terms of sustainable development is strongly influenced by these differences. According to [23], pesticide sales between 2011 and 2020 emphasise substantial regional differences in both the absolute amount of pesticides used (per hectare of agricultural land) and in total sales. Sales have increased on the markets of certain member states (for example, Latvia, Austria, Germany, France, and Hungary) and fell in others (for example, Czech Republic, Portugal,

Denmark, Romania, Belgium, Ireland, Italy, Sweden, Slovenia, Netherlands, and Cyprus). The difficulty of setting national targets in CAP national strategic plans (in this case, for the use of pesticides) is increased because there are differences in both national evolutions and absolute quantities of pesticides (per hectare of agricultural land and overall). This applies to all quantitative targets of the Green Deal currently specified at the EU level [24]. Although member states are required to adopt legally binding targets for the achievement of the overall EU targets, in regard to determining national targets, members have the flexibility to take into account their own national circumstances, including their level of pesticide use and their historical level of progress.

This article aims to serve as a starting point to reveal the socio-economic and political factors that reinforce pesticide dependence in the European Union and determine the slow pace in the transition to more sustainable agricultural practices. The analysis is narrowed to pesticide consumption since pesticides are the most widely used tool in intensive agriculture and because the European Green Deal includes more stringent reduction targets. The paper has two main objectives: (1) to identify pesticide consumption determinants in the European Union (EU) by running an empirical model based on panel data for EU27 between 2001 and 2019; and (2) to emphasize the existence of some regional patterns across EU countries and classify the EU27 member states into broad categories according to the factors that had been shown to significantly affect pesticide consumption. By achieving these objectives, cluster characteristics can be developed into recommendations to combat the persistent overuse and reliance on chemical pesticides at regional levels.

Our findings may contribute to the EU's capacity to generate policy change at the member state level and may be useful to a wider audience interested in the restraints national states face in adapting their policies to meet the Grean Deal objectives. This paper contributes to the body of knowledge in three ways. First, it provides a comprehensive analysis of the multiple factors that contribute to pesticide use in agriculture, in contrast to the majority of existing studies that discuss pesticide consumption from either micro or macro-economic perspectives. Second, the article looks at the EU market and provides, for the first time, a cluster analysis applied to the determinants of pesticide use in agriculture. Lastly, the original elements of the paper reside in the complementarity between regression and cluster analysis, with the purpose of determining common regional patterns for member states. The article is structured as follows: Section 2 presents a literature review on frequent determinants of pesticide consumption; Section 3 is dedicated to the methodological framework; Section 4 presents the findings of the regressions and the cluster analysis; and the last section brings together final observations and conclusions.

2. Review of the Scientific Literature

2.1. Determinants of the Use of Pesticides in Agriculture

The section presents a review of the literature on the impact of various economic and social factors on the widespread use of pesticides in agricultural production and provides the scientific basis for the regression analysis.

Given the multifaceted and transdisciplinary nature of pesticide dependence, it is impossible to identify a rigorous review of generally accepted and standardized variables. Although a wide range of studies have sought social, economic, and political explanations, most of these studies have focused on a small number of factors or have approached the topic from a microlevel perspective and emphasize the role of farmer decision making [25–29]. To our knowledge, no analysis has been performed at the EU level on the determinants of pesticide use in agriculture. Most studies have addressed the environmental and health impact of pesticide use or have performed a comparative analysis of National Action Plans of member states [30–32]. One general observation from the literature reviews is that, while some research investigates the factors of pesticides and fertilizers together, others study them individually. The widespread use of pesticides among intensive agricultural tools and the larger reduction target outlined in the European Green Deal regulations required a focus on analysing the determinants of pesticide consumption. Depending on data collection possibilities, we tested the impact of several factors on the use of pesticides in European agricultural practices at country level, continuing with a cluster analysis to highlight possible geographical patterns of these determinants.

As described in the Introduction, the use of pesticides reduces crop losses and contributes to an increase in overall crop yield [33,34]. The primary factors that have caused an increase in agricultural productivity have deep roots and do not all operate in the same way under comparable conditions.

In the literature, the most cited determinant for the widespread use of pesticides is the growing demand for food from the growing population at the global level, which puts greater pressure on increasing crop yields and using resources more efficiently [35]. At the European level, between 1960 and 2020, the population has decreased overall, while net migration to Europe increased during the same period [36]. At the same time, trade liberalization offers opportunities for European farmers to supply foreign markets [37], especially from Eastern Europe [38,39]. To some, this reduces the relevance of the link between the population of a country and the pressure to use various methods that improve productivity. With these constraints, we will test the impact of the population on the use of pesticides in the same country.

Demand and production for food, in general, and for healthy food, more specifically, varies between countries, depending on the country's wealth. Economic development determines an inverse U-shaped evolution of the curve that describes the use of pesticides. The least developed countries have a small consumption of chemicals in agriculture because the prices of these inputs are prohibitive; developed countries are heavy users of pesticides, while the most developed countries use them more efficiently to increase production [40,41]. Ref. [42] observed an increase in the use of pesticides in the least developed countries as their trade connectivity with developed countries improves, resulting in increased imports of pesticides and increased exports of agricultural products. Ref. [43] found an increase in the market for organic products in rich countries in western Europe as they have a higher demand for healthy food. Using GDP per capita as the independent variable, our aim is to determine whether there is a causal relationship between economic growth and pesticide use based on the references mentioned above.

From a supply perspective, the available workforce in agriculture was reduced by the urbanization process [44] (later in the case of Eastern European countries), imposing the need for higher labour productivity. The issue of labour scarcity in agriculture has been partly diminished since 1990 by the migration of low-wage labour to rural areas either from the same country or to less developed countries [45] as well as technological progresses that have improved productivity through mechanization [11]. Ref. [46] underlined a stronger effect of mechanization when labour is scarce or expensive, while pesticides are used more intensively when land is expensive. On many occasions, chemical inputs were the solution to increasing productivity in order to face increasing labour costs [38], higher land prices, and growing competition [47–49].

Another set of complementary factors with potential impact on pesticide use are land fragmentation and farm size. When land is fragmented, it becomes more difficult to use farming equipment efficiently, which decreases productivity and increases costs [50]. The results of a survey of Chinese farmers point out that the use of pesticides does not necessarily depend on farm size but rather on certain psychological aspects: in order to preserve the soil, farmers use them less frequently when they believe the land to be clean, but more frequently when they believe their chances of remaining in agriculture are limited [51]. At the same time, recent research emphasizes the possibility of increased crop yields when using traditional farming methods. For example, crop rotation favours organic farming [52] and the participation of household labour in small farms improves farm efficiency [52]. Therefore, smaller traditional farms are more suitable for organic farming, while economies of scale can be obtained when farms are larger. Once the land is introduced into the organic farming circuit, producers cannot maintain certifications as organic farmers unless they change the type or amount of pesticides they use; this production system should maintain profitability over the long term [53]. The transition to organic farming imposes certain costs and changes in farm structure [54] that are not justified in the short term. In our analysis, we use the total organic agricultural area as a factor that is expected to negatively impact the use of pesticides. Many farms indicate a high level of farm fragmentation, so we initially tested the impact of the number of producers in agriculture (as a proxy for farm size) on the use of pesticides, but it turned out to be insignificant, as we eliminated the indicator from the final regression.

Regarding the drivers behind the transition to organic farming, it can be found that this depends on the possibility of making a profit coming from two main directions: higher selling prices and support payments for organic farming [55]. Normally, when product prices grow on the market, we would expect organic crop production to grow because, although the total costs in this segment are higher, farmers would have increased opportunities to make profit. However, more farmers are stimulated to meet market opportunities when they appear [56], and they continue producing even more inorganic products, which means a higher total pesticide usage. Other authors found that low-income farmers use more inorganic fertilizers, even in unnecessary amounts [57,58], while [59] identified a negative impact of farmer income on the efficiency of pesticide use. In our analysis, we test the impact of sales prices and real income on the consumption of pesticides.

Subsidies represent a solution for farmers to either: (1) invest in new technologies [60] that boost productivity as an alternative to the use of more chemicals or (2) sell organic farm products at competitive prices regardless of the technologies used [61,62]. However, the impact of subsidies on the consumption of pesticides should be interpreted according to the conditions applied by the states that offered subsidies in a given period of time. In the 1990s, eastern European countries withdrew many subsidies previously given to farmers to buy chemicals to improve agricultural productivity [63]. The policy's withdrawal in 2013, when the European Union introduced green payments (focusing on measures such as crop diversification) in the Common Agriculture Policy, led to a substantial decrease in yield [64]. Different types of pesticide subsidies have been applied in several European countries, including lower VAT rates (e.g., Austria, Belgium, Bulgaria, Croatia, France, Ireland, Italy, Poland, Portugal, Romania and Spain) [65]. However, some authors believe that that improved market access for organic products have greater overall effects than subsidies in reducing chemical use [66]. In other cases, the total value of available subsidies is not very relevant if the conditions imposed on farmers who access them are very restrictive or the contract terms are not flexible enough [67]. Taking these limits into account, in the current article, subsidies on agricultural products are expected to have a significant influence on pesticide consumption.

Specific social characteristics also determine the choice of certain types of farming method. For example, older, more experienced, and more educated farmers would be more prone to take the hard way and implement traditional methods [68]. The general education level of the population positively influences the demand for healthy food and therefore negatively influences the use of pesticides [66].

The main determinants of pesticide application identified in the literature were summarized in Table 1.

The intricacy of multiple problems that must be addressed simultaneously, as well as the link between different strategies, highlight the necessity for research that examines pesticide use in the context of the agricultural system and on a regional scale. Multicriteria evaluation and decision support systems, in conjunction with pest monitoring programmes, can aid in the development of region-specific and long-term policies that are coordinated within an EU framework.

2.2. Cluster Analysis of Pesticide Consumption in Agriculture

A cluster analysis of agricultural systems provides categorization (and grouping) of countries, which can then serve as the basis for policymakers interested in establishing targets that clusters can strive to attain within certain time periods. This is of utmost importance since, on some occasions, management practices are not standardized, knowledge-sharing and learning from best practices are not well established, and systems frequently operate unaware and unconcerned about the performance of others around them.

Table 1. Selection of Primary Determinants in Pesticide Application.

Determinants		Implications	Correlation with Pesticide Use
Population increase (c)		Food demand growth (a) [69]	Positive
Economic development (a)	low	Prohibitive pesticides costs (a) [40,41]	Positive
	high	Food demand growth = heavy users (a) [40,41]	Positive
	very high	Focus on efficiency and growing demand for healthy food (a) [40,41]	Negative
Scarce or expensive workforce (c) [46]		Need to increase productivity (a)	Positive
Expensive land and growing competition (a) [47–49]			
Organic agricultural area		Specific certifications for organic products (a) [53]	Negative
Sales prices of crop products increase (a) [56]		Market for inorganic products becomes more profitable (a)	Positive
Real income increase (a) [57–59]		Pressure to improve input efficiency decreases (a)	Positive
Subsidies increase (a) [60,64]		Buy chemicals to improve agriculture productivity or Possibilities to invest in sustainable farming methods (a)	Ambiguous
Farmer age and education increase (b) [68]		Implementation of traditional farming methods (a)	Negative
Consumers' education increase (b) [66]		Demand for healthy food (a)	Negative

(a) economic factors, (b) social factors, (c) socio-demographic factors. Source: authors' computation based on literature review.

Cluster analysis have been used to identify patterns of energy and land use in agriculture [70], heavy metal sources in soils [71], and farmer search behaviour of various types of information, including pesticide use [72]. To our knowledge, cluster analysis has not been applied to the determinants of pesticide use in agriculture grouped at the geographical level.

3. Data and Methodology

3.1. Methods

Given the ambitious goal of the European Union to reduce the use of chemical pesticides by 50% by 2030, the primary objective of this paper is to understand the influence of a set of economic and social variables on their use in the EU by conducting several regression analyses. We have considered the evolution of pesticide consumption between 2000–2019, and estimations were performed on panel data extracted from international databases [73,74] for all 27 member states (excluding the United Kingdom). The time frame is long enough to draw meaningful conclusions that are helpful in understanding the perspectives of the European Union in the field of agriculture.

Estimates were made using seemingly unrelated regressions, a method that takes into account heteroskedasticity and correlations between errors. The empirical study initially focuses on the influence of GDP per capita and selling prices of crop production on the use of pesticides. Given the complexity of the topic and the variety of factors that influence the use of pesticides, we have expanded our model with other variables referring to the governmental support granted to agriculture and (subsidies on agricultural crops) and a social-demographic variable (population in each member state) (Equation (2)). Furthermore, given the EU's ambitious goal of improving the health of its citizens, we subsequently included an independent variable related to the organic crop area in Equation (3). Lastly, the FAO-calculated index of real income factors in agriculture was added as an explanatory variable to account for the impact of factor productivity on pesticide use [75]. Initially, we have estimated the regression by including independent variables: employment in agriculture (1000 persons), labour force participation rate in rural areas (% of total population ages 15+), number of producers in agriculture, export value index for agricultural products (2014–2016 = 100) and number of people who completed tertiary education. However, these variables had no significant impact in the regression and were not preserved in the estimations presented below, representing one of the limits of the research. With the purpose of examining the determinants of the use of chemical pesticides in the EU (27) between 2000 and 2019, we have estimated an empirical model based on panel data, gradually extending the equations with explanatory variables, specifically related to the agricultural sector or aiming at the macroeconomic and social framework, as follows:

$$Pesticides_{i,t} = a + \beta_1(GDP\ per\ capita_{i,t}) + \beta_2(Selling\ prices_{i,t}) + u_{i,t} \quad (1)$$

$$Pesticides_{i,t} = a + \beta_1(GDP\ per\ capita_{i,t}) + \beta_2(Selling\ prices_{i,t}) + \beta_3(Subsidies_{i,t}) + \beta_4(Population_{i,t}) + u_{i,t} \quad (2)$$

$$\begin{aligned} Pesticides_{i,t} = \; & a + \beta_1(GDP\ per\ capita_{i,t}) \\ & + \beta_2(Selling\ prices_{i,t}) + \beta_3(Subsidies_{i,t}) + \beta_4(Population_{i,t}) + \beta_5(Organic_{i,t}) + u_{i,t} \end{aligned} \quad (3)$$

$$\begin{aligned} Pesticides_{i,t} = \; & a + \beta_1(GDP\ per\ capita_{i,t}) \\ & + \beta_2(Selling\ prices_{i,t}) + \beta_3(Subsidies_{i,t}) + \beta_4(Population_{i,t}) \\ & + \beta_5(Organic_{i,t}) + \beta_6(Real\ income_{i,t}) + u_{i,t} \end{aligned} \quad (4)$$

where a = constant, $u_{i,t}$ = error term; t = 1, ... , T (years); i = 1, ... , N (countries)

The variables, definitions, sources, and the expected influence are presented in Table 2.

Table 2. Variables, definitions, and sources used in the first empirical model.

Variable	Definition	Source
Pesticides	Pesticide use per area of cropland (kilograms per hectare)	Our World in Data
GDP per capita	Real GDP per capita (chain-linked volumes 2010, euro per capita)	Eurostat
Selling prices	Sales prices of crop products (absolute prices, euro per 100 kg)	Eurostat
Subsidies	Subsidies on agricultural products (million euro)	Eurostat
Population	Population (total number)	Eurostat
Organic agricultural area	Hectares of organic crop area fully converted and under conversion to organic farming)	Eurostat
Real income	Index of the real income of factors in agriculture per annual work unit (2010 = 100)	Eurostat

Source: Authorial computation.

Panel data for all variables were tested for stationarity by using the Levin, Lin & Chu unit root test [76]. By applying this test, we have started from the hypothesis that the data have a unit root and are not stationary. Initially, we tested for stationarity using an individual intercept and trend, then with an individual intercept or no regressors. If the data did not show stationarity at level, we have checked for the first difference. The data were tested at level, initially including the trend and intercept in the equations. If using this variant revealed that the panel data have a unit root, we resorted to including only intercept or no regressor in the equation. Given the results obtained by applying the root tests (Prob. < 5%), we concluded that the data is stationary at level for all variables, except for the organic agricultural area (Table 3).

Table 3. Stationarity test.

Panel Unit Root Test-Levin, Lin & Chu [76]					
Variable	**Type of Test**	**t-Statistic**	**Prob.**	**Cross-Sections**	**Obs.**
Pesticides' use	level, individual intercept and trend	−5.24169	0.0000	26	494
GDP per capita	level, individual intercept and trend	−1.75557	0.0396	27	484
Selling prices	level, individual intercept and trend	−10.3856	0.0000	25	426
Subsidies	level, individual intercept and trend	−3.34717	0.0004	26	378
Population	level, individual intercept and trend	−5.43467	0.0000	27	486
Organic agricultural area	level, individual intercept and trend	−0.00809	0.4968	27	427
	level, individual intercept	1.49809	0.9329	27	427
	level, none	1.49809	0.9329	27	427
	1st difference, individual intercept and trend	−7.28873	0.0000	27	400
Real income	level, individual intercept and trend	−4.1117	0.0000	27	455

Source: Authorial computation.

3.2. Cluster Analysis

A cluster analysis was carried out based on the most relevant factors in the use of inorganic pesticides that had a significant influence in the regressions (p-value less than 10%). The analysed variables were the following: GDP per capita, population, sales prices, subsidies, organic agricultural area, index of real income of factors in agriculture, all having equal weight in the cluster formation. Cluster analysis is applied at a one-year level. The most recent year for which statistics were available was chosen (2019 in most cases, except France, where the last available data for selling prices are from 2016). Cyprus and Malta were excluded from the cluster analysis due to the lack of data on sale prices.

Since the variables were different sizes and to prevent large-scale variables from dominating the cluster formation, the data have first been normalized to have a mean of 0 and a variation of 1. We have used SAS software and applied the Ward minimum variance method, which groups observations based on the minimum distance between them (the distance being the ANOVA sum of squares). Clusters are grouped in subsequent stages at each level of the hierarchy until we obtain the optimum number of clusters, which have the maximum distance between them [77].

4. Results and Discussion

4.1. Regression Analysis

Table 4 illustrates the statistical description of the variables included in the empirical model. At first glance, the differences between the minimum and maximum values emphasize the heterogeneous evolution among the member states in the field of macroeconomic, social, and agricultural related variables. The average amount of pesticides used per hectare in the European Union was 3.1 kg, with significant differences between member states. For example, countries such as the Netherlands and Belgium had a consumption of pesticides greater than 12 kg per hectare, while the Baltic States and Bulgaria recorded values under half kilogram per hectare. In the field of economic development, the GDP per capita suggests a heterogeneous evolution among member states, with an average income of 24,279 euros for the time interval 2000–2019. In terms of agricultural sector performance, the average selling prices for agricultural products were 15 euros per 100 kg, with Italy and France recording the highest performances. Referring to the amount of subsidies granted to agriculture, the average value recorded between 2000 and 2019 was 331 million euros, with significant differences between member states. The highest amounts were given to France

(5121 million euros) and Germany (3335 million euros), while the countries that benefited the least from the subsidies were Slovakia, Denmark, and Ireland. For the entire period, the average surface aimed at organic agriculture was 337,059 hectares per member state, while in the field of real income in agriculture, the mean value of the index was 102.48 euros.

Table 4. Statistical description of the variables in the empirical model.

Variable/ Indicator	Mean	Median	Max.	Min.	Std. Dev.	Obs.
Pesticides	3.153327	2.260000	12.06000	0.240000	2.720559	520
GDP per capita	24,279.59	20,245.00	88,120.00	2990.000	16,500.82	538
Selling prices	15.05340	14.97000	33.07000	7.940000	3.881449	476
Subsidies	331.1159	34.52000	5121.500	0.010000	816.3447	439
Population	16,239,962	8,343,323	83,019,213	388,759.0	21,354,486	540
Organic agricultural area	337,059.8	167,538.0	2,354,916	1.000000	440,774.1	482
Real income	102.4768	100.0000	250.3800	33.60000	29.49577	509

Source: Authorial computation.

Given the estimated values of the coefficients presented in Table 5, the equations are as follows:

$$\begin{aligned} Pesticides_{i,t} = \; & -5.6136 + 0.2252(GDP\ per\ capita_{i,t}) \\ & +0.5727(Selling\ prices_{i,t}) + u_{i,t} \end{aligned} \tag{5}$$

$$\begin{aligned} Pesticides_{i,t} = \; & -8.9782 + 0.5787(GDP\ per\ capita_{i,t}) \\ & +0.1098(Selling\ prices_{i,t}) - 0.0024(Subsidies_{i,t}) + 0.2312(Population_{i,t}) + u_{i,t} \end{aligned} \tag{6}$$

$$\begin{aligned} Pesticides_{i,t} = \; & -9.4161 + 0.5466(GDP\ per\ capita_{i,t}) \\ & +0.2299(Selling\ prices_{i,t}) - 0.01224(Subsidies_{i,t}) + 0.4083(Population_{i,t}) \\ & -0.1948(Organic_{i,t}) + u_{i,t} \end{aligned} \tag{7}$$

$$\begin{aligned} Pesticides_{i,t} = \; & -10.8320 + 0.5097(GDP\ per\ capita_{i,t}) \\ & +0.0829(Selling\ prices_{i,t}) - 0.0067(Subsidies_{i,t}) + 0.4327(Population_{i,t}) \\ & -0.1902(Organic_{i,t}) + 0.3710(Real\ income_{i,t}) + u_{i,t} \end{aligned} \tag{8}$$

Table 5 and the estimation presented above illustrate the results of the empirical analysis aimed at identifying pesticide determinants of use in the European Union. Pesticide use was mainly influenced by the economic performance of the member states, particularly the level of GDP per capita. Consequently, an increase of one euro of GDP per capita will determine a rise of 0.5 kg of pesticide use per hectare, according to Equation (4). The increase in GDP per capita had a positive and strong influence on the use of pesticides in the European Union, confirming that many developed countries are still heavy users of pesticides. Our results are in line with [78], which also showed a positive connection between pesticide consumption, population and GDP per capita for several countries, including Europe, between 1990 and 2014. We cannot firmly contradict [40,41] which showed an inverted U-shaped evolution of pesticides along with GDP growth because we only checked a linear relationship applying a regression model for the entire time frame. Looking at the primary data that we used in the sample, one can find that several countries have reduced pesticide consumption over recent years (from 2017, 2018, or 2019) after increasing it between 2010 and 2016: Austria, Belgium, Bulgaria, Czech Republic, France, Germany, Italy, Netherlands, Poland, Portugal, and Sweden. Other countries have continued to use more pesticides until 2019: Croatia, Estonia, Latvia, Romania, Slovakia, and Spain. Although GDP per capita has continuously grown in the mentioned countries from 2010 to 2019, we can only associate a decrease in pesticide use with a higher GDP per capita for two or three years. This trend should be followed for a few more years to be able to draw more pertinent conclusions.

Table 5. Regression output.

Dependent Variable: Annual Pesticides' Use (2001–2019)				
Method: Generalized Least Squares-Seemingly Unrelated Regression				
Equation	**1**	**2**	**3**	**4**
No. Obs./ Variable	**474**	**408**	**375**	**361**
C	−5.6136 * (0.3812) (−14.7237)	−8.9782 * (0.5365) (−16.7321)	−9.4161 * (0.5459) (−17.248)	−10.8320 * (0.5862) (−18.477)
GDP per capita	0.2252 * (0.2148) (1.0485)	0.5787 * (0.0409) (14.1375)	0.5466 * (0.0495) (11.0409)	0.5097 * (0.0463) (10.9983)
Selling prices	0.5727 * (0.0377) (15.1749)	0.1098 * (0.0303) (3.6202)	0.2299 * (0.0412) (5.5710)	0.0829 *** (0.0441) (1.8799)
Subsidies		−0.0024 (0.0049) (−0.4892)	−0.0122 ** (0.0054) (−2.2562)	−0.0067 (0.0055) (−1.2197)
Population		0.2312 * (0.0227) (10.1592)	0.4083 * (0.0306) (13.3363)	0.4327 * (0.0306) (14.1349)
Organic agricultural area			−0.1948 * (0.0232) (−8.3907)	−0.1902 * (0.0242) (−7.8432)
Real income				0.3710 * (0.0507) (7.3132)
Prob (F-statistic)	0.0000	0.0000	0.0000	0.0000
R^2	0.3441	0.3316	0.4005	0.4508
Adjusted R^2	0.3413	0.3250	0.3924	0.4415
Durbin–Watson stat	1.9784	1.9254	1.8424	1.7928

Note: robust standard errors and t-statistics are in parentheses. *—*p*-value < 1%, **—*p*-value < 5%, ***—*p*-value < 10%. Source: Authorial computation.

The analysis shows that farmers tend to use more chemical pesticides as the population of EU member states increases, indicating a potential increase in food demand. The population growth of one unit generates a 0.43 kg per hectare increase in pesticides (Equation (4)). This finding confirms the relationship between the increase in demand and the response to improve productivity (in line with [69]). Consequently, European farmers still place a priority on producing a large amount of food, while the production of healthy, high-quality food (including few or no pesticides) is less attractive. The increase in demand (expressed as a higher quantity) is still a more interesting opportunity compared to the advantages of organic agricultural production. The profits obtained from organic farming may not yet be enough to determine the specialization in this area. For example, ref. [79] found a similar profit per surface of cultivated land for conventional and organic farming in certain regions in Germany.

Another factor that had a high influence on the dependent variable was real income, which determined an increase in pesticide use of 0.37 kg per hectare (Equation (4)). Moreover, having a high statistical significance in Equation (1), the resulting coefficients suggest that an increase with 1 euro of selling prices per 100 kg determines a rise in the quantity of pesticide per hectare of 0.57 kg. Therefore, when profits are higher, producers prefer to retain them rather than invest in the switch to organic farming, confirming that the behaviour observed by [56], as presented in the literature review, continues to manifest in the same way. When income increases, the preoccupation with efficiency diminishes and more pesticides are wasted, similar to what [59] found in Chinese agricultural practices.

Although not statistically significant in the second and fourth equations, subsidies granted to agricultural products tend to negatively influence the consumption of pesticides. Their growth in one unit leads to a decrease in pesticide consumption of 0.0067 kg per hectare. This result is probably due to the orientation of these funds toward organic farming. This is an interesting finding because it confirms the fact that if a certain conditionality is imposed on accessing subsidies, such as following sustainable farming methods, it would motivate farmers to embrace them and still have sufficient profit. Ref. [80] also identified direct subsidies to be effective in determining a switch to organic farming practice, but that it also caused a price decrease for both organic and conventional farming output. The decrease in the output price for organic products makes them more available to consumers, but when prices for conventional farming products are also lower, the profits for this specialization also decrease. This is when subsidies directed to organic farming compensate for the profit loss. Ref. [55] pointed out that both increased prices and subsidies for organic farming are effective methods to increase profits and encourage farmers use sustainable production methods.

The results also suggest that the increase in the area intended for organic agriculture tends to negatively influence the consumption of chemical pesticides, as the independent variable is significant in the estimations. More specifically, the expansion with one hectare of organic farming creates a decrease in pesticide use of 0.19 kg per hectare (Equation (4)). This relationship is obvious because organic farming means using fewer synthetic pesticides [81]. The question that needs to be further explored is whether an increase in the surfaces dedicated to organic agriculture determines a progressively higher reduction of chemical pesticide use as more farmers learn from the experience and as certain economies of scale or scope can be obtained. However, such an analysis can only be performed when and where sustainable agriculture is more widely spread.

The values of the coefficient of determination (R2) imply that the model explains the variation of the dependent variable in a percentage that spans from 34% (Equation (1)) to 45% (Equation (4)). To check the validity of the model, we tested the classical linear regression assumptions. The general form of the multiple linear regression is as follows:

$$Y_{i,t} = a + \beta X_{i,t} + u_{i,t}$$

where Y_i = dependent variable; $X_{i,t}$ = independent variable; $u_{i,t}$ = error term $t = 1, \ldots, T$ (time); $i = 1, \ldots, N$ (cross-sections).

First, we have verified whether there is serial independence, which assumes that the errors are distributed independently. To test the first-order correlation, we used the Durbin–Watson test. The value of around 2 confirms that there is no first-order correlation between errors, so we consider that the empirical model is valid. Subsequently, we have tested the validity of the model by looking at another assumption of the linear regression model, the multicollinearity, which implies that the explanatory variables are not correlated. To identify multicollinearity, we used the VIF test (variance inflation factors). As described in Table 6, the results of the centred VIF are around 1 for all variables. Consequently, we have concluded that there is not a highly collinear relationship between explanatory variables that could bias the estimates. We have also tested another assumption of the linear regression model—the homoscedasticity—which assumes that all error terms have the same variance, respectively $var(\varepsilon_{i,t}) = \sigma^2$ = constant for all t [73]. The histogram confirms that the residuals have a normal distribution, with a probability value above 5%. Finally, we have tested whether the explanatory variables are representative. The redundant variable test illustrates that the variable related to the population is significant, as we have rejected the null hypothesis. Moreover, we were also interested in determining if another variable initially taken into consideration for the estimation was omitted in the model. By accepting the null hypothesis (Prob. > 5%), we concluded that agriculture (% of GDP) would not be significant (Table 7).

Table 6. Variance Inflation Factors.

Variance Inflation Factors	
Variable	**Centred VIF**
GDP per capita	1.1476
Selling prices	1.1878
Subsidies	1.2151
Population	1.9623
Organic agricultural area	1.7374
Real income	1.2083

Source: Authorial computation.

Table 7. Coefficient diagnosis.

Test	Null Hypothesis	Result				Decision
Redundant variable test	Population is not significant		Value	Df	Probability	Reject the null hypothesis
		t-statistic	11.3756	354	0.0000	
		F-statistic	129.4044	(1, 354)	0.0000	
Omitted variable test	Agriculture (% of GDP) is not significant		Value	Df	Probability	Accept the null hypothesis
		t-statistic	1.6448	353	0.1010	
		F-statistic	2.7043	(1, 353)	0.1010	

Source: Authorial computation.

4.2. Cluster Analysis

The next step involved clustering the European countries based on the factors of pesticide consumption that we found in the regression models. The goal was to identify similar arrangements of these drivers so that comparable pesticide mitigation strategies could be developed. Based on the Ward method's application of the following criteria, we have obtained three clusters, as shown in Table 8. Pseudo F statistic (14.6) and cubic cluster criterion (0.28) were high showing a high separation between clusters, while pseudo T-squared was low (5.4) indicating that the variance between clusters relative to the variance within clusters is low. Figure 1 illustrates the cluster formation stages. The red line marks the stage in which the three clusters were obtained.

Table 8. The groups obtained in the cluster analysis.

Cluster Number	Countries within Each Cluster
1	Italy, Spain, France, Germany
2	Poland, Lithuania, Greece, Portugal, Slovenia, Estonia, Romania, Croatia, Latvia, Czech Republic, Hungary, Slovakia, Bulgaria
3	Ireland, Luxembourg, Finland, Denmark, Netherlands, Belgium, Sweden, Austria

Source: Authorial computation.

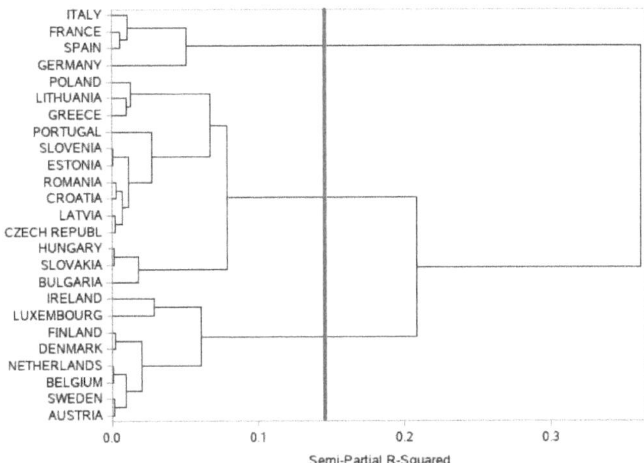

Figure 1. Hierarchical clustering tree. Source: Authorial computation.

Based on the average values that the variables take in each country (Table 9), we further comment on the main characteristics and propose a set of recommendations in each case.

Table 9. Average values of variables in each cluster.

Cluster no.	GDP per Capita	Population	Selling Prices	Subsidies	Real Income	Organic Crop Area
1	30,427.50	64,237,645.50	18.85	253.20	125.62	1,969,944.25
2	14,486.92	9,496,740.62	16.58	94.26	151.16	291,458.85
3	48,998.75	8,083,597.00	15.38	3.79	104.87	265,543.88

Source: Authorial computation.

The first cluster is the most homogeneous (semi-partial R-squared is 0.0506) and contains three of the founding member states (Italy, Germany, and France) together with Spain, which joined the EU in 1986. These are the countries with the largest population in our sample (64 million people on average) and the largest organic crop area (1.97 million hectares on average), which is not surprising given the large territory of these countries. The total subsidies for agriculture are the highest in this cluster (an average of 253.2 million euros), especially in the case of France (306.16 million euros) and Spain (301.93 million euros), although this does not translate into lower selling prices for agricultural products. Compared to other clusters, this one exhibits the largest average value for selling prices (18.85 euros per 100 kg). However, since subsidies are expressed in absolute terms and are not related to either the quantity of products or the cultivated surface, we cannot establish a clear connection between subsidies and sales prices. The average GDP per capita (30,427.50 euros) and the average real income of the factors in agriculture are at middle levels among the three clusters, indicating good possibilities to develop organic agriculture, but also putting the producers in a more comfortable situation, which brings little motivation to switch to sustainable agriculture.

Most of the newest EU members (Poland, Lithuania, Slovenia, Estonia, Latvia, Czech Republic, Hungary, Slovakia, Romania, Bulgaria, and Croatia) and two South European countries (Greece and Portugal) are included in the second cluster, which has the highest disparities, as suggested by the highest semipartial R-squared in the dendrogram (0.05060). This cluster has the smallest average GDP per capita (14,486.92 euros). Their population (9.5 million people on average) and organic crop farming area (291,459 hectares on average) are between the other clusters, but closer to the average figures in the third cluster. The level of subsidies is larger than in the third cluster (94.26 million euros on average). The average of the index of real income of factors in agriculture per annual work unit is the highest for this cluster, showing that farmers obtain high productivity and good development possibilities in agriculture. Although these countries lag in terms of economic development, which could indicate a lower demand for organic food, they could serve other developed markets through the single market as well.

The third group consists of Northern and Western European countries and, among the three clusters, it occupies a middle position in terms of homogeneity (semipartial R-squared equals 0.06094). Its members (Ireland, Luxembourg, Finland, Denmark, Netherlands, Belgium, Sweden, Austria) are characterized by having the highest GDP per capita (48,998.75 euros on average), a smaller population compared to the other clusters (8 million people on average), the lowest level of subsidies for agriculture (3.79 million euros on average), and the lowest real income in agriculture (an average index of 104.87). This occurs against a backdrop of limited agricultural specialization, which is specific to developed countries with small surfaces, low population, or scarce population in general. However, existing agricultural production, even if smaller, has the prerequisites to be turned into organic agriculture, improving the quality of the products, and addressing high income markets. Indeed, the average organic crop area (265,544 hectares) is close to the one reported by the countries in the second cluster. Austria (671,703 hectares) and Sweden (613,964 hectares) even have a higher organic crop area than any country in the second cluster. The selling prices of farm products are the smallest in the case of this cluster (an average value of 15.38 euros per 100 kg), and when combined with the small level of subsidies for agriculture and small real income, they indicate a lower profitability of agriculture currently. Although prices are low, an increase in direct subsidies might be possible to counteract this disadvantage for producers.

5. Conclusions and Recommendations

This paper consists of an analysis built up in two main stages. The first carried out a set of four seemingly unrelated regressions aimed at identifying the impact of various economic and social determinants on the consumption of pesticides in EU member countries, and the second retained the determinants with a significant impact to be used as factors in the cluster analysis. It resulted in three main clusters that place the member countries on different levels of similar conditions that determine the current level of pesticide consumption and represent barriers or opportunities to switch to sustainable agricultural practices.

Our study revealed that wealthy countries use more pesticides in agriculture, but on a downward trend over the previous two or three years, as wealthier consumers can afford healthier food. GDP per capita had the greatest impact on pesticide use (a 0.50 coefficient in the fourth equation). A larger population determines the use of more pesticides (coefficient: 0.43), establishing the link between food demand and productivity pressure.

From the supply perspective, results showed a positive and asymmetrical influence of sale prices of agricultural products (p-value < 10%) and real income of agricultural components (p-value less than 1%) on the inputs of pesticides. This emphasises the fact that improved market opportunities, expressed through favourable prices, motivate farmers to produce more, sell more and gain more profit.

As organic crop area grows, pesticide input decreases, showing that organic farming experience encourages sustainable pest control. Although conventional agriculture is still profitable, subsidies, especially those targeted at sustainable production techniques, are the only economic leverage that can push farmers to investigate alternative pest management methods and reduce synthetic pesticide use.

The cluster analysis resulted in three country clusters on which we can draw the following conclusions and recommendations.

- Research and development and sharing expertise with other market participants might improve the experience of having a large organic agricultural area for the nations with the biggest population and GDP per capita (Italy, Spain, France, Germany).

This cluster's subsidies are higher than others, but better targeting, complementary regulatory circumstances, market access, etc. can increase their efficiency.

- The second cluster (Poland, Lithuania, Greece, Portugal, Slovenia, Estonia, Romania, Croatia, Latvia, Czech Republic, Hungary, Slovakia, Bulgaria) has average pesticide use determinants. Competitive prices and large agricultural surfaces could benefit these countries. Subsidies can be increased while imposing their use in sustainable farming procedures, and there is a great deal of work to be completed in terms of regulations, market access for organic products not only on their own markets, but also on other European markets where consumers have higher incomes to spare on healthy food.
- Ireland, Luxembourg, Finland, Denmark, Netherlands, Belgium, Sweden, Austria, and Luxembourg had the lowest agricultural production and subsidies in the third cluster. More in-depth studies are needed to see for which type of crop, subsidies would not represent a waste of resources given the more difficult environmental conditions.

Originality elements for this study derive from the complementary study of determinants of pesticide consumption that considers the synergies between the agricultural, macroeconomic, and social levels, and the analysis of regional differences reflected in the cluster analysis. Consequently, our findings can contribute to the creation of targeted national sustainable production policies and the design of practical measures by providing specific quantitative information.

The analysis takes into consideration crop production as a whole without being able to differentiate between organic and conventional farming. Such a distinction would have been useful for variables such as sales prices, real income, and subsidies. In the case of some explanatory variables (subsidies on agricultural products), data for some member states were not available for the entire period of time. However, we were aware of this down side at the beginning of this study and believed that the estimation would not be biased.

Given the impossibility of capturing the multitude of factors that influence pesticide consumption, another limit of the study derives from the restrained set of parameters included in the empirical models. Future research can address the study of other social, economic and technical factors specific to the farm environment that affect the acceptance of sustainable production methods, continuing a previous work on good practices for lowering the use of pesticides and fertilizers [82].

Another important research direction would be to deepen the examination of the conditions under which subsidies or other forms of public financial support would be efficient in extending sustainable agriculture practices. Diversified farming systems that combine conventional and sustainable agriculture production methods are perhaps worth considering; this was also identified as a research direction in [83].

Eliminating pesticide use is difficult. Farmers have few pest and weed control choices after years of relying on them. Many farmers, especially those who have invested in conventional farms, cannot afford alternative pest treatment equipment and longer production time. However, a rise in organic farms, knowledge of sustainability in modern agriculture, and government-sponsored efforts are propelling the biopesticide business and pushing more farmers to adopt sustainable agricultural production.

Author Contributions: Conceptualization, V.C. and D.M.; methodology, V.C., R.G.R., A.P.A., and A.-M.H.; software, A.-M.H. and R.G.R.; validation, V.C., D.M., R.G.R., and A.P.A.; formal analysis, V.C.; investigation, A.P.A., R.G.R., and A.-M.H.; resources, A.-M.H. and R.G.R.; data curation, A.-M.H., R.G.R., A.P.A., and V.C.; writing—original draft preparation, R.G.R., A.P.A., and A.-M.H.; writing—review and editing, R.G.R., A.P.A., V.C., and D.M.; supervision, V.C. and D.M. All authors have read and agreed to the published version of the manuscript.

Funding: This research received no external funding.

Institutional Review Board Statement: Not applicable.

Informed Consent Statement: Not applicable.

Data Availability Statement: Authors may provide access to data necessary to validate their conclusions to the extent that their legitimate interests or constraints are safeguarded.

Conflicts of Interest: The authors declare no conflict of interest.

References

1. UC Davis. "What is Sustainable Agriculture?" UC Sustainable Agriculture Research and Education Program. UC Agriculture and Natural Resources. 2021. Available online: https://sarep.ucdavis.edu/sustainable-ag (accessed on 3 July 2022).
2. Pham, L.V.; Smith, C. Drivers of agricultural sustainability in developing countries: A review. *Environ. Syst. Decis.* **2014**, *34*, 326–341. [CrossRef]
3. Ericksen, P.J. Conceptualizing food systems for global environmental change research. *Glob. Environ. Change* **2008**, *18*, 234–245. [CrossRef]
4. European Environment Agency (EEA). *The European Environment State and Outlook 2020: Knowledge for Transition to a Sustainable Europe*; European Environment Agency (EEA): Coppenhagen, Denmark, 2009; p. 351. Available online: https://www.eea.europa.eu/soer/publications/soer-2020 (accessed on 18 July 2022).
5. Puma, M.J.; Bose, S.; Chon, S.Y.; Cook, B.I. Assessing the evolving fragility of the global food system. *Environ. Res. Lett.* **2015**, *10*, 024007. [CrossRef]
6. Ben Hassen, T.; El Bilali, H. Impacts of the Russia-Ukraine War on Global Food Security: Towards More Sustainable and Resilient Food Systems? *Foods* **2022**, *11*, 2301. [CrossRef]
7. Jagtap, S.; Trollman, H.; Trollman, F.; Garcia-Garcia, G.; Parra-López, C.; Duong, L.; Martindale, W.; Munekata, P.E.; Lorenzo, J.M.; Hdaifeh, A.; et al. The Russia-Ukraine Conflict: Its Implications for the Global Food Supply Chains. *Foods* **2022**, *11*, 2098. [CrossRef]
8. United Nations Environment Programme (UNEP). Environmental and Health Impacts of Pesticides and Fertilizers and Ways of Minimising Them: Summary for Policy Makers. 2021. Available online: https://www.unep.org/resources/report/environmental-and-health-impactspesticides-and-fertilizers-and-ways-minimizing (accessed on 2 December 2021).
9. US Environmental Protection Agency. "Why We Use Pesticides." EPA. 27 June 2017. Available online: https://www.epa.gov/safepestcontrol/why-we-use-pesticides (accessed on 29 June 2022).
10. Food and Agriculture Organization of the United Nations (FAO). *The Ethics of Sustainable Agricultural Intensification*; Food and Agriculture Organization of the United Nations (FAO): Rome, Italy, 2004.
11. Pingali, P. Agricultural mechanization: Adoption patterns and economic impact. *Handb. Agric. Econ.* **2007**, *3*, 2779–2805.
12. Leetmaa, S.E.; Arnade, C.; Kelch, D. A comparison of US and EU agricultural productivity with implications for EU enlargement. In *U.S. EU Food and Agriculture Comparisons*; DIANE Publishing: Collingdale, PA, USA, 2004; p. 33.
13. Vanlauwe, B.; Wendt, J.; Giller, K.E.; Corbeels, M.; Gerard, B.; Nolte, C. A fourth principle is required to define conservation agriculture in sub-Saharan Africa: The appropriate use of fertilizer to enhance crop productivity. *Field Crops Res.* **2014**, *155*, 10–13. [CrossRef]
14. Fixen, P.; Brentrup, F.; Bruulsema, T.; Garcia, F.; Norton, R.; Zingore, S. Nutrient/fertilizer use efficiency: Measurement, current situation and trends. In *Managing Water and Fertilizers for Sustainable Agricultural Intensification*; International Fertilizer Industry Association (IFA): Paris, France; International Water Management Institute (IWMI): Colombo, Sri Lanka; International Plant Nutrition Institute (IPNI): Norcross, GA, USA; International Potash Institute (IPI): Zug, Switzerland, 2015; p. 270.
15. Schreinemachers, P.; Tipraqsa, P. Agricultural pesticides and land use intensification in high, middle and low income countries. *Food Policy* **2012**, *37*, 616–626. [CrossRef]
16. Ackrill, R. *Common Agricultural Policy*; Sheffield Academic Press: Sheffield, UK, 2000.
17. Lefebvre, M.; Espinosa, M.; Gomez y Paloma, S. *The Influence of the Common Agricultural Policy on agricultural landscapes. Joint Research Centre Scientific and Policy Reports*; European Commission, Joint Research Center: Brussels, Belgium, 2012; p. 7.
18. Liu, Q.; Xu, H.; Yi, H. Impact of fertilizer on crop yield and C: N: P stoichiometry in arid and semi-arid soil. *Int. J. Environ. Res. Public Health* **2021**, *18*, 4341. [CrossRef]
19. Serebrennikov, D.; Thorne, F.; Kallas, Z.; McCarthy, S.N. Factors influencing adoption of sustainable farming practices in Europe: A systemic review of empirical literature. *Sustainability* **2020**, *12*, 9719. [CrossRef]
20. Organisation for Economic Co-operation and Development (OECD). Agricultural Policy Monitoring and Evaluation, Trade and Agriculture Directorate Committee for Agriculture, Working Party on Agricultural Policies and Markets. Available online: https://www.oecd.org/officialdocuments/publicdisplaydocumentpdf/?cote=TAD/CA/APM/WP(2021)6/FINAL&docLanguage=En (accessed on 22 July 2022).
21. Benbrook, C.; Kegley, S.; Baker, B. Organic farming lessens reliance on pesticides and promotes public health by lowering dietary risks. *Agronomy* **2021**, *11*, 1266. [CrossRef]
22. Bremmer, J.; Gonzalez Martinez, A.R.; Jongeneel, R.A.; Huiting, H.F.; Stokkers, R. Impact Assessment Study on EC 2030 Green Deal Targets for Sustainable Food Production: Effects of Farm to Fork and Biodiversity Strategy 2030 at farm, national and EU level. 2021. Available online: https://www.wur.nl/en/show/Impact-Assessment-Study-on-EC-2030.htm (accessed on 22 December 2021).

23. Eurostat. 346000 Tonnes of Pesticides Sold in 2020 in the EU 2022. Available online: https://ec.europa.eu/eurostat/web/products-eurostat-news/-/ddn-20220502-1#:~{}:text=Btween%202011%20and%202020%2C%20sales,346%20000%20tonnes%20were%20sold (accessed on 20 December 2021).
24. Guyomard, H.; Bureau, J.-C. *Research for AGRI Committee—The Green Deal and the CAP: Policy Implications to Adapt Farming Practices and to Preserve the EU's Natural Resources*; European Parliament, Policy Department for Structural and Cohesion Policies: Brussels, Belgium, 2020.
25. Staudacher, P.; Fuhrimann, S.; Farnham, A.; Mora, A.M.; Atuhaire, A.; Niwagaba, C.; Stamm, C.; Eggen, R.I.; Winkler, M.S. Comparative Analysis of Pesticide Use Determinants Among Smallholder Farmers from Costa Rica and Uganda. *Environ. Health Insights* **2020**, *14*. [CrossRef] [PubMed]
26. Rahman, S. Farm-level pesticide use in Bangladesh: Determinants and awareness. *Agric. Ecosyst. Environ.* **2003**, *95*, 241–252. [CrossRef]
27. Pan, Y.; Ren, Y.; Luning, P.A. Factors influencing Chinese farmers' proper pesticide application in agricultural products—A review. *Food Control* **2021**, *122*, 107788. [CrossRef]
28. Aminu, F.O.; Ayinde, I.A.; Sanusi, R.A.; Olaiya, A.O. Determinants of Pesticide Use in Cocoa Production in Nigeria. *Can. J. Agric. Crops* **2019**, *4*, 101–110.
29. Rahman, S.; Chima, C.D. Determinants of Pesticide Use in Food Crop Production in Southeastern Nigeria. *Agriculture* **2018**, *8*, 35. [CrossRef]
30. Helepciuc, F.-E.; Todor, A. Evaluating the effectiveness of the EU's approach to the sustainable use of pesticides. *PLoS ONE* **2021**, *16*, e0256719. [CrossRef]
31. Wossink, G.A.; Feitshans, T.A. Pesticide Policies in the European Union. *Drake J. Agric. Law* **2000**, *5*, 223.
32. DG Health and Food Safety. Overview Report on the Implementation of Member State's Measures to Achieve the Sustainable Use of Pesticides under Directive 2009/128/EC. October 2017. Available online: http://euclidipm.org/images/documents/2017-6291---Final.pdf (accessed on 29 December 2022).
33. Lechenet, M.; Dessaint, F.; Py, G.; Makowski, D.; Munier-Jolain, N. Reducing pesticide use while preserving crop productivity and profitability on arable farms. *Nat. Plants* **2017**, *3*, 17008. [CrossRef]
34. Popp, J.; Pető, K.; Nagy, J. Pesticide productivity and food security. A review. *Agron. Sustain. Dev.* **2013**, *33*, 243–255. [CrossRef]
35. Spiertz, J.H.J.; Ewert, F. Crop production and resource use to meet the growing demand for food, feed and fuel: Opportunities and constraints. *NJAS Wagening. J. Life Sci.* **2009**, *56*, 281–300. [CrossRef]
36. Eurostat. Population and Population Change Statistics. 2021. Available online: https://ec.europa.eu/eurostat/statistics-explained/index.php?title=Population_and_population_change_statistic (accessed on 15 November 2021).
37. Zolin, B. The EU and Asia: World Trade Liberalisation and the Evolution of Agricultural Product Flows. University Ca'Foscari of Venice. *Dep. Econ. Res. Pap. Ser.* **2008**, *18*, 1–22.
38. Chico, J.R.; Peña Sánchez, A.R.; Jiménez García, M. Food exports competitiveness in EU by countries. *Ekon. Manaz. Spektrum* **2017**, *11*, 25–36. [CrossRef]
39. Bojnec, Š.; Čechura, L.; Fałkowski, J.; Fertő, I. Agri-food Exports from Central-and Eastern-European Member States of the European Union are Catching Up. *EuroChoices* **2021**, *20*, 11–19. [CrossRef]
40. Celikkol Erbas, B.; Guven Solakoglu, E. In the presence of climate change, the use of fertilizers and the effect of income on agricultural emissions. *Sustainability* **2017**, *9*, 1989. [CrossRef]
41. Yu, X.; Schweikert, K.; Doluschitz, R. Investigating the environmental Kuznets curve between economic growth and chemical fertilizer surpluses in China: A provincial panel cointegration approach. *Environ. Sci. Pollut. Res.* **2021**, *29*, 18472–18494. [CrossRef]
42. Frey, R.S. The international traffic in pesticides. *Technol. Forecast. Soc. Change* **1995**, *50*, 151–169. [CrossRef]
43. Komorowska, D. Development of organic production and organic food market in Europe. Acta Scientiarum Polonorum. *Oeconomia* **2014**, *13*, 91–101.
44. Fielding, A.J. Migration and urbanization in Western Europe since 1950. *Geogr. J.* **1989**, *155*, 60–69. [CrossRef]
45. Rye, J.F.; Scott, S. International labour migration and food production in rural Europe: A review of the evidence. *Sociol. Rural.* **2018**, *58*, 928–952. [CrossRef]
46. Strijker, D. Marginal lands in Europe—Causes of decline. *Basic Appl. Ecol.* **2005**, *6*, 99–106. [CrossRef]
47. Van Meijl, H.; van Rheenen, T.; Tabeau, A.; Eickhout, B. The impact of different policy environments on agricultural land use in Europe. *Agric. Ecosyst. Environ.* **2006**, *114*, 21–38. [CrossRef]
48. Banse, M.; van Meijl, H.; Tabeau, A.; Woltjer, G.; Hellmann, F.; Verburg, P.H. Impact of EU biofuel policies on world agricultural production and land use. *Biomass Bioenergy* **2011**, *35*, 2385–2390. [CrossRef]
49. Eickhout, B.V.; van Meijl, H.; Tabeau, A.; van Rheenen, T. Economic and ecological consequences of four European land use scenarios. *Land Use Policy* **2007**, *24*, 562–575. [CrossRef]
50. Liu, X.; Zhang, W.; Zhang, M.; Ficklin, D.L.; Wang, F. Spatio-temporal variations of soil nutrients influenced by an altered land tenure system in China. *Geoderma* **2009**, *152*, 23–34. [CrossRef]
51. Zheng, W.; Luo, B.; Hu, X. The determinants of farmers' fertilizers and pesticides use behavior in China: An explanation based on label effect. *J. Clean. Prod.* **2020**, *272*, 123054. [CrossRef]

52. Ciaian, P.; Rajcaniova, M.; Guri, F.; Zhllima, E.; Shahu, E. The impact of crop rotation and land fragmentation farm productivity in Albania. *Stud. Agric. Econ.* **2018**, *120*, 116–125.
53. Reddy, B.S. Organic farming: Status, issues and prospects—A review. *Agric. Econ. Res. Rev.* **2010**, *23*, 343–358.
54. Ferreira, S.; Oliveira, F.; Gomes da Silva, F.; Teixeira, M.; Gonçalves, M.; Eugénio, R.; Damásio, H.; Gonçalves, J.M. Assessment of factors constraining organic farming expansion in lis valley, Portugal. *AgriEngineering* **2020**, *2*, 111–127. [CrossRef]
55. Nieberg, H.; Offermann, F. The profitability of organic farming in Europe. In *Organic Agriculture: Sustainability, Markets and Polices*; OECD Workshop on Organic Agriculture: Washington, DC, USA, 2003.
56. Food and Agriculture Organization of the United Nations (FAO). *Fertilizers Use by Crop in the Syrian Arab Republic*; Food and Agriculture Organization of the United Nations (FAO): Rome, Italy, 2003; Available online: https://www.fao.org/3/Y4732E/y4732e09.htm (accessed on 29 November 2021).
57. Zhou, Y.; Yang, H.; Mosler, H.J.; Abbaspour, K.C. Factors affecting farmers' decisions on fertilizer use: A case study for the Chaobai watershed in Northern China. *Cons. J. Sustain. Dev.* **2010**, *4*, 80–102.
58. Waithaka, M.M.; Thornton, P.K.; Shepherd, K.D.; Ndiwa, N.N. Factors affecting the use of fertilizers and manure by smallholders: The case of Vihiga, western Kenya. *Nutr. Cycl. Agroecosystems* **2007**, *78*, 211–224. [CrossRef]
59. Bai, X.; Zhang, T.; Tian, S. Evaluating Fertilizer Use Efficiency and Spatial Correlation of Its Determinants in China: A Geographically Weighted Regression Approach. *Int. J. Environ. Res. Public Health* **2020**, *17*, 8830. [CrossRef] [PubMed]
60. Wu, L.; Hu, K.; Lyulyov, O.; Pimonenko, T.; Hamid, I. The Impact of Government Subsidies on Technological Innovation in Agribusiness: The Case for China. *Sustainability* **2022**, *14*, 14003. [CrossRef]
61. Sane, M.; Hajek, M.; Nwaogu, C.; Purwestri, R.C. Subsidy as An Economic Instrument for Environmental Protection: A Case of Global Fertilizer Use. *Sustainability* **2021**, *13*, 9408. [CrossRef]
62. Skevas, T.; Stefanou, S.E.; Lansink, A.O. Can economic incentives encourage actual reductions in pesticide use and environmental spillovers? *Agric. Econ.* **2012**, *43*, 267–276. [CrossRef]
63. Kelch, D.R.; Osborne, S. Crop production capacity in Europe. *Agric. Sci.* **2009**, *2*, 143.
64. European Council. Timeline—History of the CAP 2022. Available online: https://www.consilium.europa.eu/en/policies/cap-introduction/timeline-history/ (accessed on 5 June 2022).
65. European Commission. A Toolbox for Reforming Environmentally Harmful Subsidies in Europe: Detailed Annexes. Final Report 2022. Available online: https://circabc.europa.eu/ui/group/c1a5a4e9-7563-4d0e-9697-68d9cd24ed34/library/7ff9e898-823f-4b06-985a-119d9e25e529/details (accessed on 22 August 2022).
66. Lohr, L.; Salomonsson, L. Conversion subsidies for organic production: Results from Sweden and lessons for the United States. *Agric. Econ.* **2000**, *22*, 133–146. [CrossRef]
67. Christensen, T.; Pedersen, A.B.; Nielsen, H.O.; Mørkbak, M.R.; Hasler, B.; Denver, S. Determinants of farmers' willingness to participate in subsidy schemes for pesticide-free buffer zones—A choice experiment study. *Ecol. Econ.* **2011**, *70*, 1558–1564. [CrossRef]
68. Alexopoulos, G.; Koutsouris, A.; Tzouramani, I. Should I stay or should I go? Factors affecting farmers' decision to convert to organic farming as well as to abandon it. In Proceedings of the 9th European IFSA Symposium 2010, Vienna, Austria, 4–7 July 2010; pp. 1083–1093.
69. Carvalho, F.P. Pesticides, environment, and food safety. *Food Energy Secur.* **2017**, *6*, 48–60. [CrossRef]
70. Ghisellini, P.; Setti, M.; Ulgiati, S. Energy and land use in worldwide agriculture: An application of life cycle energy and cluster analysis. *Environ. Dev. Sustain.* **2016**, *18*, 799–837. [CrossRef]
71. Li, J.; He, M.; Han, W.; Gu, Y. Analysis and assessment on heavy metal sources in the coastal soils developed from alluvial deposits using multivariate statistical methods. *J. Hazard. Mater.* **2009**, *164*, 976–981. [CrossRef] [PubMed]
72. Ecker, O.; Breisinger, C. *The Food Security System: A New Conceptual Framework*; International Food Policy Research Institute (IFPRI): Washington, DC, USA, 2012; Volume 1166.
73. Eurostat. 2022. Available online: https://ec.europa.eu/eurostat/data/database (accessed on 15 June 2022).
74. Our World in Data, Pesticides (Total)—Use per Area of Cropland (Kilograms per Hectare). 2022. Available online: https://ourworldindata.org/grapher/pesticide-use-per-hectare-of-cropland?tab=table (accessed on 15 June 2022).
75. Food and Agriculture Organization (FAO). Food and Agriculture Data. 2022. Available online: https://www.fao.org/faostat/en/ (accessed on 15 June 2022).
76. Levin, A.; Lin, C.F.; Chu, C.S. Unit root tests in panel data: Asymptotic and finite-sample properties. *Journal of Econometrics* **2002**, *108*, 1–24. [CrossRef]
77. Ward, J.H., Jr. Hierarchical grouping to optimize an objective function. *J. Am. Stat. Assoc.* **1963**, *58*, 236–244. [CrossRef]
78. Hedlund, J.; Longo, S.B.; York, R. Agriculture, pesticide use, and economic development: A global examination (1990–2014). *Rural Sociol.* **2020**, *85*, 519–544. [CrossRef]
79. Batáry, P.; Gallé, R.; Riesch, F.; Fischer, C.; Dormann, C.F.; Mußhoff, O.; Tscharntke, T. The former Iron Curtain still drives biodiversity–profit trade-offs in German agriculture. *Nat. Ecol. Evol.* **2017**, *1*, 1279–1284. [CrossRef]
80. Pietola, K.S.; Lansink, A.O. Farmer response to policies promoting organic farming technologies in Finland. *Eur. Rev. Agric. Econ.* **2001**, *28*, 1–15. [CrossRef]
81. Food and Agriculture Organization (FAO). Organic Agriculture and the Law, FAO Legislative Study 107. 2012. Available online: https://www.fao.org/3/i2718e/i2718e.pdf (accessed on 3 August 2022).

82. Alexoaei, A.P.; Robu, R.G.; Cojanu, V.; Miron, D.; Holobiuc, A.M. Good Practices in Reforming the Common Agricultural Policy to Support the European Green Deal–A Perspective on the Consumption of Pesticides and Fertilizers. *Amfiteatru Econ.* **2022**, *24*, 525–545. [CrossRef]
83. Rosa-Schleich, J.; Loos, J.; Mußhoff, O.; Tscharntke, T. Ecological-economic trade-offs of diversified farming systems—A review. *Ecol. Econ.* **2019**, *160*, 251–263. [CrossRef]

Article

European Integration Processes in the EU GI System—A Long-Term Review of EU Regulation for GIs

Karola Schober [1,2,*], Richard Balling [2], Tobias Chilla [1] and Hannah Lindermayer [2]

1 Institute of Geography, University of Erlangen-Nuremberg, 91058 Erlangen, Germany
2 Bavarian State Ministry for Food, Agriculture and Forestry, 80539 Munich, Germany
* Correspondence: karola.schober@stmelf.bayern.de

Abstract: *Prosciutto di Parma, Bavarian beer* and *Roquefort*—Geographical Indications (GIs) have been systematically protected at the EU level for 30 years and are now an important part of the *farm2fork strategy*. The article analyses how the integration of the EU GI system can be explained from an institution and discourse theoretical perspective and ties in with the soft spaces debate. In doing so, scalar shifts in competence from a German perspective and the role of discursive spatial relations are examined in more detail. The empirical results are based on a mix of methods that includes the evaluation of secondary statistical data, document analyses, participant observation and expert interviews. The study shows that the European Commission (EC) is increasingly acting as a spokesperson for GIs, but that regional actors are also playing a more important role in implementation and enforcement. This development is fed by the influence of the agricultural policy instrument in terms of competition, but also consumer protection and trade policy. Overall, there are three development layers: protect and systematise, legitimise and expand and open and defend. A more independent development of the EU GI system as an instrument of quality policy and for the development of rural areas could give greater weight to the sustainability-relevant, environmental policy aspects currently demanded by society.

Keywords: Geographical Indications; regional products; regional food; CAP; European integration; institutional spillover; soft spaces

Citation: Schober, K.; Balling, R.; Chilla, T.; Lindermayer, H. European Integration Processes in the EU GI System—A Long-Term Review of EU Regulation for GIs. *Sustainability* **2023**, *15*, 2666. https://doi.org/10.3390/su15032666

Academic Editors: Mariarosaria Lombardi, Erica Varese and Vera Amicarelli

Received: 27 November 2022
Revised: 26 January 2023
Accepted: 29 January 2023
Published: 1 February 2023

1. Introduction: Origin Labelling of Food in Europe

Geographical Indications (GIs) have been systematically protected at the EU level for 30 years and are now an important part of the farm2fork strategy. The three decades of policy formulation and implementation have to be understood as an incremental process. A high number of actors, institutions and interests have been interacting across the EU member states and throughout the changing periods of political priorities.

Our paper analyses this process and intends to explain the development layers. This obviously involves questions of mandate and power [1–6]: who is allowed to determine the "rules of the game"? What kind of resources are made available? What can shifting competences to one level mean for others? These questions are of particular complexity as the involved EU member states are characterised by their different food traditions and political cultures. The EU multi-level governance system is a challenging policy area [7–9], and in the GI case, it involves a series of sectoral policies such as the Common Agricultural Policy (CAP), trade and economic policy (in particular the European Single Market (ESM). We approach the complexity of these longstanding processes by means of a mixed methods approach, based on the Reflexive Grounded Theory [10].

Moreover, the protection regime as such comes along with complex spatial implications. For each product, the concrete perimeter has to be defined and also the specific ingredients or production methods that are bound to these areas. We understand the involved processes of geographical regulations as regionalisation processes [11]. This means the political negations processes of scales and perimeters come along with a certain "hardening" of spaces and their roles. Our analysis aims to explain these processes of geographical definitions, mainly based on the "soft spaces" approach [1–3,6,12].

The article reconstructs and analyses interdependencies of effects, asks about the use of discursive spatial references and sees European integration processes as being highly shapeable by regional and national actors. The article analyses the extent to which regional and national influences affect (de)stabilisation processes at the European level. In doing so, this contribution closes a gap in Europeanisation studies, especially with respect to GIs.

The structure of this paper is as follows: Section 2 summarises the state of academic debate regarding European Policy and national practices on origin labelling in a broader sense and the present understanding of processes of European integration. Section 3 clarifies the research methodology of this paper. Thereafter, in Section 4, the results are reported. A discussion regarding the research findings, its limits and future avenues is given in Section 5. Finally, Section 6 concludes this study by showing the limits of the work and a brief outlook pointing to the potential for continued independent development of the EU GI system.

2. Literature Review

The EU GI system is addressed by various scientific disciplines even if the agricultural economic perspective is of particular prominence here. This is also because the EU GI system goes back to French wine law and involved the idea of the terroir [13–17].

Not only in France but also in other European, mainly Romance countries, national protection systems with a focus on origin developed [16]. Increasing world trade questioned the efficiency of national protection systems, as many imitations appeared outside the national territory–which is why a European protection system was created in 1992 [18] (p. 1).

In Germany, GI research started in the early 2000s from a mainly agricultural economic view [19–22].

Today the economic benefit analysis of GI protection is more interdisciplinary [23,24], and also more strongly related to its collective character [21,25], trade with third countries [26–29] and its innovation potential [30]. Additionally, the GI system is discussed scientifically in the context of rural development, location development or spatial and regional development, or in terms of regional value chains [25,31–33], mostly case study-based [34–36]. The (agricultural) economic perspective mainly examined consumers' willingness to pay, price premium, market size and rural development (according to a meta-study [37]). More generally, it can be stated that information on origin leads to a purchase or additional price argument for consumers. This has been confirmed many times by country-of-origin research (first [38], more recently, e.g., [39]). With smaller-scale references, price effects in the area of origin protection have also been proven on the basis of case studies [37,40–42]. In recent years the field of social and cultural science has focused on changeability and its impact on governance [43–49]. Recently, Belletti and Marescotti (2021) have published guidelines for context-sensitive evaluation of GIs [50].

The discourse on GIs is accompanied by a legal debate [17,51–53] that reveals questions of demarcation and discusses the edges of the legal framework as well as illustrating the changeability of *hard* institutions.

At the European level and in the EU member states, various options for origin labelling have developed. Conceptually, these are distinguished between *simple, combined* and *qualified* types of origin-related designations (see Table 1) [14].

Table 1. Types of origin-related designations and their association with quality.

Types of Origin-Related Designations	Simple	Combined	Qualified
The Link between Origin and Quality	Origin	Origin Quality	Origin Quality
Examples	"Made in France", "Pork and beef completely from Hesse", "Packed in 80801 Munich", German "Regionalfenster"	Label Rouge (France), Bayerisches Bio-Siegel (Germany), Red Tractor (UK)	Prosciutto di Parma PDO (Italy), Bavarian Beer PGI (Germany)

Source: [14] p. 2ff, adapted and supplemented.

As Table 1 shows, in the case of simple origin-related designations, the message of origin stands on its own without a quality statement, but it can be ethnocentrically charged and thus interpreted as *better because of origin*. If the quality is functionally justified, it is a *qualified* indication of origin, e.g., in the sense of the EU GI system. In the case of combined origin-related designations, by definition, there does not have to be a functional link between quality and origin (typicity). In the case of combined origin-related designations, the quality is rather to be seen as an additional purchase argument that stands next to the message of origin [14]. For reasons of state aid law, the quality must be in the foreground; the origin is secondary and in principle can be combined flexibly.

Origin-related designations can also be categorised according to whether they are obligatory or optional from the producer's point of view and whether they were initiated by the state or privately (Table 2).

Table 2. Examples of origin labelling: classification proposal.

	Obligatory	Optional
Public Law	Special regulations for meat according to Regulation (EU) No 1169/2011, national regulations for certain products (e.g., for milk in Italy and France)	GIs according to Regulation (EU) No 1151/2012, EU-notified national quality (and origin labelling) systems
Private Law	none	Regionalfenster (in Germany)

Source: own representation.

In principle, origin labelling is voluntary for companies in the EU. However, the regulation regarding food information for consumers (EU) No 1169/2011 contains a whole series of obligatory special regulations, the main aim of which is to guarantee food safety (e.g., in the sensitive beef sector; keyword: BSE) [54]. In addition, within the framework of the marketing standards (Regulation (EU) No 589/2008 [55]), the indication of origin by means of a code is obligatory for eggs (e.g., 1-DE-1234567 = free-range farming method, country of origin Germany, identity of the laying farm with house number). The name, company name or address of the producer is not considered an indication of origin but is mandatory (Regulation (EU) No 1169/2011 Art. 2(2)(g)). These obligatory regulations are driven by consumer policy; they are therefore intended to promote traceability and thus food safety. The *farm2fork strategy* was also born out of a similar idea; approaches can already be found in the beginnings of the so-called *General Food Law* (Regulation (EC) No 178/2002 [56]).

Regardless of whether the indication of origin is voluntary or obligatory, the following applies: the consumer must not be misled (Regulation (EU) No 1169/2011 Art. 7(1)(a)). As soon as the product suggests a certain origin (and this impression could also be created by the use of symbols typical for the country such as the national flag, the national colours and famous buildings), the actual origin must be indicated, so to speak, as a clarifying or de-localising indication (Art. 26(2)(a)).

In contrast, (national) state activities aimed at highlighting a specific origin are in principle critically discussed at the European level due to their internal market-distorting character (TFEU Art. 28, Art. 30, Art. 107(1)). France was able to introduce national mandatory origin labelling for milk in 2017 as part of a two-year test phase, which was extended unilaterally, followed by seven other countries (including Italy, Portugal and Spain) [57,58]. However, the European Court of Justice (ECJ) clarified on 1 October 2020 (Case C-485/18) that when introducing mandatory origin labelling, there must also be a demonstrable link between quality and origin; in France's specific case, the national origin labelling requirement was therefore anti-competitive. Nevertheless, France has enforced mandatory origin labelling for certain meat dishes in community catering in 2022 and Sweden has also launched a similar initiative [59,60]. This also indicates that the discussion on whether origin labelling should be regulated nationally or at the European level has recently gained momentum again.

State programmes related to origin fall into the area of voluntary measures. They also provide the opportunity for state support for certification, controls and consumer communication. However, as state aid for certain locally bound producers, these are considered to have a market-distorting effect under European law (TFEU Art. 28, Art. 30, Art. 107(1)). These state programmes are only accepted by the EC if there is an explanation under state aid law for the corresponding market intervention. The limits are currently set by the Guidelines for State Aid in the Agriculture and Forestry Sector and in Rural Areas 2014–2020 (2014/C 204/01) and the Agricultural Block Exemption Regulation (Regulation (EU) No 702/2014 [61]).

In addition, the origin may only be secondary to the quality message. This means that, in EU logic, state programmes of origin are considered, rather, as quality programmes, i.e., quality marks or seals with a subordinate message of origin in the labelling sense. They are supported or promoted by the state and are offered to companies as voluntary labelling options (e.g., Label Rouge (France), AMA-Qualitätszeichen (Austria), Bayerisches Bio-Siegel (Germany)). In Germany, in addition to other regional quality programmes, there is the Regionalfenster (regional window), a state-initiated but privately funded system of origin labelling. This provides a standardised space to indicate optional indications of origin (without reference to quality)—a voluntary declaration field [62].

In contrast to all these options of state (quality and) origin labelling, which are generally viewed rather critically by European legislation, the promotion of GIs (PGI/PDO) is recognised and worthy of support at the European level: European resources, primarily from the agricultural budget, but also from the Structural Funds, flow into financial instruments (such as certification, sales and regional promotion). This tension between competitive and primarily agricultural policy goals raises the question of integration processes in the GI system.

This article shows how the GI system with its functional terroir logic, originally a French-influenced policy upload, was integrated at the EU level and has since been further developed and strengthened vis-à-vis other national systems as well as international trademark law [63] (p. 151ff).

The EU GI system is part of overarching Europeanisation and European integration processes [64]. Basically, Europeanisation is seen as the successful adoption of European values at the sub-European level, which is expressed above all in the implementation of European regulations and programmes in the member states [8] (p. 38ff). Europeanisation is thus the consequence of European integration at the sub-European level [8] (p. 38). European integration is the strengthening of institutional units at the European level [8] (p. 38ff)—in principle independent of their democratic legitimacy. A large number of authors deal with integration dynamics (e.g., [65–69], but the opposing concept of disintegration has also been studied (e.g., [70,71]).

These interwoven processes of European integration and Europeanisation can also be seen as a necessary product (and perhaps a historical intermediate outcome) of globalisation [72] (p. 9). The underlying assumption is that they happen, or better, are made, procedurally and revolve around the allocation of power, competences and resources [73] (p. 1).

For the analytical reconstruction of vertical processes of change, the article draws on institutional and discourse theory considerations. This is based on a broad understanding of institutions: i.e., in the development of European GI institutions, not only the organisational units (such as Directorates-General and associations) or the regulatory framework were considered, but also the practices of actors (e.g., activities in committees).

The institutional economic game metaphor (according to [74]) serves to describe and explain these processes. In an ideal-typical view, formal institutions can be seen as the rules of the game (e.g., the EU GI regulation), while the game moves as actions (e.g., issue-related cooperation) are to be assigned to informal institutions [75] (p. vii). In practice, the processes of rulemaking and rule application are interwoven. In the legal literature, the basic idea of the processual shaping of structures is also found, especially clearly in Georg Jellinek's (1900) idea of the "normative power of the factual" [76]. There, the changeability of rules on the basis of social reality is also emphasised and the enforcement mechanisms are illuminated (described in more detail in the anthology by [77]).

The sociological institutional theory further suggests focusing on four allocation problems: "competences", "resources", "legitimacy" and "control" [5] (p. 61ff). This model, originating from the 1990s, is accentuated for the present analysis of the EU GI system in the direction of process orientation and emphasis on interaction possibilities in order to emphasise relationalities more strongly. "Interaction" is cited as an additional allocation category because the ability to develop and use networks is seen in innovation sociology as a key element for institutional change processes [78,79].

The basic assumption of this multidimensional analysis model is that institutions developed in a particular space have an effect on that space's organisational stability. Stabilisation processes can even lead to the sphere of influence of this spatial unit expanding beyond its administrative boundaries in individual dimensions (e.g., enforcement of European standards within the framework of free trade agreements).

In this ideal typical view, the legitimisation of these allocation processes by local people is seen as the foundation for these same processes in the *Europeanisation game*, as institutions are seen as the product of social negotiation processes [5] (p. 62).

"Discourse coalitions" [80] (p. 217) are characterised by the fact that a group of actors uses a particular set of narrative strands during a certain period of time. However, the approach does not assume that the supporting actors of discourse coalitions necessarily have to agree to do so [80] (p. 12) or act intentionally (cf. [81] (p. 14)), but rather that it is sufficient that a common work of meaning is carried out in the public debate and shared narrative patterns are claimed [80] (p. 12).

The concept of path dependencies (understood as developmental important pre-structuring, [82]) is suitable for examining significant institutional changes over time. In the analysis of GI processes, it serves to explore the context of action more closely, while avoiding the danger of limiting oneself to trivial "history matters" [82]. The concept of path dependency, first used by the economic mathematician William Brian ARTHUR (1989), has also been used in recent decades to explain processes of social transformation [83,84]—a debate in which the EU GI system has been discussed more recently [85,86].

In this work, the institutional theory approaches find a connection to the geographical debate through the "soft spaces" approach [1–3,6,12]. The idea of examining spatial changes through institutional distribution mechanisms is based, for example, on rescaling studies from spatial and environmental planning [9,87].

Following the (post-)structuralist understanding of space or scale, such vertical processes of change are also socially constructed and changeable over time, become relevant only through their relations and are socially and politically controversial [88].

Soft spaces thus exist along and alongside territorially ordered spaces and scales ("hard spaces" [6]) and are to be understood as a "shorthand for the wide range of governance bodies and strategy-making processes and implementation practices" [6] (p. xxi). The concept is deliberately relatively broad in order to maintain the explorative potential of this "increasingly important research area" [6]. The increasing importance is also due to the fact that throughout Europe "new, non-statutory or informal spaces [...] can be found in a variety of circumstances and with diverse aims and rationales" [6].

Allmendinger et al., however, see these new spaces as legitimised less through elected representatives and more through a heterogeneous group of political, social and market actors [6] (p. 4). Cohen and McCarthy (2015) also emphasise that decisions are often shifted to a level of scale that is only indirectly legitimised democratically and even see such rescaling processes as an expression of neoliberal restructuring [89].

In their study of the Baltic Sea macroregion, Metzger and Schmitt (2012) identify as an important developmental step the fact that the EC was able to act "as a designated spokesperson for the interests of the wider Baltic Sea Region" [4] (p. 265). They see the establishment of a spokesperson role as an indication of the stabilisation of a spatial unit. They also emphasise the high importance of intersubjective recognition of this quasi-imagined order (similar to [90] (p. 58)).

Similarly, the fading-in and fading-out of spatial categories in political discourse can play a decisive role in the political process [6]. Thus, political processes can get by without spatial references in some phases ("spatial tabooing" [6] (p. 2713). Often, it is precisely this fading out of spatial references at the European level that is a discursive enforcement strategy at a certain point in time. At a later point in time, spatial references become the subject of discussion again, especially when spatial conflicts of use arise ("reterritorialization" [6]).

3. Data and Methods

The data originating from the underlying research project were collected and analysed mainly qualitatively and by applying the "Reflexive Grounded Theory" [10]. As is typical for such a research design, the research process was not considered in a linear way, but conceptual considerations and findings from the empirical data collection were put into a reciprocal relationship.

This recursive approach, which actively includes the researcher's point of view, is a hermeneutic principle and should ensure quality [10]. The inclusion of the researcher's point of view also results in the partial focus on the German perspective, e.g., when it comes to explaining scalar shifts. Different methods were thus combined with the aim of complementary compensation [91,92]. The resulting mix of methods is as follows:

1. Analysis of secondary statistical data, primarily eAmbrosia [93].
2. Broad document analysis (according to Flick 2012 [92]), especially legal bases and legal evaluations.
3. Participant observation (according to Mayring 2002 [94], 26 events and expert discussions) and research diary (FTB; according to Breuer et al. 2017 [10]). The events took place in the national context of Germany (15 events) or with a stronger European perspective in Brussels (11 events).
4. A total of 49 guided, anonymised expert interviews (according to Meuser/Nagel 1991 [95]) at European, national and regional levels with an average duration of 70 min. The interviewees were persons who have been responsible for the conception or implementation of national origin labelling systems in a member state or region or who have been involved in the integration process of the EU GI system (lobbyists, GI experts, (former) representatives of the EC). The interview guide used was increasingly fine-tuned in the selection process of the study, but also modified with regard to the respective person to be interviewed and their scalar and thematic location.

The empirical data were collected between January 2019 and October 2020. Over the equivalent of around two months (29 April to 17 May 2019, 21 October to 25 October 2019 and 27 January to 28 February 2020), many expert interviews and participant observations took place in Brussels. The intensive study of the legal matter took place mainly in the first half of 2019. All available documents that contained primary or secondary collected data were analysed using the qualitative data and text analysis software MaxQDA. Rule-based techniques and procedures were used to find, construct and elaborate analysis steps during the research process [10].

The corpus was considered relatively complete or meaningful if the data collections on which the results were based did not suggest any new thematic phenomena [96] (p. 130). Basically, the aim was to elaborate a specific vocabulary for reconstructing and making transparent the focused field of action "Institutionalisation Processes and Spatialities in the EU GI System". Such an inductive–deductive approach is not new and can be described as "interpretative-structuralist-pragmatic analysis" [97] or "methodological hybrid" [97]. The more recent developments, i.e., those of the last two years, are also discussed against this background.

Overall, the result of this procedure can best be described as an "object-based theory" [94], i.e., as a "middle-range theory" [95], in other words, its explanatory power is limited to the object of study. For details of this study see [63].

4. Results: Stabilisation of the EU GI System over Time

The stabilisation processes of the GI system can be divided into three development phases that lead to overlapping "layers", namely (1) protect and systematise, (2) legitimise and expand and (3) open and defend. These phases are shown in Figure 1 next to the increasing number of GIs from EU Member States (by product group).

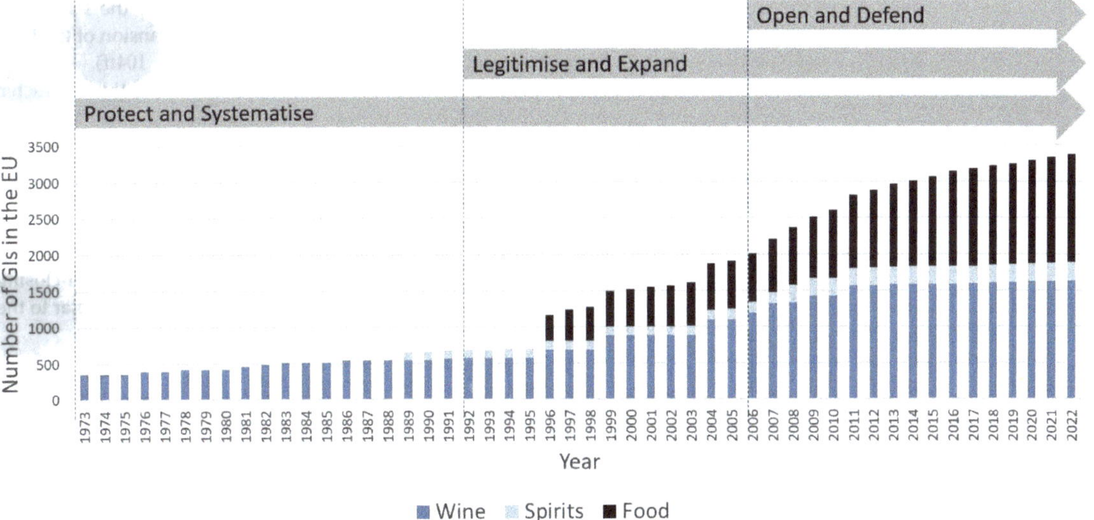

Figure 1. Phases in the development of EU GIs into Rural Intellectual Property Rights (IPRs). Source: own presentation. Data source: eAmbrosia (as of 28 December 2022).

Table S1 "Milestones in the development of GI rules for agricultural products and foodstuffs" shows the relevant elements of the development of the GI rules from their introduction (Regulation (EEC) No 2081/1992 [98]) to the current amending regulation under CAP 2023 to 2027 (Regulation (EU) No 2021/2117 [99]) or proposal for the coming reform of the regulation.

4.1. Layer 1: Protect and Systematise

The focus on *protection and systematisation* is characterised by the institutional establishment of the possibility of protection for GIs. The adoption of Regulation (EEC) No 2081/1992 (entry into force: 24 July 1993) was the decisive step for a systematised protection. A GI system had already been discussed by the twelve member states of the European Economic Community (EEC) since the mid-1980s within the CAP framework. In 1992, it was then introduced for the first time with its own regulation for the relatively broad field of agricultural products and foodstuffs.

Various path dependencies played a role in the enforcement of the first GI Regulation, such as:

- The traditionally relatively strong regulation of agricultural policy
- Its stronger orientation towards quality instead of quantity since the 1980s
- The European goal of realising a free internal market
- The relatively strong role of the national states in the process ("regulatory committee procedure")
- The not yet developed but in a certain way competing trademark law
- The strengthening ECJ case law
- The existing rules on European level for wines, aromatised wine products and spirits

Especially the historically grown protection rights for wines (since 1970), aromatised wine products (since 1991) and spirits (since 1989) [100] (p. 351) played an important role in the introduction process. These old designation protection rights were originally national protection systems that were now recognised at the European level (expert interviews). Other arguments were made with regard to the increasing consumer demand for quality products as well as the increasing intra-European trade, which required the same standards. These framework conditions laid the foundation for the later expansion of the EC's role as spokesperson for regional products. For more details, see [63] (p. 104ff).

In the 1990s, the GI system was discussed, due to its market intervention character, on the one hand as a contradiction to the declared goal of a free European single market (a rather market liberal argumentation) and on the other hand precisely as a means for the functioning of such a single market (a rather Keynesian argumentation). In other words, the main issue was how to implement the European single market with as uniform standards as possible and how much regulation was needed to achieve this.

The relatively high scope of protection [21] (p. 1031) and the relatively strong enforcement mechanisms (private sector as well as state) make the EU GI system a market-closing mechanism. However, this was contrasted with more competitive elements familiar to the trademark protection process:

- Creation of product specifications
- Registration procedure
- Existence of a public register

It is undisputed that during the introduction process, the member states were oriented towards the so-called *Romanic* concept of origin protection [22,53,100] meaning basically the protection of qualified type of origin-related designations (terroir concept). With Regulation (EEC) No 2081/1992, however, two concepts were established, which can be summarised as a narrow (PDO) and a broad (PGI) terroir concept [63] (pp. 88ff, 127ff).

The narrow, French PDO understanding is based more on the assumption of an objective spatial structural connection, while the broad, German PGI understanding is based more on the recognition of (inter-)subjective traceability (consumer reputation; [97]). The original proposal for a regulation at the end of 1991/beginning of 1992 envisaged one category—namely PDO.

Consequently, many German designations would not have been eligible for protection at the community level [97], [101] (p. 4). The existing protection of German designations abroad would also have been uncertain (ibid.). Therefore, Germany rejected this proposal and insisted on a regulation that was also open to German designations [102]

(p. 194). In this context, it had been important to convince other delegations—especially France—of the German position [102]. Thus, the PGI concept was implemented, which also functioned for the designations recognised under German law and was rather oriented towards competition policy [102].

In addition to the introduction of the PGI concept, Germany's national representatives also managed to successfully introduce their interests in four other areas: procedures regarding recognition and control, the relationship with trademark protection, the relationship with national protection systems, as well as start-up aids [63] (p. 110ff).

However, the PDO concept, or the underlying assumption of a causal link between a certain geographical area and product quality, was not free of criticism either [103] (p. 43), [104].

Together, the PDO and PGI concepts form a Europeanised understanding of terroir, which reflects the view of the EEC member states at the time with their different legal systems ([63], p. 117ff). With the first GI regulation, it was no longer exclusively recognised that a spatial structural link to a certain food quality existed in an objectifiable way (PDO), but it was sufficient that it was recognised intersubjectively (PGI).

This last point was developed in subsequent years. In the ECJ ruling of 10 November 1992 (Case C-3/91, Turrón de Alicante ruling), the PGI concept was strengthened [15] (p. 67). The stronger position of the consumer is also reflected in the fact that in the case of PGIs the *general business practice*, which in the past was based on manufacturer and expert assessments, is continuously being replaced by the consumer's perception (reputation; recognition in the market; for more details see [63] (pp. 147ff, 171f)).

One point that seems to be in need of legal clarification is the question of geographical references in consumer/trade perception. Specifically, it is a question of whether the country of origin or the country of destination is decisive in the claim for injunctive relief (similarly also [105] (p. 590)). For example, the German Federal Supreme Court ruled on 12 December 2019 in the *Culatello di Parma* case that it was not the perception in the country of origin of the product (Italy) that mattered, but the perception of the German consumer as part of *the European consumer* [105] (p. 589). In this case, the (consumer) perception in the country of destination was decisive.

According to experts, however, there is a tendency in borderline cases to be guided by the consumer's view of the country of origin of the product.

4.2. *Layer 2: Legitimise and Expand*

After the many attempts to exert political influence during the introduction process (see also [15] (p. 85)), the adopted regulation needed legal classification and clarification. The ECJ played an important role here. It stabilised the system as a whole and thus strengthened it vis-à-vis trademark law (see [106] (p. 292)). Essentially, the ECJ case law of the last decades has had the following effects [63] (p. 151f):

- Clarification of the importance of the EU GI system as an independently introduced area in EU law
- Clarification and broad interpretation of the introduced rules on violations of protection with regard to references and other misleading practices
- Extension of less defined, possibly generic food names
- Further definition of (quality-influencing) production steps based on existing practice

Elements of competition policy were already established with the introduction of the first GI regulation in 1992, but the goal of safeguarding intellectual property rights was only explicitly addressed for the first time in Regulation (EU) No 1151/2012, Recital 19 [107]. This is not surprising, as the topic of intellectual property as a whole only gained momentum late: the basis for the *Common IPR Strategy* was laid for the first time in 2000 in Article 17(2) of the Charter of Fundamental Rights of the European Union.

The fact that GIs, which were originally conceived as designation rights, are being developed more and more in the direction of a brand or trademark is illustrated, for example, by the gradual obligatory use of the logo on the product (see also [108] (p. 211)) and, more recently, on advertising material (Regulation (EU) No 2021/2117 Art. 2 [99]).

How agricultural, trade or competition policies are affected can also be seen from the EU institutions involved in the narrower sense: the Directorate General for the Internal Market, Industry, Entrepreneurship and SMEs (DG GROW) is responsible for the enforcement of intellectual property rights. Agricultural designations are an exception due to their history, so DG AGRI has exclusive competence for GIs at the EU level. In the area of enforcement, however, the two DGs have recently been cooperating more intensively, especially in the development of the *GIview* database, resulting in the integration of GIs into the EUIPO trademark register in a separate section [63] (p. 174ff). This illustrates the influence of trademark law and competition policy.

While experts nowadays repeatedly emphasise the benefits of the relatively strong, trademark-like enforcement mechanism, some also stress the importance of the special role of GIs in the IPR world. They describe GI protection as a "Rural Intellectual Property Right", "protection of farmers' and producers' IPR" and "a specific collective right assigned to a region". Overall, it can be stated that from today's perspective, an agricultural policy framework for the protection of certain origin-related designations was created by the nation-states in the 1990s and established with strong competition policy enforcement mechanisms and this was expanded and manifested by the EC in the following decades.

Already during the introduction process, consumer interests were argued for. This was mainly about satisfying the demand (for regional or traditional products) and providing guidance (Regulation (EEC) No 2081/1992, recitals [98]). Overall, the demand argument is seen in a more differentiated way today:

Regulation (EU) No 1151/2012 mentions further socio-political aspects such as the preservation of cultural heritage (Recital 1) as well as resource and animal protection (Recital 23). With the last consolidation of the Food Regulation (amending Regulation (EU) No 2021/2117), the topic of "sustainability, including the economic, social, environmental and climate sustainability" (Recital 1) was introduced due to the overarching CAP objectives [99]. Consequently, products can now also be protected on the basis of their potential contribution to sustainable development (Regulation (EU) No 1151/2012 Art. 1(2)(b)) [107].

Consumer protection arguments are also used at the enforcement level. For example, GIs as state-certified and controlled products at the production level are also part of the *farm2fork strategy* [109] (p. 12). The *farm2fork strategy* originating from the Directorate General for Health and Food Safety (DG SANTE) aims to make food systems fair, healthy and environmentally friendly [109]. In this context, information on origin is also given importance in the sense of consumer clarity with respect to market control [109] (p. 13). Ensuring traceability and avoiding food fraud can also be seen as "typical value chain thinking" (expert interview). The relevant legal foundations are the Food Information Regulation (Regulation (EU) No 1169/2011 [54]) and the Official Controls Regulation (Regulation (EU) No 2017/625 [110]).

For a long time, food control was primarily concerned with the prevention of health risks, but in recent years consumer deception regarding origin has been seen much more strongly as a separate offence of food crime (*food fraud*; expert interview, [111]). For this reason, national reference centres have been set up within the EU and annual control audits have been carried out within the framework of consumer protection. Furthermore, new European fields of action include the introduction of the so-called *GI ID Cards* to improve market controls and the establishment of the cross-product area database *GIview* [63] (p. 174ff).

Furthermore, the EC promotes advertising campaigns for GI products within the framework of EU promotion policy under Regulation (EU) No 1144/2014 [112]. EU promotion policy traditionally has an export focus, but also raises awareness of the system within the EU. It does not focus exclusively, but specifically on GIs (expert interviews). In 2005, more than 30% of the EU promotion policy budget went to food GIs (excluding wine and spirits; [113] (p. 189)). Overall, the funds for EU promotion policy have been greatly expanded in the last 15 years or so, although they have been declining over the last three years (from EUR 27.6 million in 2006 to EUR 185.9 million in 2023) [109,114].

Highlighting the so-called *Union message* is seen as the central element of EU co-financed campaigns (FTB, participant observation). This is also where the obligatory logo Enjoy! It's from Europe! (Figure 2) was introduced.

Figure 2. Example of an EU co-financed promotion policy programme on Instagram using the logo Enjoy! It's from Europe! Source: Instagram, European Sustainable Wines.

The compulsory use of the EU logo for co-financed promotion activities by the producer associations, also within the EU, illustrates an increasing role of the sender or spokesperson of the EU for regional culinary matters.

Furthermore, the present analysis of GI rules concerning the scope of application shows an expansion in four dimensions (see also [15] (p. 94ff):

(a) Extension of CAP competence:

The object of Regulation (EU) No 1151/2012 refers—as a CAP policy instrument—in principle only to agricultural products (so-called *Annex I products* (TFEU)) [107]. However, Heine already noted that Regulation (EEC) No 2081/1992 also covered foodstuffs, which the EEC Treaty did not define as an agricultural policy competence [115] (p. 96f). The decisive factor were the economic significance and the inclusion of special cases of the Member States [104,115,116].

(b) Extension to include product areas without existing GI rules of their own:

The product groups that can be found in the annex to the GI regulation have been continuously expanded. The comparative analysis of the reformed regulations shows that only one product group was deleted (natural mineral and spring water) and 16 product groups (e.g., pasta, salt and chocolate) were included (own analysis).

(c) Addition of product groups with existing, own GI rules:

Since around the mid-2000s, harmonisation processes have been increasing in regulatory areas of products that traditionally have their own rules of origin protection (aromatised wine products, wine, spirits). In a way, this can be seen as a broadening of the scope. When the experts talk about *harmonisation of the GI regulations*, they are mostly talking about procedural alignment with Regulation (EU) No 1151/2012, in particular concerning product specifications, the registration procedure and the public register (see Supplementary). While there is a general tendency towards alignment with the Food Regulation (EU) No 1151/2012, in some cases sectoral regulations are also taken into account: For example, the reform proposals of the EC with regard to the topic of "goods in transit" are oriented towards regulations for spirits (Regulation (EU) No 2019/787 [117]). Harmonisation processes became very clear in the recent complete integration of aromatised wine products into Regulation (EU) No 1151/2012. DG AGRI plans to further harmonise and simplify the GI system across sectors (expert interviews). Since 2019, a uniform GI regulation has also been under discussion (expert interviews). In March 2022, the Commission presented a first draft of a regulation that merges the regulations previously spread across several legal bases [118]. This has since been updated in May [119].

(d) Non-restriction through vague definitions and the avoidance of definitions:

Sometimes the avoidance of definitions or the preservation of interpretative leeway is also conducive to the expansion of competence. For example, this is the case when dealing with so-called generic terms that cannot be protected [22,51,115,120] and deletions, both of which potentially devalue a term forever (expert interviews, [15]).

Taking a closer look at the harmonisation processes, the cross-sectoral (agricultural products and food, wines and spirits) comparative analysis of the GI rules showed that spatially relevant categories were hardly harmonised (see Table 3).

Table 3 shows that the protection concept is in principle based on the functional argument of typicality in all three product areas. However, while for food and agricultural products as well as wine very similar PDO and PGI concepts have been introduced, for spirits the more loosely justified link based on a broader GI concept (TRIPS) applies. However, the fact that the broader spirit GI concept and the PGI concept are to be interpreted in a relatively similar way is indicated by the possibility of labelling protected spirits with the PGI logo as well.

When considering the possible territorial scope, it becomes clear that in the case of foodstuffs and wines, this can only extend to entire states in exceptional cases—and in fact, this almost never happens. In the case of spirits, the reference to states is not only traditionally possible, but also common practice. However, the typicality of the product, which refers, for example, to the special soil and climate conditions, is more difficult to prove in the case of larger areas such as nation-states with potentially heterogeneous characteristics (expert interviews).

The comparative analysis with regard to regional raw material binding shows that different rules exist. A relevant point seems to be, in the case of PGIs a raw material linkage must be justified by its specificity (expert interviews, [119]).

With regard to the geographical definition of production stages or steps of further processing, there are also different formulations in the three regulations. The geographical limitation of the value chain must be factually justified with specificity. This becomes clear at the product or regional level in the application or later court proceedings. For example, the Schutzverband der Nürnberger Bratwurst e. V. (Nuremberg Bratwurst Protection Association) successfully argued that packaging at the place of production contributes to maintaining quality, while in the wine sector, it is discussed under which conditions bottling is a quality-determining aspect (expert interviews).

Table 3. Spatial references in three * basic GI regulations.

	Agricultural Products and Foodstuffs	Wine	Spirits
Regulation	(EU) No 1151/2012	(EU) No 1308/2013	(EU) No 2019/787
Protection schemes	PDO, PGI	PDO, PGI	GI
Justification of the link to quality and origin	PDO: quality or characteristics "essentially" or "exclusively" attributable to "geographical environment" PGI: quality, reputation or other characteristic is "essentially attributable to its geographical origin" (Art. 5)	PDO: quality or characteristics "essentially" or "exclusively" attributable to "geographical environment" PGI: specific quality, reputation or other characteristics attributable to that geographical origin (Art. 93)	"quality, reputation or other characteristic of that spirit drink is essentially attributable to its geographical origin" (Art. 3)
Territorial Scope	PDO: place, region, exception: country PGI: place, region, country (Art. 5)	PDO/PGI: region, specific place, exception: country (Art. 93)	Locality, region or country (Art. 3)
Regional raw material binding	PDO: 100 % ** PGI.: not explicitly regulated ***	PDO: 100 % PGI: 85 % of grapes (Art. 93)	not explicitly regulated
Regional processing	PDO: all production steps PGI: "at least one of the production steps" (Art. 5)	PDO: "all the operations involved, from the harvesting of the grapes to the completion of the wine-making processes" PGI: "its production takes place in that geographical area" (Art. 93)	"the production steps which give the spirit drink the quality, reputation or other characteristic that is essentially attributable to its geographical origin, take place in the relevant geographical area" (Art. 35)

* Due to the integration of the regulation for aromatised wine products and its relatively small scope of application, this was excluded from the analysis [107,117,121]. ** For animal products incl. feed (if technically not possible: up to 50 % from outside; Regulation (EU) No 664/2014 [122]). *** Can result from the justification of typicality (reputation). Source: own representation

4.3. Layer 3: Open and Defend

Overall, the reforms of the regulations also show the increasing influence of trade policy. Regulation (EEC) No 2081/1992 states in the recitals that "provision should be made for trade with third countries" [98], but HEINE (1993) already stated that third countries had to virtually adopt the community system [115] (p. 103). As a consequence, there was a WTO dispute settlement procedure, which led to a greater opening of the system through Regulation (EC) No 510/2006 (USA/Canada v. EU; expert interviews), [120,123]. This second reform thus opened the third phase, which is characterised by system opening and defence.

At the system level, GI law is opposed to trademark law. The American GI critic K. William WATSON even describes these processes as a "battle" that takes place on different "fronts" [124] (p. 8). On the one hand, the parties are concerned with defending their own territory, but on the other hand, they are also concerned with *exporting* their own protection concept. However, this is not only about the directly negotiating partners, but rather about displacing other parties in certain markets or securing export markets (so also [124]).

The overall trade policy influence on the EU GI system can be clearly seen through its international expansion and accompanying opening and defence processes, which are described in more detail in the following paragraphs. As summarised by an EC representative at an event, the internationalisation of GIs has three objectives: the defence of intellectual property, the idea of creating value locally as well as protecting it globally and helping producers to better reach international markets (participant observation).

Overall, the EC's spokesperson role vis-à-vis third countries is particularly evident in five fields of action: FTAs and special GI agreements, multilateral agreements, development cooperation, system enforcement and export promotion.

The EU has 34 agreements concluded or in force with 47 states, from which 1588 designations have been protected within the EU, and 16 ongoing negotiations (as of 13 May 2019; [125], participant observation). In these negotiations, EC representatives act as spokespersons for EU GIs. The aim is to "bring GI law to the world" (expert interview).

EC members repeatedly emphasise the sensitivity of the issue in free trade agreements, describing it, for example, as a "deal breaker", a trigger for "dramas"—although GIs generally only make up a "very small part in very comprehensive free trade agreements". Because of this sensitivity, negotiations are often pushed into the final phase, where several points of contention end up. Thus, GIs sometimes become "bargaining chips" that are used to achieve other goals, "completely different, unrelated topics" such as export quotas. In order not to fall victim to such packaged solutions, agreements that deal with the issue separately, as in the case of China, would be preferable. This would make the issue workable, GI experts would negotiate and the larger free trade agreement would be relieved (expert interviews).

During recent years, the EU has gone through a learning process in negotiations (expert interview, FTB). The strategy of the EC staff seems to be successful: due to the free trade agreements with Canada, Japan and Singapore, a GI system has been established there (expert interview, participant observation).

When protecting designations in the context of negotiations with third countries, the question of orientation towards the place of origin or destination—the geographical reference—also plays a role. In addition to the product's reputation, which is particularly relevant in the European market, other aspects (e.g., language and writing system) can be decisive for the value of the protection (e.g., Japanese 神戸牛 for *Kobe beef*). Normally, in negotiations with third countries, the original names are protected in the national language. In the case of CETA, the protection communities of *Prosecco* and *Bavarian beer* have therefore opted for trademark protection (expert interviews).

The formerly most important agreement for GIs at the multilateral level, the so-called TRIPS Agreement (WTO has been stalled since 1995).

These days the Lisbon Agreement for the Protection of Appellations of Origin and their International Registration is seen as the more relevant multilateral agreement for European interests (expert interviews).

The negotiation process started in 2008 (expert interview) and was supplemented by the Geneva Act, which entered into force on 26 February 2020. It required five members to do so (in comparison TRIPS listed one hundred and sixty-four signatories as of 7 February 2020, [126], expert interviews). The EU's objectives during the negotiations were largely achieved and referred to

- The modernisation of the legal framework
- The admission of supranational confederations of states
- The extension of the scope of application to include PGIs (expert interviews)

On further consideration, by opening up or expanding the system, not only can producers of traditional foods from Europe better reach international markets, but also producers from regions in non-EU countries can better place their products on the European market. The latter was, after all, the impetus for the second reform of GI law in the sense of opening up the system. However, third-country products recognised by the EU within the framework of free trade or multilateral agreements cannot bear the EU logos (expert interview).

GI protection can certainly open up new market opportunities and thus have a positive impact on regional development, as the example of *Café de Valdesia PDO* shows [127]. Here, registration as a PDO contributed to strategic quality work, more intensive cooperation along the value chain, better visibility and image enhancement of the product, new investments in local structures and sustainable regional development [127].

With regard to development policy aspects, there is also GI cooperation with, for example, the African Union. Currently, on the initiative of the African Union, potentially protectable products are being identified together with local actors, the FAO and the EUIPO, and the basis for registration is being created (expert interview). Attempts are also being made to call on funding from the EU's development policy (expert interview).

A total of 217 products from third countries are currently listed in eAmbrosia (as of 8 November 2022), i.e., in addition to the designations recognised in political agreements. Moreover, the fact that 47 new applications are also in progress shows not only that direct registration can be interesting in the context of development policy, but that the full adoption of the EU regulatory system is useful for many an international producer group.

Furthermore, enforcement or control mechanisms concerning GIs have been strengthened at the EU level with competition and consumer protection policy instruments, as already mentioned. However, an interest in trade policy also reveals new European fields of action in the area of control. Thus, cross-border issues affecting trade with third countries, such as *goods in transit* and *abuse of names on the internet*, became more important. They have since been incorporated into the revised Regulation (EU) No 1151/2012 and illustrate the increasing institutionalisation of control-related fields of action at the EU level and thus the strengthening of protection.

5. Discussion

The basis of the EU's spokesperson role for GI was gradually laid through old protection rights and manifested with the introduction of the first GI regulation in 1992. Since the introduction of systematised GI rules at the EU level some 30 years ago, the system has become more institutionalised in all allocation dimensions considered (competences, control, resources, interaction and legitimacy). The role of the EC as a spokesperson for regional culinary matters has been expanded in and outside the EU, e.g., in free trade, consumer protection and competition policy. DG AGRI is the key actor because it has exclusive competence at the European level. Enabling conditions (path dependencies) and common interests across policy fields (discourse coalitions) were crucial for this development.

With regard to European integration, understood as an interplay of territorial and relational processes [6], it can be stated that it is discussed to which extent national or European competence exists for the regulation of origin-related designations in a broader sense, which means that there is probably "pooled-territoriality" [6] (p. 2705).

The results illustrate that the complex processes at the EU level are characterised above all by sectoral as well as multiscale influences. Moreover, although the EU GI system has never lost its agricultural policy focus, it has been increasingly influenced by other policy fields in recent decades and has developed above all in the direction of *IPR*.

The increase in the European level's scope for action is clearly evident in the area of enforcement and control. Here, on the one hand, competition policy institutions played a more important role; on the other hand, consumer policy instruments have also been expanded. They are intended to improve the strength of enforcement at the regional level, on the one hand by strengthening GI producer groups as rights holders, and on the other hand by strengthening (nationally) state-coordinated market control.

These central points raise in particular the questions of what role "spillover effects" (first [128]) played in the integration of the EU GI system, to what extent spatial references were used discursively and what potentials arise from strengthening the regional level or the functionally delimited speciality regions (soft spaces).

In its role as a regulator for consumer protection-related market controls, the EC—or more specifically DG SANTE—has also increased the enforcement power of the system in the intra-European market. DG AGRI has also increased enforcement in international trade (e.g., on the internet). These actions concerning controls arise from the discourse coalition of agricultural, consumer protection and competition policy interests and the recognition of EU GIs as *Rural IPR*. This seems to be a spillover effect, which basically describes the institutional radiating from one limited policy field into other fields.

The strengthening and harmonisation of the system within Europe since the mid-2000s can also be seen in connection with global integration processes and its assertion against trademark law. It is not clear to what extent competitive logics are due to impulses from global or intra-European trade. It seems logical, however, that the EC is better able to enter into international negotiations and conduct advertising campaigns with a unified, coherent system than with a sectorally fragmented one. These processes can also be seen as an expression of the attempt to speak with one voice. At the same time, these harmonisation processes are accompanied by the loss of special roles (in more detail [63] (p. 165f)). The moving context could affect the attractiveness of GIs and hence rural development [129].

The tendency towards "supra-territoriality" [6] (p. 2705) can also be seen in the increasing spokesperson activities of the EC for GIs in relation to the restrictions under EU state aid law for simple and combined origin-related designations. On the other hand, the role of the EC as a spokesperson for origin-specific culinary affairs also depends on how the possibilities of *national* origin labelling change. These are currently the subject of intense discussion. For example, Germany, Austria and Sweden are currently discussing mandatory origin labelling in certain areas (e.g., food retail, out-of-home catering) or for certain products (e.g., pork), while in France and Italy, this has already been implemented and expanded.

The comparative analysis of directly spatially relevant categories across the sectors of *agricultural products and food, spirits* and *wine* has shown that these have not yet been harmonised or, in other words, discursively excluded ("discursive deterritorialization" [8]). The harmonisation largely concerns the procedures. Spatial references (especially raw material binding and local consumer expectations) usually come back into the discussion at a later point in time, namely at the regional level or in concrete disputes ("discursive reterritorialisation" [8]). The classification in the quality programmes with Regulation (EU) No 1151/2012 can also be interpreted as a deterritorialisation process because the aspect of origin is pushed into the background.

In the political process, spatial references in the GI system are often still used today as a category of distinction (*Europe of North and South*), and in the national context of Germany, mainly as an argument for defending the system (*Romanic System; System of the Others*; for more details: [63] (p. 241ff)). However, Germany has already helped to shape the system under the mentioned five important points during the introduction process, including the PGI concept.

Another point where spatial references could play a decisive and also economically relevant role is that of consumer perception. However, it has not yet been legally clarified whether the perception of consumers in the country of origin or in the country of protection is decisive for protection in the case of a lawsuit [63,105].

Furthermore, speciality regions, meaning regional territorial demarcations for certain PDO/PGI/GI products, can also be operationalised as soft spaces with their European framework and their functional founding logic. The functional link between certain spatial conditions at a certain point in time and a specific product quality is the core of the terroir principle on which the idea of GI protection is based.

Firstly, this link is expressed in the consumers' spatial appreciation, which is the basis for protection, and which is shown in the market through the act of purchase or a willingness to pay more. Secondly, there are the functional arguments that have to be put forward by the producer groups in the context of the protection application and which relate, for example, to soil and climate conditions and to the space-specific reputation of the product among consumers.

What distinguishes speciality regions from other soft spaces such as macro-regions is that they are currently unlimited in time. They are also less informal spatial concepts [130] (p. 83), but rather strongly formalised through their legal definition.

However, like other soft spaces, speciality regions can be seen as "testing grounds" [6] (p. 4) and new cooperation spaces. The GI system is relevant as a sustainability innovation in that it can reduce the risk of the (global) market's focus on quantity and price in order to protect traditional foods. Spatial characteristics that are justified by traditional—or more recently also explicitly sustainable—production methods (e.g., preservation of diversity, environmental protection, animal welfare) are relevant to specification.

With the strengthening of the regional level (especially producer groups as rights holders, but also regional enforcement of market control in the sense of consumer protection), regional actors are also given more responsibility. At the level of the producer groups, there is still a need for clarification regarding the organisational forms and the necessity of recognition [63] (p. 172f), [131,131].

The example of Germany, however, clearly shows that where other systems are already established—which, unlike the GI system, are based, for example, on state-established quality criteria and administrative boundaries—the enforcement of the GI policy is more difficult than in other member states [63] (p. 258ff).

The support of governments and administrations can play a key role in the legitimisation process of the EU GI system in national and regional contexts [46,63,132,133], but there is also evidence that the timing of EU accession influences the chances of protection [119] and [63] (p. 157f). GIs can be an instrument for regional agricultural policy to promote a quality strategy, alongside national quality programmes and the promotion of regional collective and certification marks [63,134].

6. Conclusions

Looking at the limitations of the study, it can be stated that since the study refers to European integration processes and therefore focused on intra-European experts and sources, no conclusive statements can be made about the influence of international or multilateral institutions.

Furthermore, a more political science perspective on the described spillover effect could provide further insights for the integration debate.

Moreover, in order to explore the actual connectivity to structures (e.g., legal frameworks, associations) and interests in implementation processes, a comparison of regional practices based on themes would be valuable in the context of border studies, where the national context has recently been seen more as an explanatory factor for governance structures [135]. The need for further development of GI-specific measurement methods with the help of greater consideration of the respective context has also been stated from a more legal perspective [132] (p. 56ff). The "simplified notion of reputation" [132] (p. 59) oriented towards trademark law, which is based on sales figures or consumer surveys at a certain point in time, does not do justice to the required continuity or the "ongoing vitality" [132] (p. 56) of cultural developments.

Particularly demanding, therefore, but also relevant, seems to be the institutional integration of sustainability goals, which contains a clearly future-oriented time component. Further scale-sensitive studies that understand speciality regions even more strongly as cooperation areas could provide better insights at this point.

It remains to be seen whether the EC, in designing the EU GI system, as suggested by Regulation (EU) No 1151/2012, consolidated on 8 July 2022, and the new EC proposal for a unified GI regulation of 2 May 2022, will be more strongly oriented towards competition policy mechanisms and administered, for example, as pure IP law at EUIPO, or whether it will be further developed as an independent system. The latter would have the potential to give greater weight to cultural embedding and other aspects relevant to sustainability and specific to the region, such as those currently demanded by society, and not to reduce them to *purely legal issues*.

At the regional level, the EU GI system is a benefit for consumers, producers and regional development. It offers clearly defined, publicly accessible, officially controlled and clearly communicated origin criteria—in a uniform system across Europe. Too much focus on trademark mechanisms seems to contribute little to the *sui generis* legal field of EU GI, the preservation of cultural culinary arts and the further development towards sustainability. In any case, if the goal of protecting regional specificities is to be carried into the future, harmonisation *automatisms* should always be the object of critical reflection.

Supplementary Materials: The following supporting information can be downloaded at: https://www.mdpi.com/article/10.3390/su15032666/s1, Table S1: Milestones in the development of GI rules for agricultural products and foodstuffs.

Author Contributions: Conceptualisation, investigation, methodology, writing—original draft preparation, project administration, K.S.; writing—review and editing, H.L., R.B. and T.C.; supervision, T.C. and R.B. All authors have read and agreed to the published version of the manuscript.

Funding: This research was funded by German Academic Scholarship Foundation (Studienstiftung des deutschen Volkes e. V.) We also acknowledge financial support by Deutsche Forschungsgemeinschaft and Friedrich-Alexander-Universität Erlangen-Nürnberg within the funding programme "Open Access Publication Funding".

Informed Consent Statement: Informed consent was obtained from all subjects involved in the study.

Data Availability Statement: Publicly available datasets were analyzed in this study. This data can be found here: https://ec.europa.eu/info/food-farming-fisheries/food-safety-and-quality/certification/quality-labels/geographical-indications-register/ (accessed on 28 December 2022).

Acknowledgments: We thank all the interviewees who contributed their time and knowledge to this study.

Conflicts of Interest: The authors declare no conflict of interest. The funders had no role in the design of the study; in the collection, analyses, or interpretation of data; in the writing of the manuscript; or in the decision to publish the results.

References

1. Allmendinger, P.; Haughton, G. Soft Spaces, Fuzzy Boundaries, and Metagovernance: The New Spatial Planning in the Thames Gateway. *Environ. Plan A* **2009**, *41*, 617–633. [CrossRef]
2. Haughton, G.; Allmendinger, P.; Counsell, D.; Vigar, G. *The New Spatial Planning: Territorial Management with Soft Spaces and Fuzzy Boundaries*; Routledge: London, UK; New York, NY, USA, 2009; ISBN 978-0-415-48335-3.
3. Haughton, G.; Allmendinger, P.; Oosterlynck, S. Spaces of Neoliberal Experimentation: Soft Spaces, Postpolitics, and Neoliberal Governmentality. *Environ. Plan A* **2013**, *45*, 217–234. [CrossRef]
4. Metzger, J.; Schmitt, P. When Soft Spaces Harden: The EU Strategy for the Baltic Sea Region. *Environ. Plan A* **2012**, *44*, 263–280. [CrossRef]
5. Lepsius, M. *Rainer Interessen, Ideen, Institutionen*; Abingdon: Nashville, TN, USA, 2013; ISBN 978-3-322-94352-1.
6. Allmendinger, P.; Chilla, T.; Sielker, F. Europeanizing Territoriality—Towards Soft Spaces? *Environ. Plan A* **2014**, *46*, 2703–2717. [CrossRef]
7. Wong, R.Y.-P. The Europeanization of Foreign Policy. In *International Relations and the European Union*; Hill, C., Smith, M., Eds.; Oxford University Press: Oxford, UK, 2005; pp. 134–153.
8. Chilla, T. *Punkt, Linie, Fläche—Territorialisierte Europäisierung*; Peter Lang: Bern, The Switzerland, 2013; ISBN 978-3-653-03087-7.
9. Becker, S.; Naumann, M. Rescaling Energy? Räumliche Neuordnungen in Der Deutschen Energiewende. *Geogr. Helv.* **2017**, *72*, 329–339. [CrossRef]
10. Breuer, F.; Muckel, P.; Dieris, B. *Reflexive Grounded Theory: Eine Einführung für die Forschungspraxis*, 3rd ed.; Springer: Wiesbaden, Germany, 2018; ISBN 978-3-658-15421-9.
11. Chilla, T. Die flexible Region. In *Die Region—Eine Begriffserkundung*; Ermann, U., Höfner, M., Hostniker, S., Preininger, E.M., Simic, D., Eds.; Transcript: Bielefeld, Germany, 2022; pp. 81–90. ISBN 978-3-8376-6010-4.
12. Haughton, G.; Allmendinger, P. Soft Spaces' in Planning. *Town Ctry. Plan.* **2007**, *76*, 306–308.
13. Barjolle, D.; Boisseaux, S.; Dufour, M. *Le Lien Au Terroir: Bilan des Travaux de Recherche*; Ecole polytechnique fédérale de Zurich, Institut d'économie rurale: Zürich, Switzerland, 1998.
14. Becker, T. *Zur Bedeutung geschützter Herkunftsangaben*; University of Hohenheim: Stuttgart, Germany, 2006. [CrossRef]
15. León Ramírez, C.A. *Der Schutz von geographischen Angaben und Ursprungsbezeichnungen für Agrarerzeugnisse und Lebensmittel nach der Verordnung (EG) Nr. 510/2006*; Nomos: Baden-Baden, Germany, 2007; ISBN 978-3-8329-2797-4.
16. Calboli, I. Geographical Indications between Trade, Development, Culture, and Marketing: Framing a Fair(er) System of Protection in the Global Economy? In *Geographical Indications at the Crossroads of Trade, Development, and Culture*; Calboli, I., Ng-Loy, W.L., Eds.; Cambridge University Press: Cambridge, UK, 2017; pp. 3–35. ISBN 978-1-316-71100-2.

17. Zappalaglio, A. The Debate Between the European Parliament and the Commission on the Definition of Protected Designation of Origin: Why the Parliament Is Right. *IIC* **2019**, *50*, 595–610. [CrossRef]
18. O'Connor and Company; Insight Consulting. Geographical Indications and TRIPS: 10 Years Later . . . A Roadmap for GI Holders to Get Protection in Other WTO Members. 2007. Available online: https://trade.ec.europa.eu/doclib/docs/2007/june/tradoc_135088.pdf (accessed on 21 January 2023).
19. Grienberger, R. *Die Herkunftsangabe als Marketinginstrument: Fallstudien aus Italien und Spanien*; Fraund: Mainz, Germany, 2000.
20. Thiedig, F.; Sylvander, B.; Thiedig, F.; Sylvander, B. Welcome to the Club?—An Economical Approach to Geographical Indications in the European Union. *Ger. J. Agric. Econ.* **2000**, *49*, 428–437. [CrossRef]
21. Menapace, L.; Moschini, G.C. Strength of Protection for Geographical Indications: Promotion Incentives and Welfare Effects. *Am. J. Agric. Econ.* **2014**, *96*, 1030–1048. [CrossRef]
22. Thiedig, F. *Spezialitäten mit geographischer Herkunftsangabe: Marketing, Rechtlicher Rahmen und Fallstudien*; Peter Lang: Frankfurt am Main, Germany, 2004; ISBN 978-3-631-52540-1.
23. Ermann, U.; Langthaler, E.; Penker, M.; Schermer, M. *Agro-Food Studies: Eine Einführung*; utb: Stuttgart, Germany, 2017; ISBN 978-3-8385-4830-2.
24. Chilla, T.; Fink, B.; Balling, R.; Reitmeier, S.; Schober, K. The EU Food Label 'Protected Geographical Indication': Economic Implications and Their Spatial Dimension. *Sustainability* **2020**, *12*, 5503. [CrossRef]
25. Fernández-Barcala, M.; González-Díaz, M.; Raynaud, E. Contrasting the Governance of Supply Chains with and without Geographical Indications: Complementarity between Levels. *SCM* **2017**, *22*, 305–320. [CrossRef]
26. Barham, E. Translating Terroir: The Global Challenge of French AOC Labeling. *J. Rural. Stud.* **2003**, *19*, 127–138. [CrossRef]
27. Bowen, S. Development from Within? The Potential for Geographical Indications in the Global South. *J. World Intellect. Prop.* **2010**, *13*, 231–252. [CrossRef]
28. Curzi, D.; Huysmans, M. The Impact of Protecting EU Geographical Indications in Trade Agreements. *AAEA* **2022**, *104*, 364–384. [CrossRef]
29. Marie-Vivien, D. *The Protection of Geographical Indications in India: A New Perspective on the French and European Experience*; SAGE: Los Angeles, CA, USA, 2021; ISBN 978-93-5328-856-3.
30. Gocci, A.; Luetge, C.; Vakoufaris, H. Between Tradition and Sustainable Innovation: Empirical Evidence for the Role of Geographical Indications. *IBR* **2020**, *13*, 101. [CrossRef]
31. Quiñones-Ruiz, X.F.; Penker, M.; Belletti, G.; Marescotti, A.; Scaramuzzi, S.; Barzini, E.; Pircher, M.; Leitgeb, F.; Samper-Gartner, L.F. Insights into the Black Box of Collective Efforts for the Registration of Geographical Indications. *Land Use Policy* **2016**, *57*, 103–116. [CrossRef]
32. Van Caenegem, W.; Cleary, J.A.; Drahos, P. Pride and Profit: Geographical Indications as Regional Development Tools in Australia. *J. Soc. Policy* **2014**, *16*, 90–114.
33. Voss, J.; Spiller, A. Das EU-System zum Schutz geographischer Herkunftsangaben und Ursprungsbezeichnungen. In *Ernährung, Kultur, Lebensqualität—Wege regionaler Nachhaltigkeit*; Antoni-Komar, I., Pfriem, R., Raabe, T., Spiller, A., Eds.; Metropolis: Oldenburg, Germany, 2008; ISBN 978-3-89518-634-9.
34. Arfini, F.; Antonioli, F.; Cozzi, E.; Donati, M.; Guareschi, M.; Mancini, M.C.; Veneziani, M. Sustainability, Innovation and Rural Development: The Case of Parmigiano-Reggiano PDO. *Sustainability* **2019**, *11*, 4978. [CrossRef]
35. Arfini, F.; Bellassen, V. *Sustainability of European Food Quality Schemes: Multi-Performance, Structure, and Governance of PDO, PGI, and Organic Agri-Food Systems*; Springer International Publishing: Cham, Switzerland, 2019; ISBN 978-3-030-27507-5.
36. Bätzing, W. Nutzungskonflikte zwischen Teichwirtschaft, Naturschutz und Freizeitinteressen im Aischgrund. Probleme und Potenziale bei der Aufwertung des "Aischgründer Karpfens" zum Qualitätsregionalprodukt. *Mitt. der fränkischen geogr. Ges.* **2013**, *59*, 81–100. Available online: https://www.archiv.geographie.uni-erlangen.de/wp-content/uploads/publ_wba/wba_publ_270.pdf (accessed on 21 January 2023).
37. Jantyik, L.; Török, Á. Estimating the Market Share and Price Premium of GI Foods—The Case of the Hungarian Food Discounters. *Sustainability* **2020**, *12*, 1094. [CrossRef]
38. Schooler, R.D. Product Bias in the Central American Common Market. *J. Mark. Res.* **1965**, *2*, 394–397. [CrossRef]
39. European Commission. *Joint Research Centre. Empirical Testing of the Impact on Consumer Choice Resulting from Differences in the Composition of Seemingly Identical Branded Products*; Publications Office: Luxembourg, 2020. Available online: https://data.europa.eu/doi/10.2760/497543 (accessed on 21 January 2023).
40. Capelli, M.G.; Menozzi, D.; Arfini, F. Consumer willingness to pay for food quality labels: Evaluating the prosciutto di parma PDO quality differentiation strategy. In Proceedings of the International Congress of the European Association of Agricultural Economists, Ljubljana, Slovenia, 26–29 August 2014. [CrossRef]
41. Garavaglia, C.; Mariani, P. How Much Do Consumers Value Protected Designation of Origin Certifications? Estimates of Willingness to Pay for PDO Dry-Cured Ham in Italy. *Agribusiness* **2017**, *33*, 403–423. [CrossRef]
42. Menapace, L.; Colson, G.J.; Grebitus, C.; Facendola, M. *Consumer Preferences for Country-Of-Origin, Geographical Indication, and Protected Designation of Origin Labels, Economics Working Papers (2002–2016), 147, 2009*; Iowa State University: Ames, IA, USA, 2009. Available online: https://core.ac.uk/download/pdf/128975881.pdf (accessed on 21 January 2023).

43. Coombe, R.J.; Ives, S.; Huizenga, D. Geographical Indications: The Promise, Perils and Politics of Protecting Place-Based Products. In *Sage Handbook on Intellectual Property*; David, M., Halbert, D., Eds.; Sage Publications: Thousand Oaks, CA, USA, 2014; pp. 207–223. [CrossRef]
44. Coombe, R.J.; Malik, S.A. Rethinking the Work of Geographical Indications in Asia: Addressing Hidden Geographies of Gendered Labor. In *Geographical Indications at the Crossroads of Trade, Development, and Culture*; Calboli, I., Ng-Loy, W.L., Eds.; Cambridge University Press: Cambridge, UK, 2017; pp. 87–121. ISBN 978-1-316-71100-2.
45. May, S.; Tschofen, B. Regionale Spezialitäten als globales Gut. Inwertsetzungen geografischer Herkunft und distinguierender Konsum. *Z. Für Agrargesch. Und Agrarsoziol.* **2016**, *64*, 61–75. [CrossRef]
46. May, S. *Ausgezeichnet!: Zur Konstituierung kulturellen Eigentums durch geografische Herkunftsangaben*; Göttingen Studies in Cultural Property; Göttingen University Press: Göttingen, Germany, 2016; ISBN 978-3-86395-289-1.
47. May, S.; Sidali, K.L.; Spiller, A.; Tschofen, B. *Taste | Power | Tradition: Geographical Indications as Cultural Property*; Göttingen Studies in Cultural Property; Göttingen University Press: Göttingen, Germany, 2017; ISBN 978-3-86395-208-2.
48. Welz, G. Europäische Produkte. Nahrungskulturelles Erbe und EU-Politik. Am Beispiel der Republik Zypern. In *Prädikat "Heritage". Wertschöpfungen aus kulturellen Ressourcen*; Hemme, D., Tauschek, M., Bendix, R., Eds.; LIT: Berlin, Germany, 2007; pp. 323–335.
49. Welz, G. Contested Origins: Food Heritage and the European Union's Quality Label Program. *Food Cult. Soc.* **2013**, *16*, 265–279. [CrossRef]
50. Belletti, G.; Marescotti, A. *Evaluating Geographical Indications—Guide to Tailor Evaluations for the Development and Improvement of Geographical Indications*; FAO: Rome, Italy, 2021; ISBN 978-92-5-134869-7.
51. Hacker, F.; Ströbele, P. *Markengesetz: Kommentar*, 12th ed.; Carl Heymanns: Köln, Germany, 2018; ISBN 978-3-452-28553-9.
52. Knaak, R. Geographical Indications and Their Relationship with Trade Marks in EU Law. *IIC* **2015**, *46*, 843–867. [CrossRef]
53. Loschelder, M. Geografische Herkunftsangaben—Absatzförderung oder erzwungene Transparenz. *GRUR* **2016**, *4*, 339–346.
54. REGULATION (EU) No 1169/2011 OF THE EUROPEAN PARLIAMENT AND OF THE COUNCIL of 25 October 2011 on the Provision of Food Information to Consumers, Amending Regulations (EC) No 1924/2006 and (EC) No 1925/2006 of the European Parliament and of the Council, and Repealing Commission Directive 87/250/EEC, Council Directive 90/496/EEC, Commission Directive 1999/10/EC, Directive 2000/13/EC of the European Parliament and of the Council, Commission Directives 2002/67/EC and 2008/5/EC and Commission Regulation (EC) No 608/2004. Available online: https://eur-lex.europa.eu/legal-content/EN/ALL/?uri=CELEX%3A32011R1169 (accessed on 22 December 2022).
55. COMMISSION REGULATION (EC) No 589/2008 of 23 June 2008 Laying Down Detailed Rules for Implementing Council Regulation (EC) No 1234/2007 as Regards Marketing Standards for Eggs. Available online: https://eur-lex.europa.eu/legal-content/EN/TXT/?uri=CELEX%3A32008R0589 (accessed on 22 December 2022).
56. REGULATION (EC) No 178/2002 of the European Parliament and of the Council of 28 January 2002 Laying Down the General Principles and Requirements of Food Law, Establishing the European Food Safety Authority and Laying Down Procedures in Matters of Food Safety. Available online: https://eur-lex.europa.eu/legal-content/EN/ALL/?uri=celex%3A32002R0178 (accessed on 22 December 2022).
57. Lehmann, N.; EU-Lebensmittelkennzeichnung. EuGH Setzt Nationalen Herkunftsangaben für Milch enge Grenzen. *Agrarheute* **2020**. Available online: https://www.agrarheute.com/management/recht/eugh-setzt-nationalen-herkunftsangaben-fuer-milch-enge-grenzen-573527 (accessed on 21 January 2023).
58. Agrarzeitung. Herkunftsangaben Werden zum Ärgernis. Available online: https://www.agrarzeitung.de/nachrichten/wirtschaft/Herkunftsangaben-werden-zum-Aergernis-67912 (accessed on 21 January 2023).
59. Michel, J. Frankreich Weitet Nationale Herkunftskennzeichnung auf Gastronomie aus. *Agrarheute* **2022**. Available online: https://www.agrarheute.com/politik/frankreich-weitet-nationale-herkunftskennzeichnung-gastronomie-590754 (accessed on 21 January 2023).
60. DGS (Magazin für Geflügelwirtschaft). Herkunftskennzeichnung Wird Umgesetzt. Available online: https://www.dgs-magazin.de/themen/themen-a-z/article-7210351-194087/herkunftskennzeichnung-wird-umgesetzt-.html (accessed on 21 January 2023).
61. COMMISSION REGULATION (EU) No 702/2014 of 25 June 2014 Declaring Certain Categories of Aid in the Agricultural and Forestry Sectors and in Rural Areas Compatible with the Internal Market in Application of Articles 107 and 108 of the Treaty on the Functioning of the European Union. Available online: https://eur-lex.europa.eu/legal-content/EN/TXT/?uri=CELEX%3A32014R0702 (accessed on 22 December 2022).
62. Regionalfenster. Kennzeichnung für regionale Produkte. Available online: https://www.regionalfenster.de/ (accessed on 21 January 2023).
63. Schober, K. Regionale Produkte in Europa. Raumbezogene Institutionalisierungsprozesse beim Europäischen Herkunftsschutz. Doctoral Thesis, FAU Erlangen-Nuremberg, Erlangen, Germany, 2021. Available online: https://www.urn:nbn:de:bvb:29-opus4-175055 (accessed on 21 January 2023).
64. Wagener, H.-J.; Eger, T. *Europäische Integration: Wirtschaft und Recht, Geschichte und Politik*, 3rd ed.; Vahlen: München, Germany, 2014; ISBN 978-3-8006-4761-3.
65. Chilla, T.; Evrard, E. Spatial integration revisited—New insights for cross-border and transnational contexts. In *Science in support of European Territorial Development and Cohesion: Second ESPON 2013 Scientific Report*; ESPON 2013 Programme, Ed.: Luxembourg, 2013; pp. 44–49.

66. Chilla, T.; Streifeneder, T. Interrelational space? The spatial logic of the macro-regional strategy for the alps and its potentials. *Eur. Plan. Stud.* **2018**, *26*, 2470–2489. [CrossRef]
67. Dühr, S.; Colomb, C.; Nadin, V. *European Spatial Planning and Territorial Cooperation*; Routledge: London, UK, 2010. [CrossRef]
68. Mendoza, J.E.; Dupeyron, B. Economic Integration, Emerging Fields and Cross-Border Governance: The Case of San Diego–Tijuana. *J. Borderl. Stud.* **2020**, *35*, 55–74. [CrossRef]
69. Nugent, N. *The Government and Politics of the European Community*, 2nd ed.; Comparative Government and politics; Macmillan Education: London, UK, 1991; ISBN 978-0-333-55799-0.
70. Jones, E. Towards a Theory of Disintegration. *J. Eur. Public Policy* **2018**, *25*, 440–451. [CrossRef]
71. Durand, F.; Decoville, A.; Knippschild, R. Everything All Right at the Internal EU Borders? The Ambivalent Effects of Cross-Border Integration and the Rise of Euroscepticism. *Geopolitics* **2020**, *25*, 587–608. [CrossRef]
72. Nitschke, P. *Gemeinsame Werte in Europa? Stärken und Schwächen im normativen Selbstverständnis der Europäischen Integration*; Nomos: Baden-Baden, Germany, 2019; ISBN 978-3-8452-9162-8.
73. Brasche, U. *Europäische Integration: Wirtschaft, Euro-Krise, Erweiterung und Perspektiven*, 4th ed.; De Gruyter Oldenbourg: Berlin, Germany, 2017; ISBN 978-3-11-049547-8.
74. North, D. *Institutions, Institutional Change and Economic Performance*; Cambridge University Press: Cambridge, UK, 1990. [CrossRef]
75. Erlei, M.; Leschke, M.; Sauerland, D. *Institutionenökonomik*, 3rd ed.; Schäffer-Poeschel: Stuttgart, Germany, 2016; ISBN 978-3-7910-3526-0.
76. Jellinek, G. *Allgemeine Staatslehre*; Springer: Berlin/Heidelberg, Germany, 1929.
77. Anter, A. *Die Normative Kraft des Faktischen: Das Staatsverständnis Georg Jellineks*, 2nd ed.; Nomos: Baden-Baden, Germany, 2020; ISBN 978-3-8487-5919-4.
78. Rammert, W. Technik und Innovationen: Kerninstitutionen der modernen Wirtschaft. In *TUTS Working Papers*; Technische Universität Berlin: Berlin, Germany, 2015. Available online: https://nbn-resolving.org/urn:nbn:de:0168-ssoar-12355 (accessed on 21 January 2023).
79. Hasselkuss, M. Transformative Soziale Innovation durch Netzwerk. Das Beispiel 'Bildung für nachhaltige Entwicklung'. In *Wuppertaler Forschungsschriften 2018*; Oekom: München, Germany, 2018. Available online: https://nbn-resolving.org/urn:nbn:de:bsz:wup4-opus-71081 (accessed on 21 January 2023).
80. Ochoa, C.S.; Hugendubel, M. *Umstrittene Faktenlage: Eine Diskursanalyse der Öffentlichen Diskussion um Sozioökonomische Ungleichheit in Deutschland*; Hans-Böckler-Stiftung: Düsseldorf, Germany, 2019. Available online: https://urn:nbn:de:101:1-2019102313200692525178 (accessed on 21 January 2023).
81. Schmid, H.B.; Schweikard, D.P. Einleitung: Kollektive Intentionalität. Begriff, Geschichte, Probleme. In *Kollektive Identität. Eine Debatte über die Grundlagen des Sozialen*; Schmid, H.B., Schweikard, D.P., Eds.; Suhrkamp: Frankfurt am Main, Germany, 2009; pp. 11–65.
82. Beyer, J. Pfadabhängigkeit ist nicht gleich Pfadabhängigkeit! Wider den impliziten Konservatismus eines gängigen Konzepts/Not All Path Dependence Is Alike—A Critique of the "Implicit Conservatism" of a Common Concept. *Z. Soziol.* **2005**, *34*, 5–21. [CrossRef]
83. Kropp, C. Urban Food Movements and Their Transformative Capacities. *IJSAF* **2018**, *24*, 413–430. [CrossRef]
84. Sung, B.; Park, S.-D. Who Drives the Transition to a Renewable-Energy Economy? Multi-Actor Perspective on Social Innovation. *Sustainability* **2018**, *10*, 448. [CrossRef]
85. Belmin, R.; Casabianca, F.; Meynard, J.-M. Contribution of Transition Theory to the Study of Geographical Indications. *Environ. Innov. Soc. Transit.* **2018**, *27*, 32–47. [CrossRef]
86. Vandecandelaere, E.; Samper, L.F.; Rey, A.; Daza, A.; Mejía, P.; Tartanac, F.; Vittori, M. The Geographical Indication Pathway to Sustainability: A Framework to Assess and Monitor the Contributions of Geographical Indications to Sustainability through a Participatory Process. *Sustainability* **2021**, *13*, 7535. [CrossRef]
87. Sielker, F.; Stead, D. Revisiting the Question of Scale and Rescaling in EU Macro-Regional Strategies. In *Macro-Regional Integration—New Scales, Spaces and Governance for Europe?* Sielker, F., Ed.; Dissertation; FAU Erlangen-Nuremberg: Berlin, Germany, 2017; pp. 112–133. Available online: https://nbn-resolving.org/urn:nbn:de:bvb:29-opus4-85171 (accessed on 21 January 2023).
88. Köhler, B. Die Materialität von Rescaling-Prozessen. Zum Verhältnis von Politics of Scale und Political Ecology. In *Politics of Scale. Räume der Globalisierung und Perspektiven Emanzipatorischer Politik*; Wissen, M., Röttger, B., Heeg, S., Eds.; Westfälisches Dampfboot: Münster, Germany, 2007; pp. 208–223.
89. Cohen, A.; McCarthy, J. Reviewing Rescaling: Strengthening the Case for Environmental Considerations. *Prog. Hum. Geogr.* **2015**, *39*, 3–25. [CrossRef]
90. Schluchter, W. Interessen, Ideen, Institutionen: Schlüsselbegriffe einer an Max Weber orientierten Soziologie. In *Soziale Konstellation und Historische Perspektive*; Sigmund, S., Albert, G., Bienfait, A., Stachura, M., Eds.; VS Verlag für Sozialwissenschaften: Wiesbaden, Germany, 2008; pp. 57–80. ISBN 978-3-531-15852-5.
91. Flick, U.; Von Kardoff, E.; Steinke, I. *A Companion to Qualitative Research*; SAGE: Los Angeles, CA, USA, 2004; ISBN 978-0-7619-7375-1.
92. Flick, U. *Qualitative Sozialforschung. Eine Einführung*, 5th ed.; Rowohlt: Hamburg, Germany, 2012.

93. eAmbrosia The EU Geographical Indications Register. Available online: https://ec.europa.eu/info/food-farming-fisheries/food-safety-and-quality/certification/quality-labels/geographical-indications-register/ (accessed on 28 December 2022).
94. Mayring, P. *Einführung in Die Qualitative Sozialforschung: Eine Anleitung zu Qualitativem Denken*, 5th ed.; Beltz: Weinheim Basel, Germany, 2002; ISBN 978-3-407-29093-9.
95. Meuser, M.; Nagel, U. ExpertInneninterviews—Vielfach erprobt, wenig bedacht: Ein Beitrag zur qualitativen Methodendiskussion. In *Qualitativ-empirische Sozialforschung: Konzepte, Methoden, Analysen*; Garz, D., Kraimer, K., Eds.; Westdt. Verl.: Opladen, Germany, 1991; pp. 441–471. ISBN 3-531-12289-4.
96. Jäger, S. *Kritische Diskursanalyse. Eine Einführung*, 6th ed.; Unrast: Münster, Germany, 2012.
97. Scharl, P. *Der Geopolitische Diskurs um Die Gründung Einer US-Amerikanischen International Law Enforcement Academy in Costa Rica. Eine Analyse Nationaler Interessen, Raumbezogener Diskursiver Instrumente und Ihrer Verankerung in Einem "Terrain of Resistance"*; Universität Passau: Passau, Germany, 2008. Available online: Urn:nbn:de:bvb:739-opus-12695 (accessed on 21 January 2023).
98. COUNCIL REGULATION (EEC) No 2081/92 of 14 July 1992 on the Protection of Geographical Indications and Designations of Origin for Agricultural Products and Foodstuffs. Available online: https://eur-lex.europa.eu/legal-content/EN/TXT/?uri=CELEX%3A31992R2081 (accessed on 22 December 2022).
99. REGULATION (EU) 2021/2117 OF THE EUROPEAN PARLIAMENT AND OF THE COUNCIL of 2 December 2021 Amending Regulations (EU) No 1308/2013 Establishing a Common Organisation of the Markets in Agricultural Products, (EU) No 1151/2012 on Quality Schemes for Agricultural Products and Foodstuffs, (EU) No 251/2014 on the Definition, Description, Presentation, Labelling and the Protection of Geographical Indications of Aromatised Wine Products and (EU) No 228/2013 Laying Down Specific Measures for Agriculture in the Outermost Regions of the Union. Available online: https://eur-lex.europa.eu/legal-content/EN/TXT/?uri=uriserv:OJ.L_.2021.435.01.0262.01.ENG (accessed on 22 December 2022).
100. Nathon, N. The Protection of Geographical Indications for Agricultural Products in the European Union. In *Cambridge Handbook of Intellectual Property in Central and Eastern Europe*; Sundara Rajan, M.T., Ed.; Cambridge University Press: Cambridge, UK, 2019; pp. 349–364. ISBN 978-1-316-66125-3.
101. Deutscher Bundesrat. Beschluss des Bundesrates Zum Vorschlag Einer Verordnung (EWG) des Rates zum Schutz Geographischer Angaben und Ursprungsbezeichnungen bei Agrarerzeugnissen und Lebensmitteln, SEK(90) 2415 endg., Ratsdok. 10837/90, Drucksache 83/91 (Beschluss) vom 26 April 1991. 1991. Available online: https://dserver.bundestag.de/btd/12/013/1201369.pdf (accessed on 31 January 2023).
102. Von Mühlendahl, A. Der Schutz geographischer Herkunftsangaben in der Europäischen Gemeinschaft nach der Verordnung Nr. 2081/92 vom 24. Juli 1992. *ZLR* **1993**, *1–2*, 187–200.
103. Ballarini, G.D.O.C. Mehr "Schatten" Als "Licht". *Parma Cap. Aliment.* **1989**, *22*, 43–46.
104. Glaus, U. *Die Geographische Herkunftsangabe als Kennzeichen*; Dissertation; Helbig & Lichtenhahn: Basel, Switzerland; Frankfurt am Main, Germany, 1996.
105. Hacker, F. Maßgeblichkeit, Grenzen und Perspektiven des "europäischen Verbrauchers". Zugleich Besprechung von BGH "Culatello di Parma". *GRUR* **2020**, 587–590.
106. Monteverde, P. Enforcement of Geographical Indications. *JIPLP* **2012**, *7*, 291–297. [CrossRef]
107. REGULATION (EU) No 1151/2012 OF THE EUROPEAN PARLIAMENT AND OF THE COUNCIL of 21 November 2012 on Quality Schemes for Agricultural Products and Foodstuffs. Available online: https://eur-lex.europa.eu/legal-content/EN/TXT/?uri=CELEX%3A32012R1151 (accessed on 22 December 2022).
108. Omsels, H.-J. Die Verordnung (EG) 1151/2012 über Qualitätsregelungen für Agrarerzeugnisse und Lebensmittel. *MarkenR* **2013**, *6*, 209–214.
109. European Commission. Communication from the Commission to the European Parliament, the Council, the European Economic and Social Committee and the Committee of the Regions. A Farm to Fork Strategy for a Fair, Healthy and Environmentally-Friendly Food System. COM (2020) 381 Final. 2020. Available online: https://eur-lex.europa.eu/EN/legal-content/summary/farm-to-fork-strategy-for-a-fair-healthy-and-environmentally-friendly-food-system.html (accessed on 21 January 2023).
110. REGULATION (EU) 2017/625 OF THE EUROPEAN PARLIAMENT AND OF THE COUNCIL of 15 March 2017 on Official Controls and Other Official Activities Performed to Ensure the Application of Food and Feed Law, Rules on Animal Health and Welfare, Plant Health and Plant Protection Products, Amending Regulations (EC) No 999/2001, (EC) No 396/2005, (EC) No 1069/2009, (EC) No 1107/2009, (EU) No 1151/2012, (EU) No 652/2014, (EU) 2016/429 and (EU) 2016/2031 of the European Parliament and of the Council, Council Regulations (EC) No 1/2005 and (EC) No 1099/2009 and Council Directives 98/58/EC, 1999/74/EC, 2007/43/EC, 2008/119/EC and 2008/120/EC, and Repealing Regulations (EC) No 854/2004 and (EC) No 882/2004 of the European Parliament and of the Council, Council Directives 89/608/EEC, 89/662/EEC, 90/425/EEC, 91/496/EEC, 96/23/EC, 96/93/EC and 97/78/EC and Council Decision 92/438/EEC (Official Controls Regulation). Available online: https://eur-lex.europa.eu/legal-content/EN/TXT/?uri=celex%3A32017R0625 (accessed on 22 December 2022).
111. BMEL (Bundesministerium für Ernährung und Landwirtschaft). Neue Regeln für Die Amtlichen Lebensmittel- und Futtermittelkontrollen. 2019. Available online: https://www.bmel.de/DE/themen/verbraucherschutz/lebensmittelsicherheit/kontrolle-und-risikomanagement/kontrollverordnung.html (accessed on 21 January 2023).

112. REGULATION (EU) No 1144/2014 OF THE EUROPEAN PARLIAMENT AND OF THE COUNCIL of 22 October 2014 on Information Provision and Promotion Measures Concerning Agricultural Products Implemented in the Internal Market and in Third Countries and Repealing Council Regulation (EC) No 3/2008. Available online: https://eur-lex.europa.eu/legal-content/en/TXT/?uri=OJ%3AJOL_2014_317_R_0004 (accessed on 22 December 2022).
113. Spiller, A.; Voss, J.; Deimel, M. Das EU-System zum Schutz geographischer Herkunftsangaben und Ursprungsbezeichnungen: Eine vergleichende Studie zur Effektivität des Instruments zur Förderung des ländlichen Raums und Implikationen für die deutsche Agrarförderung. In *Zur Wettbewerbsfähigkeit der Deutschen Agrarwirtschaft—Politische, Institutionelle und Betriebliche Herausforderungen*; Rentenbank, L., Ed.; Landwirtschaftliche Rentenbank: Frankfurt am Main, Germany, 2007; Volume 2, pp. 187–232.
114. European Commission. Promotion of EU Farm Products. 2022. Available online: https://agriculture.ec.europa.eu/common-agricultural-policy/market-measures/promotion-eu-farm-products_en (accessed on 28 December 2022).
115. Heine, J.F. Das neue gemeinschaftliche System zum Schutz geographischer Bezeichnungen. *GRUR* **1993**, 96–103.
116. Beier, F.-K.; Knaak, R. Der Schutz geographischer Herkunftsangaben in der Europäischen Gemeinschaft—Die neueste Entwicklung. *GRUR Int.* **1993**, 602–610.
117. REGULATION (EU) 2019/787 OF THE EUROPEAN PARLIAMENT AND OF THE COUNCIL of 17 April 2019 on the Definition, Description, Presentation and Labelling of Spirit Drinks, the Use of the Names of Spirit Drinks in the Presentation and Labelling of Other Foodstuffs, the Protection of Geographical Indications for Spirit Drinks, the Use of Ethyl Alcohol and Distillates of Agricultural Origin in Alcoholic Beverages, and Repealing Regulation (EC) No 110/2008. Available online: https://eur-lex.europa.eu/legal-content/EN/TXT/?uri=CELEX%3A32019R0787 (accessed on 22 December 2022).
118. European Commission. Proposal for a Regulation of the European Parliament and of the Council on European Union Geographical Indications for Wine, Spirit Drinks and Agricultural Products, and Quality Schemes for Agricultural Products, Amending Regulations (EU) No 1308/2013, (EU) 2017/1001 and (EU) 2019/787 and Repealing Regulation (EU) No 1151/2012, COM(2022) 134 Final, 31.03.2022. 2022. Available online: Https://eur-lex.europa.eu/legal-content/EN/TXT/?uri=CELEX%3A52022PC0134 (accessed on 21 January 2023).
119. European Commission. Proposal for a Regulation of the European Parliament and of the Council on European Union Geographical Indications for Wine, Spirit Drinks and Agricultural Products, and Quality Schemes for Agricultural Products, Amending Regulations (EU) No 1308/2013, (EU) 2017/1001 and (EU) 2019/787 and Repealing Regulation (EU) No 1151/2012, COM (2022) 134 Final/2, 02.05.2022. 2022. Available online: https://eur-lex.europa.eu/legal-content/EN/TXT/?uri=CELEX%3A52022PC0134R%2801%29 (accessed on 21 January 2023).
120. Profeta, A.; Balling, R.; Schoene, V.; Wirsig, A. The Protection of Origins for Agricultural Products and Foods in Europe: Status Quo, Problems and Policy Recommendations for the Green Book. *J. World Intellect. Prop.* **2009**, *12*, 622–648. [CrossRef]
121. REGULATION (EU) No 1308/2013 OF THE EUROPEAN PARLIAMENT AND OF THE COUNCIL of 17 December 2013 Establishing a Common Organisation of the Markets in Agricultural Products and Repealing Council Regulations (EEC) No 922/72, (EEC) No 234/79, (EC) No 1037/2001 and (EC) No 1234/2007. Available online: https://eur-lex.europa.eu/legal-content/EN/TXT/?uri=celex%3A32013R1308 (accessed on 22 December 2022).
122. COMMISSION DELEGATED REGULATION (EU) No 664/2014 of 18 December 2013 supplementing Regulation (EU) No 1151/2012 of the European Parliament and of the Council with regard to the establishment of the Union Symbols for Protected Designations of Origin, Protected Geographical Indications and Traditional Specialities Guaranteed and with Regard to Certain Rules on Sourcing, Certain Procedural Rules and Certain Additional Transitional Rules. Available online: https://eur-lex.europa.eu/legal-content/EN/TXT/?uri=CELEX%3A32014R0664 (accessed on 22 December 2022).
123. COUNCIL REGULATION (EC) No 510/2006 of 20 March 2006 on the Protection of Geographical Indications and Designations of Origin for Agricultural Products and Foodstuffs. Available online: https://eur-lex.europa.eu/legal-content/EN/ALL/?uri=CELEX%3A32006R0510 (accessed on 22 December 2022).
124. Watson, K.W. Reign of Terroir. How to Resist Europe's Efforts to Control Common Food Names as Geographical Indications. *Policy Anal.* **2016**, *787*, 1–15.
125. Tome, B. Geographical Indications: What's in It for the Agrifood Sector? In Proceedings of the "Regional Specialities and Geographical Indications" at the Representation of the State North Rhine-Westphalia to the EU, Brussels, Belgium, 14 May 2019.
126. World Trade Organization. Members and Observers. 2020. Available online: https://www.wto.org/English/Thewto_e/Whatis_e/Tif_e/Org6_e.Htm (accessed on 7 February 2020).
127. Schober, K. Geographische Herkunftsangaben als Patent der Region? In Proceedings of the "Räumliche Manifestierungen von globalen Wertketten (Dannenberg/Franz)", German Congress for Geography, Kiel, Germany, 26 September 2019.
128. Haas, E. *The Uniting of Europe*; Stevens: London, UK, 1958.
129. Marie-Vivien, D.; Bérard, L.; Boutonnet, J.-P.; Casabianca, F. Are French Geographical Indications Losing Their Soul? Analyzing Recent Developments in the Governance of the Link to the Origin in France. *World Dev.* **2017**, *98*, 25–34. [CrossRef]
130. Sielker, F. Soft Borders als neues Raumkonzept in der EU? Das Beispiel der makroregionalen Kooperationen. In *Nimm's Sportlich—Planung als Hindernislauf*; Grotheer, S., Schwöbel, A., Stepper, M., Eds.; ARL: Hannover, Germany, 2014; pp. 79–94. ISBN 978-3-88838-389-2.
131. Busse, C. Gedanken zum Verhältnis von Geoschutzgemeinschaften zu anerkannten Agrarorganisationen. *ZLR* **2018**, 486–501.

132. Gangjee, D.S. From Geography to History: Geographical Indications and the Reputational Link. In *Geographical Indications at the Crossroads of Trade, Development, and Culture*; Calboli, I., Ng-Loy, W.L., Eds.; Cambridge University Press: Cambridge, UK, 2017; pp. 36–60. ISBN 978-1-316-71100-2.
133. Balling, R.; Schober, K.; Lindermayer, H. Governance of GIs and the Role of Regional and Local Public Actors: Bavaria. In Proceedings of the International Conference on GIs: "Worldwide Perspectives on Geographical Indications", Montpellier, France, 5–8 July 2022. Available online: sciencesconf.org:gi2021:341668 (accessed on 21 January 2023).
134. Balling, R. *Gemeinschaftsmarketing für Lebensmittel*; Marketing der Agrar- und Ernährungswirtschaft; 13, Wiss.-Verl; Vauk: Kiel, Germany, 1997.
135. Chilla, T. The Domestic Dimension of Cross-Border Governance: Patterns of Coordination and Cooperation. *RuR* **2022**. [CrossRef]

 *sustainability*

Article

Skill Needs for Sustainable Agri-Food and Forestry Sectors (II): Insights of a European Survey

Ana Ramalho Ribeiro [1,2,*], Billy Goodburn [3], Luis Mayor [1], Line F. Lindner [1], Christoph F. Knöbl [1], Jacques Trienekens [4], Daniel Rossi [5], Francesca Sanna [6], Remigio Berruto [6] and Patrizia Busato [6]

1 ISEKI-Food Association, Impacthub Vienna, Lindengasse 56, 1070 Vienna, Austria
2 Research Centre for Natural Resources, Environment and Society (CERNAS), Coimbra Agriculture School, Bencanta, 3045-601 Coimbra, Portugal
3 Irish Co-Operative Organisation Society, The Plunkett House, 84 Merrion Square, DO2 T882 Dublin, Ireland
4 Business Management & Organisation Group, Social Sciences Department, Wageningen University, Hollandseweg 1, 6706KN Wageningen, The Netherlands
5 Confagricoltura, Corso Vittorio Emanuele II, 101, 00186 Roma, Italy
6 Department of Agriculture, Forestry and Food Science (DISAFA), Università di Torino, Largo Paolo Braccini, 2, Grugliasco, 10095 Torino, Italy
* Correspondence: ana.ramalho@iseki-food.net

Abstract: The agri-food and forestry sectors are in transition towards more sustainable, green, and innovative systems tackling several challenges posed by globalization, governance, and consumers' demands. This transition to novel processes, markets, and businesses requires skills and competences to prepare the new generations and upskill the actual workforce. The purpose of this paper was to assess the skills and knowledge needs of future professionals in the agri-food and forestry sectors, from European stakeholders' perspectives, by using a European questionnaire. Overall, respondents highlighted the importance of improving sustainability and soft and digital skills. In particular, food safety management and control; quality management and assurance of processes and product; efficient use of resources and organization; and planning, visioning, and strategic thinking skills ranked higher. In almost all countries, respondents had the perception that neither formal nor non-formal training covered training needs, though formal training was more suited to address education requirements. Both for organizations and individuals, it is far more relevant to have skills to perform than to have training recognition. The outcomes also provide findings that can be used to help develop updated curricula that meet the sector's needs.

Keywords: skills survey; skills training; sustainability; bioeconomy; agri-food sector; forestry sector

Citation: Ramalho Ribeiro, A.; Goodburn, B.; Mayor, L.; Lindner, L.F.; Knöbl, C.F.; Trienekens, J.; Rossi, D.; Sanna, F.; Berruto, R.; Busato, P. Skill Needs for Sustainable Agri-Food and Forestry Sectors (II): Insights of a European Survey. *Sustainability* **2023**, *15*, 4115. https://doi.org/10.3390/su15054115

Academic Editors: Mariarosaria Lombardi, Vera Amicarelli and Erica Varese

Received: 24 November 2022
Revised: 17 February 2023
Accepted: 18 February 2023
Published: 24 February 2023

1. Introduction

The European agri-food and forestry sectors are facing diverse challenges due to the impacts of climate change, war, rising energy prices, and economic uncertainty along with low incremental crop productivity [1–3]. Such vulnerabilities are stressed by an increasing demand for food and feed, rising environmental concerns, and climatic changes that generate more uncertainties [4]. The Farm-to-Fork strategy [5], a key element of the European Green Deal [6], aims to achieve an innovative and sustainable food system, targeting improvements in the whole food chain from production and processing to consumption and food waste management. In the sectors of agriculture, forestry, and bioeconomy, focus is on mitigating the effects of climate change to reduce the loss of biodiversity and shifting to a neutral or even positive environmental impact [7]. Furthermore, the bioeconomy, boosted by the European Green Deal, is now becoming a substantial element for development and growth in Europe. The bioeconomy concept harnesses both the valorization of natural resources and human manpower in a sustainable way [8]. The transition to long-term sustainable farming and food systems entails complex processes that require a structured approach, including reshaping education methods [9]. Dedicated bioeconomy training must be driven by these emerging needs to prepare the workforce with new skills. Besides being widespread in everyday life, digital tools and smart technology have been rapidly evolving in all processes across agri-food and forestry sectors, clearly shifting production with automation processes, communication, and new business management to a new level of complexity, innovation, and efficiency [10]. New products, greener processes, and complex food chains and business models demand new skills and knowledge to successfully thrive in a competitive sector [11]. Market globalization requires linking the various actors of the value chain by making use of the new circular economy model and successfully tackling many of the current challenges [12].

The evolution in different economic sectors and markets leads the search and discussion to define adequate and matching skills [13]. In 2015–2016, the European Commission launched a survey entitled "European Digital Skills Survey" to identify, among other topics, the digital skills required by employers in the workplace [14]. Several studies and reports have addressed the main skills and competences required to drive the agri-food and forestry sectors towards a more sustainable path. In the farming sector, research identified skills to engage professionals in more sustainable agriculture [11,15]. In the food sector, the transition to the highly digital and technological processes of Food Industry 4.0 is happening, and insights about the skills and new professions needed were assessed by academics and industry [10,16]. Surveys conducted in the forestry sector assessed if knowledge and skills needed for contemporary forestry careers are being provided by degree programs [17,18]. Paralleled in the labor market is the digitalization of the economy as a major transformation of how people work, representing challenges and opportunities in all production sectors. Complementary to the identification of skills gaps is the need for further engagement between education institutions and industry to design and deliver adequate training programs and to foster the development of these sectors [19,20].

The ERASMUS + "FIELDS" project (acronym for: addressing the current and future skill needs for sustainability, digitalization, and the bioeconomy in agriculture: European skills agenda and strategy) has been designed to identify the gaps and define the needs of knowledge and skills for the agri-food and forestry sectors addressing these above-mentioned emerging challenges. The FIELDS project aims to develop an integrated view of a sectoral strategy at the European level by considering four main areas of skill trends: sustainability, digitalization, bioeconomy, and soft skills and entrepreneurship. Within the scope of the FIELDS project, it was critical to have a broader understanding focused on stakeholders' views on knowledge and competences needed in the future. The survey is part of more comprehensive empirical research on skills needs including other activities such as focus groups [21] and trends and scenario analyses [22].

As a first step, nine focus groups were undertaken in different European countries to identify skills and training needs in the agri-food and forestry sectors. The outcomes of the focus groups [21] were used to design an online survey as follow-up research.

This study describes the results of the online survey, implemented as a broader exploratory tool to collect information about skills and training needs, as well as business and entrepreneurship trends, for the future of the agri-food and forestry sectors. The survey aimed to engage stakeholders from different areas of operation and to gain information on their understanding regarding skills needs and gaps in each sector. It also aimed to provide insight into the range of perceived future skills and training needs. The information provided by the survey may be used to help develop fit-for-purpose training courses for the areas of sustainability, digitalization, and bioeconomy. A coordinated strategy is needed to empower the workforce with new professional skills necessary to cope with emerging challenges and technologies of the agri-food and forestry sectors.

Due to the exploratory nature of this survey, the results are described and discussed in tandem and are herein organized by the following sections: the respondents' demographic characterization and their working context; skills needed by addressed areas (sustainability, bioeconomy—sector-specific skills, digitalization, soft skills, and business–entrepreneurship); perception of training availability by country; and business insights.

2. Materials and Methods

2.1. Description of Survey

Both current and future skills needs were first identified in previously performed FIELDS focus group sessions [21]. In these focus groups, participants were asked to select the 10 most important overall skills from skill lists that covered the topics of sustainability, digitalization, bioeconomy, soft skills, and business-entrepreneurship skills [21]. The five most-selected skills per category in this ranking exercise were included in the survey questionnaire. The web-based survey was developed and designed in English using the online SurveyMonkey® tool, which allowed the survey to be translated into nine different languages: Dutch, English, Finnish, French, German, Greek, Italian, Portuguese, Slovenian, and Spanish. It gave an overall introduction to the participants and allowed for both full and partial completion of the questions with an opt-out break built into the survey before the business trends section. In the introduction, participants were informed about the total confidentiality of the information provided to only be used in an aggregate manner for the purpose of identifying underlying trends and demands across European member states, in full compliance with the General Data Protection Regulation (EU) 2016/679. The survey comprised 10 sections with 31 questions (Appendix A) and was estimated to take no more than 15 min to complete. The questions with more relevant insights were selected to be analyzed and presented for discussion. Therefore, the survey sections were grouped as follows:

(1) Demographic profiles of participants, with particular focus on the country of work, gender, age, and organizational insights, including the working sectors and the size of their organizations.
(2) Current skill needs in the categories of sustainability, digitalization, bioeconomy—sector-specific (agriculture, forestry, and food industry), and soft and business–entrepreneurship skills. Questions about future skills needs in a 5–10 year range were also asked.
(3) Countries' particular training needs and the importance of training recognition.
(4) Business insights, including questions about business trends and challenges, current business models, and business strategy skills and tools. This final section was optional to complete.

2.2. Dissemination Campaign

The aim was to have input from stakeholders working in the agri-food sector, including the views of industry (workers, managers, cooperatives), academics, and policy entities. The core platform for dissemination was via direct email contacts with industry stakeholders through project partners' email contacts, as well as through the project website and social media (Facebook®, Twitter®, and LinkedIn®) and other accounts owned by project partners, using the snowball sampling technique [23]. The survey was shared among all 31 FIELDS partner organizations across 12 participating countries and was also disseminated in partner organizations' newsletters (ISEKI newsletter Dec 2020) and through other Erasmus+ and EU project contacts. The slogan "have your say in the future skills needs of the agri-food and forestry Sectors" was used to engage stakeholders. Dissemination was also done via direct (partner) newsletters as well as through their webpages. Other Erasmus+ and EU projects also disseminated the survey via their own platforms and social media channels. Several direct contact email reminders were sent out via the project partners and particular attention was given to countries where it was determined there were insufficient responses captured. The survey was available online between 1 December 2020 and 22 January 2021.

2.3. Data Analysis

The survey collected 517 answers; however, 123 were excluded because they only provided data on demographic questions. As a result, the considered sample size had 394 participants. Answers were exported from the SurveyMonkey® website to an Excel file for further analysis. The survey was set up with general and sector-specific questions and participants could select which questions to answer. This option led to a varying number of answers by question, which is specified in each presented figure or table. Before data analysis, the survey sections were reorganized for better clarity in the results presentation, and some less relevant sections are not discussed in this work. Graphical representation in the results section clustered respondents who answered categories "Very important" and "Absolutely essential" as "Very important," and categories "Not important at all" and "Of little importance" as "Of little importance," combining high- and low-scale scores. This data treatment was made for the sake of simplicity in the outcomes' presentation. The categories "Multiple sectors" and "Multiarea" grouped all respondents working in more than one sector or area of activity. The category "Other" within sectors comprised policy operators, educators, researchers, and service providers. In the area of operation, the category "Other" included researchers, associations, public representatives, and service providers. In the demographic profile analysis, the age categories were reduced from eight to six (<20 years; 20–29; 30–39; 40–49; 50–59; over 60). The results of the open question about skills needed in the future in 5 to 10 years were counted and we considered the number of answers by category. Regarding training needs by country, data for analysis included only countries with more than 15 answers. The section regarding business insights was optional, with 91 participants who selected to continue the survey and complete this section. The results from the open question "What do you see as the top three challenges facing your business over the next 2–3 years?" were grouped under seven topics according to the researchers' assessment, namely: "sustainability & climate," "bioeconomy & technical issues," "economic & investment & markets," "digitalization," "human resources," "soft skills," and "training & specialization." For this question, respondents (67) had the option to identify three different challenges.

Descriptive analysis regarding skills' importance was performed using cross-tabulations to analyze frequencies and associations between skills and sectors of activity. For the questions related to sustainability, digitalization, soft skills, and business and entrepreneurship skills, mean scores were calculated for the five-point Likert scale. The Kruskal–Wallis test, a nonparametric approach, was used to compare groups, followed by Dunn's post-hoc test when statistical differences were found (Appendix B Table A1) [24]. Statistical significance was tested at 0.05 probability level. Statistical tests were performed using IBM SPSS®25 software.

3. Results and Discussion

The survey collected opinions and views of professionals and other stakeholders about future skills needs in different sectors. A main goal was to reach a large audience among sectors and countries, conveying a broader view of the path to follow in upcoming years. The five surveyed and discussed skills in each category are the five most important skills obtained from the focus groups study [21].

3.1. Demographics and Organizational Insights

The demographic profile of respondents participating in the survey is presented in Table 1, totaling 394 responses. The survey gathered participants from 23 countries within the European Union, and some from the European Economic Area (EEA) (Figure 1). However, there were four countries (Spain, Italy, Ireland, and Austria) with higher inputs, corresponding to 53% of total participants. The majority of respondents were between 40 and 60 years old and 59% were male; this gender difference was reflected in all sectors except for the food industry, where more women contributed to the survey. The distribution by sector of activity was quite uneven: half of respondents worked in the Agricultural sector,

14% in the Food Industry, 10% worked in Multiple sectors (more than one considered sector of activity), and 16% in Other sectors (such as operators, educators, researchers, or service providers). The Bio-Based industries and the Forestry sectors were the least represented; the small sample size counted only 3% and 5% of the total responses. By area of operation, education providers and advisors were half of the respondents while the other half was distributed as farmers (10%), cooperatives (8.4%), agri-food companies (7.4%), and foresters-forest industries (1.5%). "Other" for area of operation included several professional areas, such as social partners, regulators, policy makers, trade associations, other industry sectors, researchers, and technicians. An overview of the size of organizations shows a fair balance between all categories included in the questionnaire, although most of the answers can be included in the range of small and medium enterprises. A large share of the respondents did not include information on organization size. The majority of respondents who answered represented micro-enterprises and SMEs while only 10% of participants were from large companies. The differences observed regarding sector of activity in the demographic profile will be further discussed in the results section.

Table 1. Description of socio-demographic characteristics of respondents in total and distributed by sector of activity.

	Total Respondents	Agriculture	Bio-Based Industries	Food Industry	Forestry	Multiple Sectors [1]	Other [2]
Number of participants	394 (100%)	201 (51%)	11 (3%)	55 (14%)	21 (5%)	41 (10%)	65 (16%)
Age (years)							
Less than 20	3 (1%)	2	0	2	0	2	0
20–29	39 (10%)	11	0	9	5	15	6
30–39	48 (12%)	12	18.2	22	14	5	8
40–49	122 (31%)	26	63.6	31	24	39	37
50–59	113 (29%)	27	18.2	27	43	27	34
over 60	66 (17%)	21	0	9	14	12	15
Gender							
Female	154 (39%)	35	36	56	24	32	46
Male	231 (59%)	64	64	36	76	63	51
Prefer not to say	9 (2%)	0.5	0	7.3	0	5	3
Area of Operation							
Advisor	45 (11.4%)	12	9	7	10	17	9
Education Provider	151 (38.3%)	32	55	36	57	22	60
Agri-Food Companies	29 (7.4%)	4	9	31	0	7	0
Co-operatives	33 (8.4%)	11	9	5	10	10	2
Farmer	40 (10.2%)	17	0	0	0	15	0
Forest Industries and Foresters	6 (1.5%)	0	0	0	24	2	0
Multiarea [3]	30 (7.6%)	6	9	15	0	22	2
Other [4]	60 (15.2%)	18	9	5	0	5	28
Organization size (persons)							
0–9	61 (15.5%)	21.9	9.1	7.3	9.5	19.5	3.1
10–49	50 (12.7%)	12.4	18.2	14.5	9.5	26.8	3.1
50–250	34 (8.6%)	10	0	12.7	4.8	9.8	3.1
250+	39 (9.9%)	7.5	0	18.2	19	14.6	6.2
No answer	210 (53.3%)	48.3	72.7	47.3	57.1	29.3	84.6

Data represented in %, except for data in column "total respondents" and line "number of participants" where data are presented as frequency and percentage. [1] Multiple sectors: grouped all respondents working in more than one sector; [2] Other sectors: comprised policy operators, educators, researchers, and service providers; [3] Multiarea: grouped all respondents working in more than one area of operation; [4] Other areas of operation included researchers, associations, public representatives, and service providers.

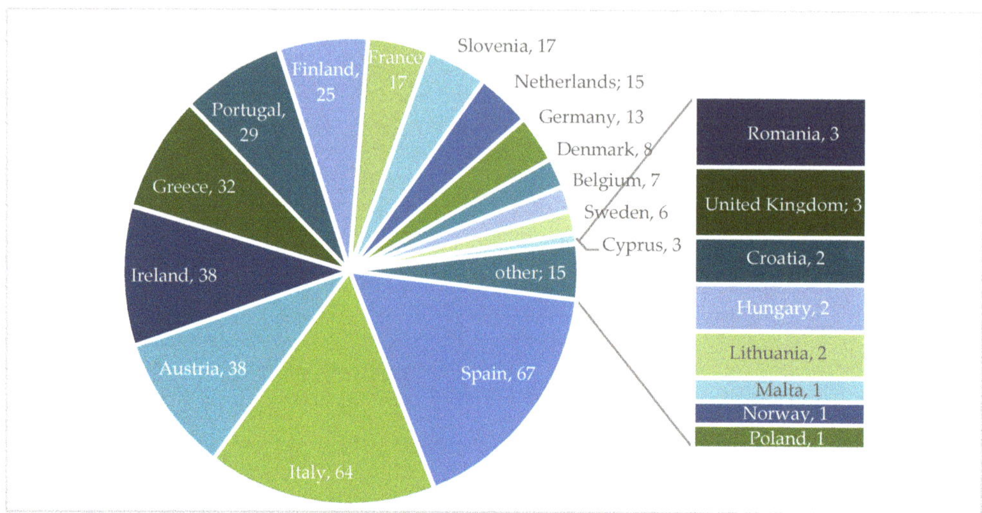

Figure 1. Number of respondents by country.

3.2. *Identified Current and Future Skills Needs*

3.2.1. Sustainability Skills

A sustainable food system encompasses environmental, health, and social benefits, as well as fairer economic gains [6]. Some studies address sustainability skills as the competences to thrive in an evolving agri-food sector, such as coping with unexpected events and adapting to new developments [25] or learning (critical thinking or communication) and life skills (flexibility and leadership) [26]. Within this study, these were considered soft skills and discussed in another section. In this study, sustainability skills were addressed as "green skills" important in the awareness of sustainable production, the mitigation of climate change, the reuse and recycling of resources and materials, and the use of renewable energy sources [15]. In this survey, the importance of five sustainability skills identified as relevant for the agri-food and forestry sectors previously in FIELDS focus groups [21] were evaluated by respondents. The results of this question are shown in Figure 2. Though all skills were important, "efficient use of resources and logistics," "mitigation and adaptation to climate change," and "good agricultural practices" were found the most important for respondents. "Efficient use of resources and logistics" was found to be important for all sectors. Skills related to "mitigation and adaptation to climate change" were identified as statistically more important by Agricultural, Food, Multiple, Forestry, and Other (Table A1) respondents' sectors. "Good agricultural practices" were found significantly more important for most of the respondents from the Agriculture, Other, and Multiple sectors (Table A1). "Soil nutrient and health management" were significantly less important to Food and Bio-Based industries (Table A1), and skills related to "by-products and co-products valorization" were those with lower shares of respondents finding the skills very important.

The agri-food industry clearly recognizes the importance of protecting and making good use of natural resources and the impact of climate change in disrupting supplies and processes as the main challenges to tackle towards a more sustainable future [27]. Agri-food and forestry activities contribute to climate change, but at the same time are dependent on natural resources and more vulnerable to its effects [12]. Previous studies suggest that technological developments to mitigate climate change effects require trained skilled workers [9]. Similar skills were previously identified in a sustainability transition context, referring to the importance of agri-food workers having global awareness of climate change impact, carbon emission reduction, water resources, and ecosystems management, but they should also have the skills to put in practice strategies related to renewable energies, by-products valorization, and more efficient production [11,15]. As indicated, to achieve the transition to sustainable production systems, it is required to train workers presently displaying poor or moderate skills levels with new competences that enable them to effectively promote sustainable agriculture through formal or life-long learning [28]. This upskilling will pave the way to apply better and more efficient processes with reduced impact on the environment and biodiversity.

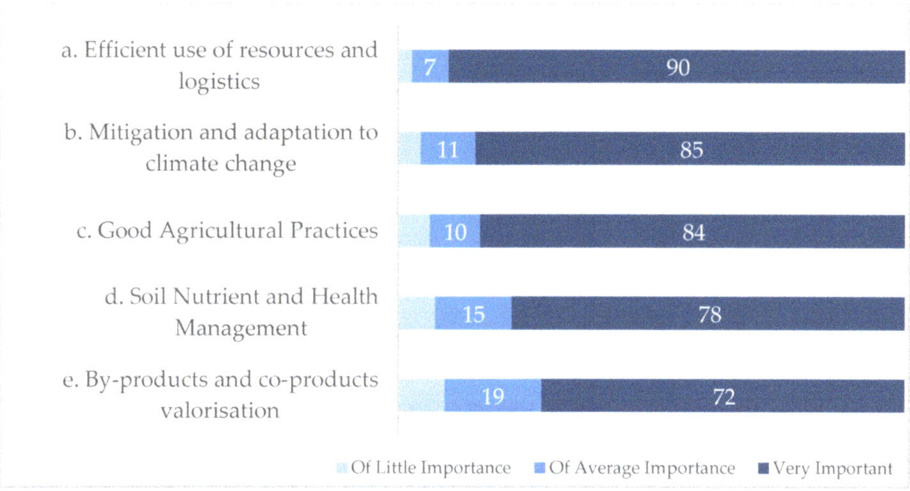

Figure 2. Categorization of the five selected sustainability skills by importance. Values in the graph represent the percentage of answers for each skill.

3.2.2. Digitalization Skills

Digital transformation occurs in everyday life and skills needed are applicable to all sectors. Information and communication technologies applied to agriculture have improved productivity, supply chains, strategic decisions, and control [29]. Digitalization is the main driver of Industry 4.0 development, comprising advanced technologies including automation of processes, use of robotics, and Internet of Things. These advances are now gaining *momentum* in the agri-food sector [30,31]. Big data is considered to be a key opportunity for the future development of agriculture since it increases the variety and velocity of data collection, enabling various tools and services to be implemented [32]. How to use and interpret the collected raw data is still a challenge. This concern is reflected in the high demand for skills related to "data handling and analysis" (Figure 3). The five selected skills from the FG outcomes were assessed in the survey questionnaire and results are presented in Figure 3. A high number of respondents also considered skills related to "everyday use of digital technology to communicate" particularly important. Likewise, a recent study on future skills required in the food industry for the transition to Industry 4.0 identified basic digital skills, data analysis, and use of complex digital communication skills as some of the essential skills in seven professional profiles of the food industry [10,33]. "Field operations management systems" and "farm management information systems (FMIS)" were also considered more important by the Agriculture and Multiple Sector respondents, as these are specific skills for agricultural practices (Table A1). These are digital operational practices developed to reduce operational and production costs with less environmental impact [34] and several obstacles related to their implementation have been identified, including insufficient farmers with adequate skills [35]. The importance of "e-commerce and e-marketing" scored lower for all sectors but still was found relevant for the respondents. E-commerce in the agri-food sector is promising and may help direct sales, shortening supply chains; however, it is still a challenging process and dependent on several factors, such as agricultural prices, logistics time, product quality, credibility, spending habits, and profitability [36,37]. Furthermore, in the agri-food sector, non-digital channels are preferably used, with this difference being even higher in rural areas [38]. It was highlighted that simple digital platforms are gaining position in communication, marketing, and business relationships in the agriculture market, particularly after the COVID-19 crisis [33], with digital technologies presenting several opportunities including the potential to reduce trade and transaction costs [39]. The inclusion of e-commerce and marketing skills in farmers' training has been proposed [40] and may help to develop food business, mainly in rural areas [36].

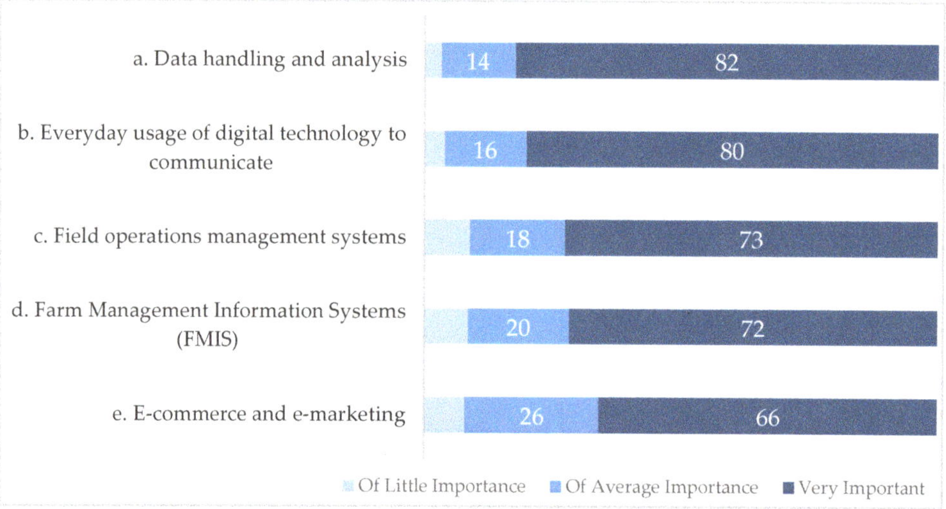

Figure 3. Categorization of five selected digitalization skills by importance. Values represent the percentage of answers for each skill.

The literature referring to the transition to Industry 4.0 for food companies describes diverse adoption patterns and technological demands [30,31], highlighting also that a skilled workforce is needed to fully exploit this technological potential [16]. Despite the numerous advantages of adopting innovative technologies in the agri-food and forestry sectors, there are challenges and risks, such as associated costs and the existence of appropriate training to support digital transition [41]. A guide for the food industry to meet the future skill requirements emerging from Industry 4.0 has recently been published [10]. The five selected digital skills in this work were also identified by Akyazi and colleagues [10] as very relevant to implementing the transition to a more digitalized food industry.

3.2.3. Bioeconomy—Sector-Specific Skills

In this survey, bioeconomy skills have been considered as those that are sector-specific for agriculture (Figure 4), food industry (Figure 5), and forestry (Figure 6) activities [8].

Regarding agriculture skills (Figure 4), the results of the survey clearly showed the importance of having a strategic and management vision to perform the activity. Skills related to "planning and coordinating production" and "calculating, handling and managing risks" were considered very important for the Agriculture sector respondents. The implementation of organizational tools in small farms, as a case study, was observed to improve productivity, product quality, and work environment [42]. "Performing farming operations" is becoming increasingly related to automated systems that reduce time and production costs and increase profitability [34]. Results showed that respondents found it more important to develop skills related to planning and coordinating production together with managing risks compared with those related to innovations in products and production, such as shifting to organic or growing new crops and developing new bioproducts more interesting from a bioeconomy point of view. In fact, using a management approach and organizational tools with a focus on planning and monitoring were suggested to increase profitability and contribute to a more agroecological and sustainable environment [27,42].

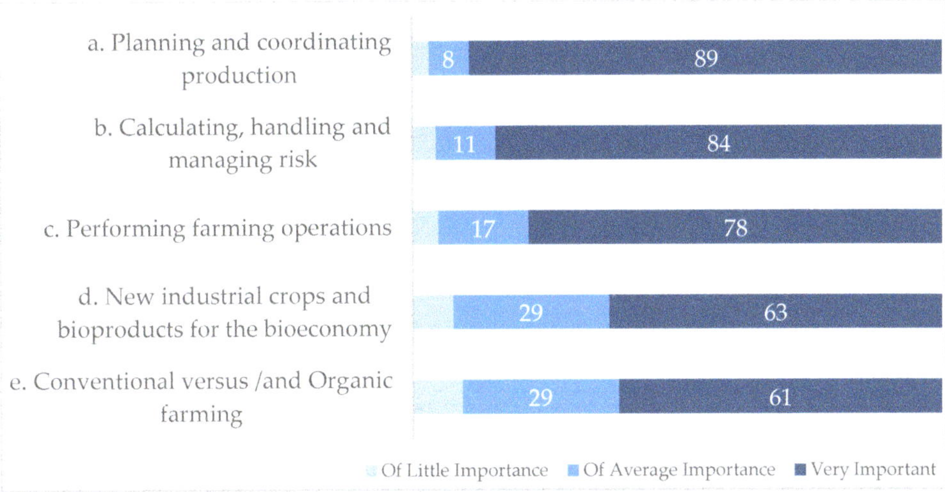

Figure 4. Categorization of five selected Bioeconomy—Agriculture skills by importance. Values represent the percentage of answers for each skill.

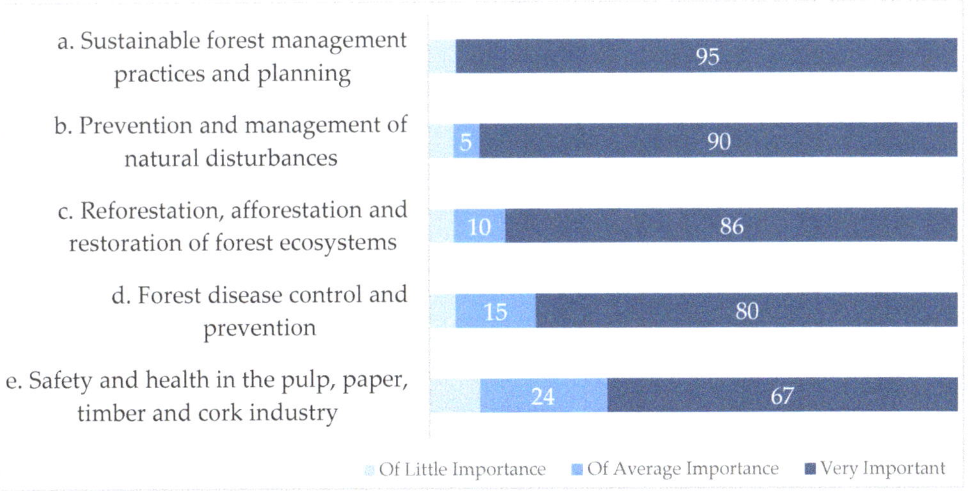

Figure 5. Categorization of five selected Bioeconomy—Forestry skills by importance. Values represent the percentage of answers for each skill.

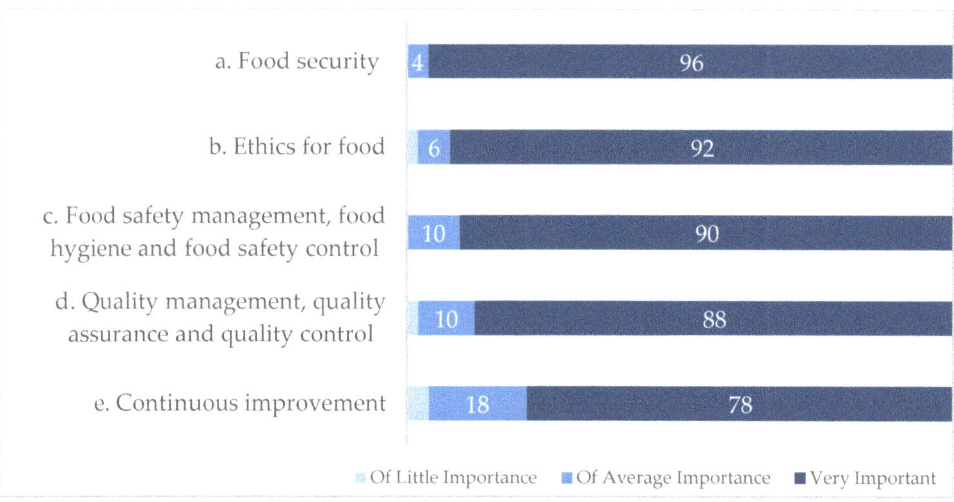

Figure 6. Categorization of five selected Bioeconomy—Food Industry skills by importance. Values represent the percentage of answers for each skill.

Regarding forestry skills (Figure 5), "sustainable forest management practices and planning," together with "prevention and management of natural disturbances," were considered the most important bioeconomy skills for the forestry sector and forest industries respondents. "Reforestation, afforestation and the restoration of forest ecosystems" were also indicated as very important. A recent online survey in the England and Wales forestry sector identified a long list of future skills needs, headed by practical skills and agroforestry and silvicultural knowledge. Other skills, such as "forest management and design" and "carbon and climate resilience," were also demanded, as well as more trained officers on plant health and diseases [18]. Another survey administered to forestry employers and students also highlighted the need for curricula improvement of land-management skills and suggested the continuous improvement of training programs to face unprecedented environmental challenges from forest disease, wildfire, drought, and population growth [17]. Different studies refer to the same skills needs as the ones here identified, which demands an effort to improve workforce abilities in this sector considering the current climate change scenario.

Regarding the food industry sector, skills related to "food security" and "ethics for food" were those found most important. These topics and their relation were referred to as one of the greatest dilemmas of our time [43]. How will nutritious food be available and provided for all, in a sustainable and safe manner, considering also complex public health problems, such as undernutrition, obesity, and micronutrient deficiencies [43]? The influence of climate change in food production is also immediately related to food systems and access to affordable, healthy food [44]. The challenges faced by food industry stakeholders are vast and include animal welfare, transparency, social justice, healthy food, and environmental issues, creating dilemmas when solutions are opposed [45]. Food industry workers need to be empowered with knowledge and decision-making tools to assist them in the ethical decision-making process [46].

"Quality management, quality assurance and quality control" and "food safety management, food hygiene and food safety controls" were also considered core skills in the food sector. The food industry is highly regulated and very compliance-driven to assure food quality for consumers. Food safety is a major concern in the sector, comprising significant challenges such as longer food chains, novel ingredient sources, new processing technologies, and higher consumer demands for fresh, low additive, and natural foods [46]. Scientific and technological advances that significantly impact food products and improve processes demand technical skills and knowledge to guarantee ingredients' safe application, processes control, and management along the food chain [47]. A European study to assess skills development of food professionals described similar important training activities more focused on current critical skills for the industry: product development, food legislation and control, food safety management, food hygiene and food safety control, quality management, quality assurance and quality control, and consumer and nutritional sciences [48].

3.2.4. Soft Skills

Soft skills are a set of positive attributes and competencies that can improve productivity and enhance relationships and are critical for performance in the workplace [49]. Like digital skills, soft skills are transversal and necessary in all sectors of activity. In general, respondents (between 89% and 91% depending on the skill) rated soft skills as very important for their work (Figure 7). Interestingly, there were only minor differences among sectors. The skills "being resilient, adaptable, and proactive" and "organization, planning, visioning and strategic thinking" were significantly more important for the Multiple, Food, Agriculture, and Other sectors. Creative and critical thinking, strategic planning, communication, networking, adaptability, and continuous learning are some of the skills identified to promote a transition to sustainable agriculture production [11,28], food industry and forestry sector development [10,16,50], and leadership in agriculture and natural resources activities [51]. A survey administered to students and employers identified communicating effectively and behaving professionally and ethically as the most important skills in forestry education [17]. In the food industry, a skill needs survey administered to food science and technology professionals and employers found that seven out of ten of the most required skills were soft skills; both groups agreed that communication was the strongest non-food skill [48]. In another survey for employers, the most significant skill gaps found were in the areas of personal attributes and attitudes [52]. In the agriculture sector, communication skills were among the required skills for young farmers [53], whereas communication, facilitating, and networking skill needs were found in agronomists' workers [28]. A shift from the traditional curricula centered on technical skills towards a curriculum equally balanced between technical, personal, and soft skills has been suggested for the forestry [18,54,55], food [10,48], and agriculture sectors [10,45]. The importance of soft skills in the workplace has long been recognized, and although these competences are generally considered to be acquired throughout life, several authors suggest that higher education institutions should be more active in promoting soft skills training in agri-food and forestry students [28,56,57].

3.2.5. Business and Entrepreneurship Skills

The business and entrepreneurship skills selected for this survey were found very important in general for the majority of respondents (Figure 8). The best qualifications were obtained by skills related to having business and management skills to consolidate a business and to find new business models or new value chains. The highest qualification was the need for "collaboration and cooperation across all sectors of the food chain." Comparing between activity sectors, higher importance scores were observed for all analyzed sectors but were significantly less important for the Forestry sector (Table A1). Furthermore, "interdisciplinary knowledge to assess the whole value chain" was highly evaluated by participants mainly working in Multiple sectors, probably because these stakeholders have a wider understanding across areas of activity for business growth, and significantly less relevant for the Bio-Based industry respondents. Collaboration along the supply chain and maintaining sustainable relations by fostering bonds were revealed to be key sources of value creation [58]. Enterprises are perceiving the importance of collaborations in their core activity and the advantages for their business competitiveness [59,60]. Skills that followed were "business planning/model and strategic management," scoring higher for all sectors, and again significantly less important for the Bio-Based industry participants (Table A1). Planning and management are found to be essential skills for business development. A study performed with small family farm businesses found that managers who focused on extensive planning and controlling perceived their business as successful [61]. However, developing and creating strategic plans was pointed out by others to be time-consuming but also difficult due to market uncertainty [62]. Skills related to "new value chains/business models" were generally regarded as important by all sectors. A well-defined business development strategy, shaped by unique features of each firm, was considered very significant to gain competitive advantage in both existing and new markets in the agri-food sector in a Northern Ireland survey [19]. However, lack of management skills, mainly in small firms, limits their innovative capability for growth [63]. These results are in line with McElwee [64], who described networking, innovation, teamwork, leadership, and business monitoring as very important skills in rural and farm entrepreneurship. Agri-food and forestry entrepreneurs are often demanded to have the technical skills to create a valuable product but also the competences to run a business. Therefore, a set of different skills involved in entrepreneurship requires a combination of theory and practice [65]. In small business, each farmer's characteristics and motivations are important drivers in the influence of their entrepreneurial activity, creating diversity in farm management, heterogeneity in value-creating strategies, and resiliency of farm systems [66]. As a general remark, new agri-entrepreneurs were shown to have fewer resources and capabilities, and in particular lower entrepreneurial skills and social capabilities, than entrepreneurs from other activities [67].

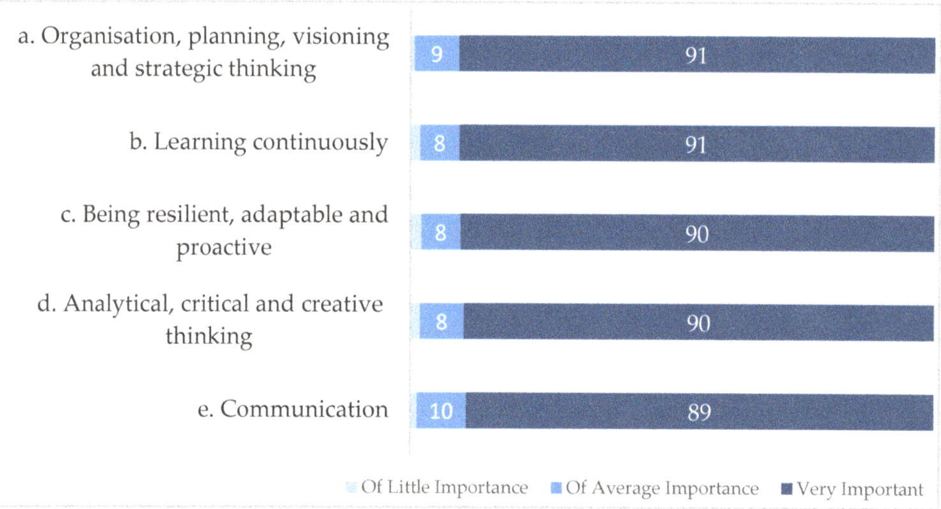

Figure 7. Categorization of five selected soft skills by importance. Values represent the percentage of answers for each skill.

3.2.6. Skills Needs in the Near Future

The participants' understandings of skills needs in the near future (5–10 years) are represented in Figure 9. Sector-specific skills were less significant in the future from respondents' point of view because they were considered more sector-specific. Digitalization skills were those found more important in the future, followed by sustainability skills, business–entrepreneurship, and soft skills. In the literature, these skills are found very relevant in the surveyed agriculture [11], food [16], and forestry [17] sectors. Lack of social competences as soft skills may limit workers to technical positions rather than filling managerial and leadership vacancies [55]. Advances in digitalization technologies are constant, meaning a continuous demand for upskilling. Digital transformation is shaping all aspects of the agri-food and forestry sectors, such as trade logistics, and distribution [39], smart farming and robotics production [31], marketing and communication, and contributing to achieving the Sustainable Development Goal [41,68].

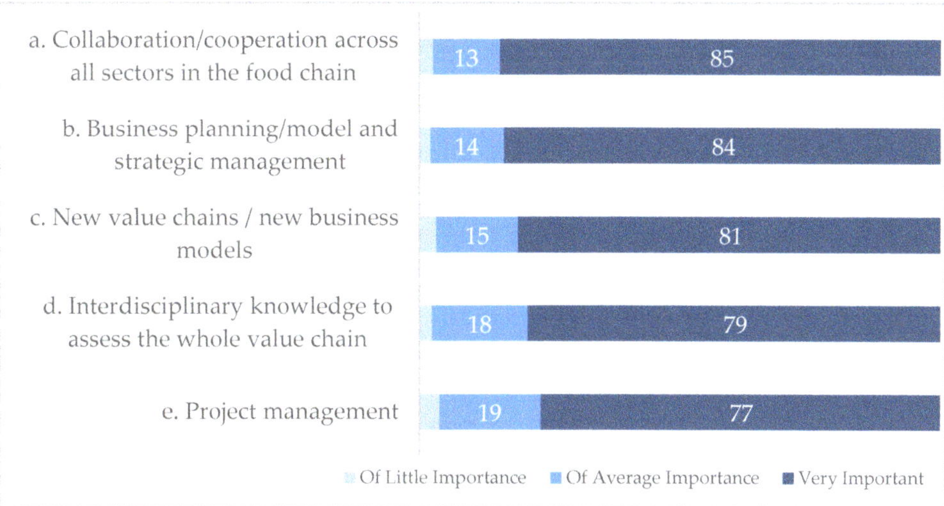

Figure 8. Categorization of five selected business and entrepreneurship skills by importance. Values represent the percentage of answers for each skill.

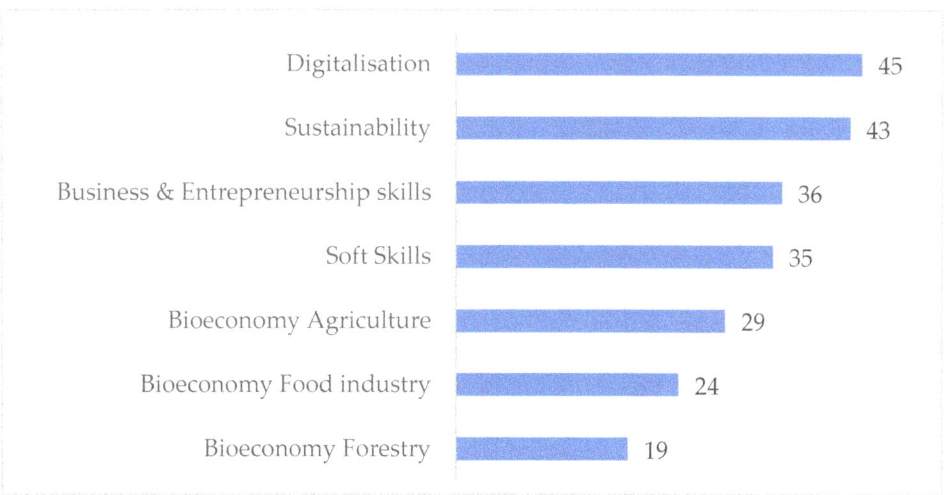

Figure 9. Skill categories required in the future for agri-food and forestry professionals (next 5 to 10 years). Values represent the percentage of answers for each skill category related to total answers.

The challenge for present and future workers in the agri-food and forestry industries is to acquire an assorted set of competences including digital and technical skills, communication and soft skills, and to efficiently manage decisions.

3.3. Countries' Training Needs and the Importance of Training Recognition

This study also intended to assess participants' views regarding the suitability of training systems, formal and non-formal, to cover existing training needs in each country (Figure 10). A considerable number of respondents were unaware of currently available training (including mostly non-formal but also about formal training) in their countries. Still, respondents were more aware of the formal education provided. Generally, formal training was considered more adequate to meet needs than non-formal, except for French respondents. For Austria, Finland, France, Ireland, the Netherlands, and Portugal, respondents clearly did not believe that the existing training systems, formal or non-formal, covered the country's needs. In contrast, in Spain, Slovenia, Greece, and Italy (only non-formal), systems were considered to cover training needs, formal and non-formal, by more than half of respondents. Respondents from Austria and Finland were more skeptical, and more than 80% of respondents considered training needs not covered by formal or non-formal training systems. Performing the analysis by sector of activity (Figure 11), participants consistently considered formal training more suitable to covering training needs compared with non-formal training. Furthermore, there is more uncertainty about existing training for non-formal systems, except for Bio-Based industry respondents. Though agricultural training varies largely throughout Europe, on average, only 8.5% of farmers have received formal agricultural training, and 70% have only practical experience [69]. Therefore, farmers' training seems to be an unresolved matter and is essential for the acquisition of skills in an ever-evolving sector [69]. Universities are viewed as essential to fostering the development of agriculture by having the ability to develop efficient training based on the latest research, to continuously evolve, and to provide education in different formats to support lifelong learning [9]. Despite these efforts, universities and training centers seem to face difficulties with providing needed education due to a lack of competent instructors and effective curricula [70]. Education institutions need to overcome conventional knowledge systems [71] and develop a new educational perspective by integrating formal and informal knowledge, scientific with technical subjects [72], and a broader understanding of challenges and opportunities in order to promote more sustainable agriculture [11].

When assessing the importance of training recognition and/or having the skills to perform the task after training (Figure 12), consideration was given at both organization and at trainee levels. Interestingly, organizations seemed to have more interest in both assessed aspects (formal qualification and having the skills to perform the task) when compared with the importance for the individual. This fact was observed across all studied sectors. It is also clearly observed that it is more important for both organizations and trainees to have the skill to perform the task than the recognition of training through formal qualifications. In the agri-food sector, the skill to perform is linked to innovation, and higher-skilled workers improve productivity by using innovative technologies at a faster rate [19], which is important. Organizations invest in upskilling employees' skills; however, around 50% of this training is "on-the-job" and provided by the firms, and only 24% are trained by a nationally recognized qualification [19].

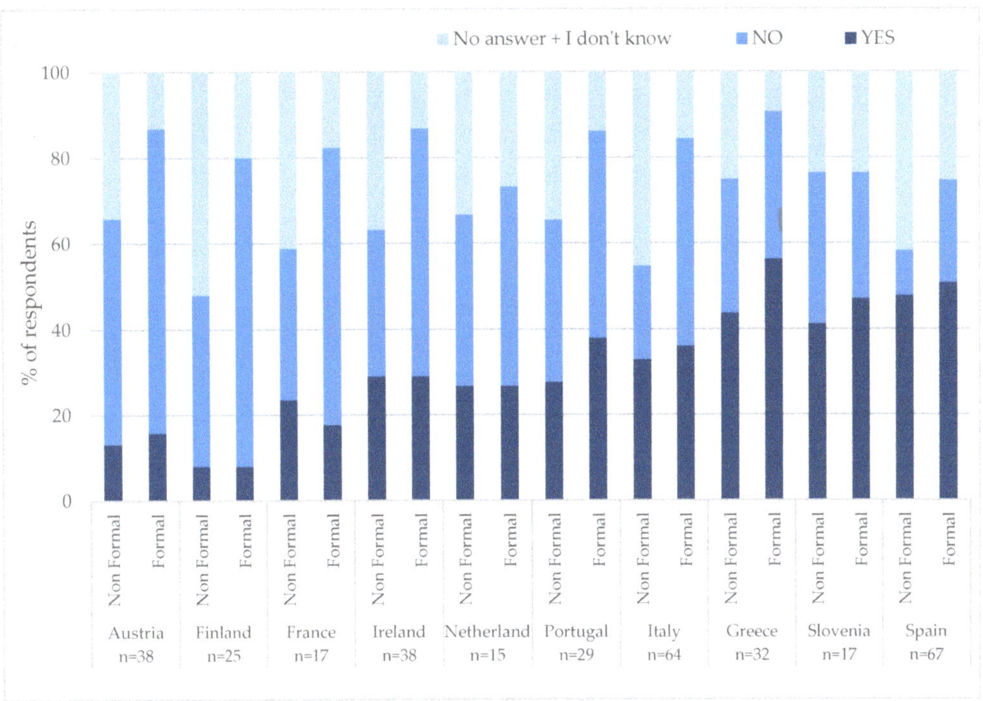

Figure 10. Perception of the suitability of training systems (formal and non-formal) to cover existing training needs by country. Bars represent the percentage of answers for each country and training system. N is the sample size.

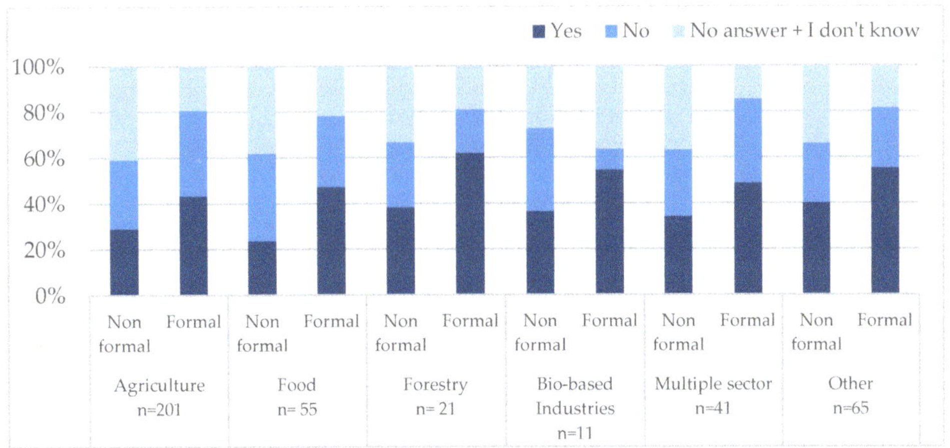

Figure 11. Perception of the suitability of training systems (formal and non-formal) to cover existing training needs by sector. Values represent the percentage of answers for each sector and training system. N is the sample size by sector.

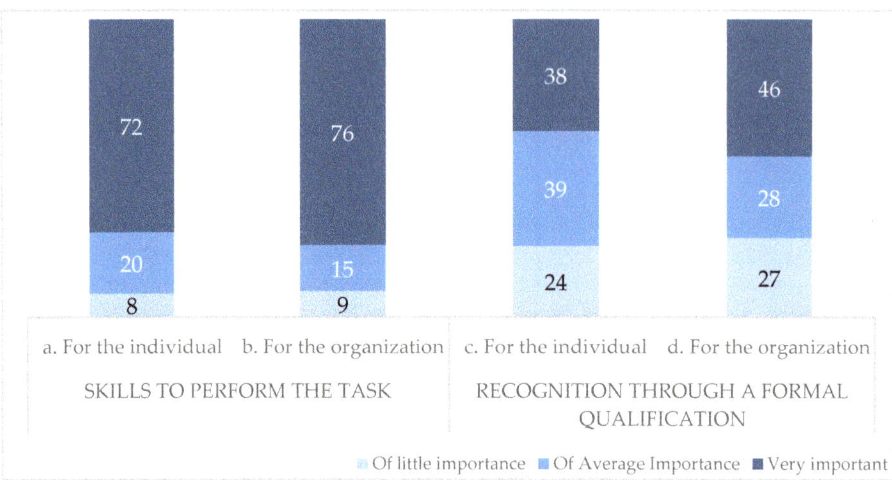

Figure 12. Results of the question on the importance of "having the skills to perform a task" after a training activity and the "recognition for a qualification" answered by trainees and organizations. Number of respondents by question: a. 176, b. 253, c. 197, d. 309. Values represent the percentage of answers for each question.

3.4. Business Trends

The questionnaire section related to business trends was optional and 91 participants agreed to proceed. More than half of the respondents were from the Agriculture sector (49), followed by Food, Other (both 13), and Multiple (12) sectors, and only some few were from Forestry (3) and Bio-Based industries (1). Within the business strategy context, consideration was given to the business operating models, the strategic business focus, and the required business strategy skills. This section first addressed the type of business models participants are operating in. The majority of respondents operated their core business model as business-to-business (B2B—56%), some were operating business-to-consumer (B2C—27%), and less (Other—7%) had a combination of the two models, or were cooperatives or research or consultancy institutions. Regarding sectors, and comparing both business models questioned, the business-to-business model was more adopted in all sectors compared with business-to-consumer: Agriculture (23 vs. 8 answers), Food (5 vs. 2 answers), Multiple (5 equal 5 answers), Forestry (3 vs. 0 answers), and Bio-Based industries (1 vs. 0 answers).

In Figure 13, the core strategic focus of the participants' business model is presented. Findings showed sustainability, innovation, business growth, and increased competitiveness as the main selected drivers for business development. Focuses on digital transformation and work to secure business also featured high on the strategic business focus. The transition to sustainable agri-food and forestry systems is closely linked with technical innovation [41] and digital transformation [73], as discussed in previous sections. Moreover, innovative business models supported by digital tools may foster the agri-food sector, value supply chains, and boost sustainability and employability [74].

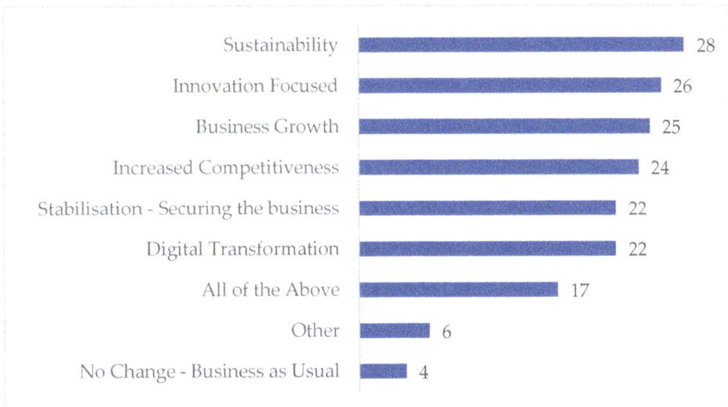

Figure 13. The strategic focus of business. Values represent the number of answers by question.

The three most relevant strategic business skills to support business models (Figure 14) were related to "business strategy, development, implementation, and analysis," "business continuity planning," "business planning/modelling and strategic management," and "recognition and realizing business opportunities." These were followed by "change management," "providing leadership," and "data analytics." "Growth mindset" and "collaboration and co-operation across sectors of the food chain" were also valued. Agriculture and Food Industry were the two sectors with more answers to this question. The agriculture sector highlighted "business strategy, development, implementation and analysis," and the Food Industry sector highlighted "new value chains and business models" as main business strategy skills. These findings showed an increased interest in new business models and management-related skills.

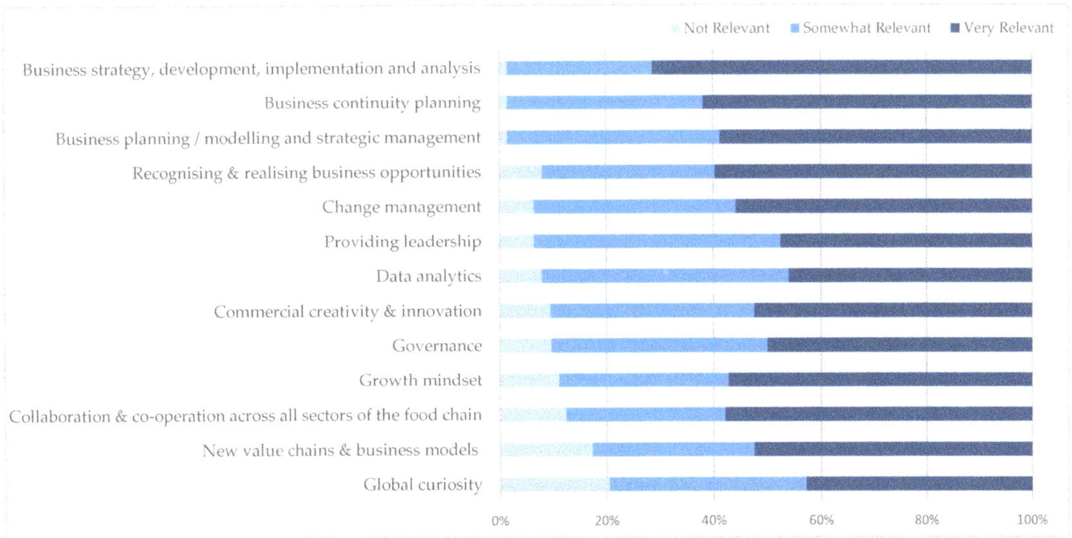

Figure 14. Answers to the question: How relevant are the following business strategy skills to your Core Business Model? Number of respondents varied between 61 and 64. Values represent the percentage of answers.

4. Conclusions

The identification of existing and emerging skills needs in bioeconomy, sustainability, and digitalization is of paramount importance to the development of a strategic approach that will bolster the European agri-food and forestry sectors in the transition to sustainable production in the long term. Survey outcomes on skills needs are in concordance with those found in the FIELDS focus groups. The skills considered most important in focus groups were found in general very important also by survey respondents, with some differences depending on the working sector and skill. Though some differences between each sector and specific skills were observed, on the main, the evaluated set of skills were considered relevant and may feature the agri-food and forestry workforce upskilling and reskilling training. A highlight should be given to the relevance of soft skills for all sectors, with all skills ranking very important for more than 90% of respondents, including skills related to "data handling analysis" and "everyday usage of digital technology to communicate" among those better evaluated by respondents. Among business and entrepreneurial skills, those considered most important were "collaboration across all sectors of the value chain" and "business planning and strategic management." For sector-specific skills, the skill with higher ranking for Agriculture was "planning and coordinating production," for Forestry it was "sustainable forest management practices and planning," and "food security" for the Food sector. According to participants' prospects of required skills in 5 to 10 years, these were mainly digital and sustainability skills. Using a trainer perspective over all the skills analyzed in this study, it is possible to remark that regardless of the category or sector assessed, "planning," "management," and "control" are key activities associated with the most-demanded skills in the survey, setting a focus to train students and workers in these competences.

There is a general perception among respondents that formal training better responds to training needs compared with non-formal training, but also that existing training systems do not cover actual skill needs. The results showed that a cross-sectoral approach developed to train a set of skills, including sustainability, digital skills, soft skills, and business skills, was identified by stakeholders as a way to tackle the agri-food and forestry sectors' challenges. Increasing technical skill levels will promote innovation and digital advances, which, harnessed with soft skills training, will create experts with sound human and technical competences who will be able to improve productivity, face business challenges, and create new markets to support and develop sustainable and solider agri-food and forestry sectors.

This study presented some limitations, mainly related to the fact that samples are not representative of studied sectors (small sample size) and not equally represented, being also quite unbalanced. This study is an exploratory study complemented by other activities on skill needs conducted in the FIELDS project, such as focus groups and scenario analysis. In the future, work skills needs will be analyzed from the perspective of the three activities.

Author Contributions: Conceptualization: R.B., P.B. and D.R.; methodology and validation: L.M., B.G. and L.F.L.; formal analysis: L.M., B.G., L.F.L., C.F.K. and A.R.R.; data curation: L.M., B.G., L.F.L. and A.R.R.; writing—original draft preparation, A.R.R., L.M. and L.F.L.; writing—review and editing: F.S., R.B., P.B., D.R., J.T. and B.G.; supervision: L.M. and L.F.L.; project administration: R.B., F.S. and P.B.; funding acquisition: R.B. and D.R. All authors have read and agreed to the published version of the manuscript.

Funding: This research was co-funded by the ERASMUS+ programme of the European Union, grant agreement number 612664-EPP-1-2019-1-IT-EPPKA2-SSA-B. This publication reflects only the authors' views, and the European Union is not responsible for any use that may be made of the information it contains.

Informed Consent Statement: Informed consent was obtained from all subjects involved in this study. This study was conducted in accordance with the Declaration of Helsinki, and the protocol was approved by the High Steering Committee of the FIELDS ERASMUS+ project (612664-EPP-1-2019-1-IT-EPPKA2-SSA-B).

Data Availability Statement: The data presented in this study are available on request from the corresponding author. The data are not publicly available due to privacy.

Acknowledgments: The authors would like to express their gratitude to the FIELDS project partners who contributed to the conducting and reporting of the survey.

Conflicts of Interest: The authors declare no conflict of interest. The funders had no role in the design of the study; in the collection, analyses, or interpretation of data; in the writing of the manuscript; or in the decision to publish the results.

Appendix A. Questionnaire

https://zenodo.org/record/7245370#.Y1eyR3bMJD8 (accessed on 17 February 2023).

Appendix B

Table A1. Categorization of selected skills by sector of activity.

	Sample (n)	Agriculture	Bio-Based Industries	Food	Forestry	Multiple Sectors	Other	*p* Value
Sustainability Skills								
Mitigation and adaptation to climate change	363	4.28^{ab}	3.90^{a}	3.92^{ab}	4.35^{b}	4.51^{b}	4.41^{ab}	0.004
Efficient use of resources and logistics	366	4.31	3.80	4.35	4.43	4.46	4.38	0.391
By-products and co-products valorisation	362	3.81	3.92	3.70	3.67	4.22	3.97	0.111
Good agricultural practices	355	4.46^{c}	3.76^{a}	3.48^{ab}	3.30^{ab}	4.59^{c}	4.07^{bc}	<0.001
Soil nutrient and health management	358	4.27^{b}	3.47^{a}	3.75^{a}	3.45^{ab}	4.26^{b}	3.98^{ab}	<0.001
Digitalization Skills								
Everyday usage of digital technology to communicate	367	4.20	3.55	4.19	4.00	4.24	4.28	0.320
Data handling and analysis	355	4.23^{ab}	3.60^{a}	4.18^{ab}	4.05^{ab}	4.49^{b}	4.00^{ab}	0.026
E-commerce and e-marketing	360	3.69	3.10	3.90	3.95	3.91	3.90	0.141
Farm Management Information Systems (FMIS)	357	4.08^{b}	3.30^{a}	3.39^{ab}	3.29^{a}	4.14^{b}	3.76^{ab}	<0.001
Field operations management systems	364	4.01^{b}	3.18^{a}	3.48^{ab}	3.62^{ab}	4.14^{b}	3.90^{ab}	0.010
Soft Skills								
Communication	368	4.31	3.80	4.44	4.24	4.57	4.35	0.072
Analytical, critical, and creative thinking	365	4.28	4.00	4.21	4.33	4.53	4.52	0.133
Being resilient, adaptable, and proactive	369	4.34^{bc}	3.60^{a}	4.34^{bc}	3.89^{ab}	4.56^{c}	4.45^{bc}	0.001
Organisation, planning, visioning, and strategic thinking	363	4.42^{bc}	3.78^{a}	4.37^{bc}	4.10^{ab}	4.66^{c}	4.33^{bc}	0.005
Learning continuously	358	4.37	4.00	4.31	4.43	4.53	4.39	0.706
Business and Entrepreneurship Skills								
Business planning/model and strategic management	359	4.26^{ab}	3.67^{a}	4.02^{ab}	3.90^{ab}	4.49^{b}	4.08^{ab}	0.006
New value chains/new business models	355	4.06	3.67	4.10	4.05	4.31	4.10	0.589
Collaboration/cooperation across all sectors in the food chain	364	4.31^{b}	3.91^{ab}	4.26^{ab}	3.68^{a}	4.52^{b}	4.29^{ab}	0.024
Interdisciplinary knowledge to assess the whole value chain	363	4.04^{ab}	3.60^{a}	4.06^{ab}	4.06^{ab}	4.46^{b}	4.17^{ab}	0.031
Project management	358	4.03	3.44	3.96	3.80	4.29	3.97	0.085

Data represent mean values of answers rated by importance. Superscript letters within rows represent statistical differences between median values ($p < 0.05$).

References

1. European Commission; Directorate-General for Research and Innovation; Froidmont-Görtz, I.; Faure, U.; Gajdzinska, M.; Haentjens, W.; Krommer, J.; Lizaso, M.; Lutzeyer, H.; Mangan, C.; et al. *Food 2030 Pathways for Action: Research and Innovation Policy as a Driver for Sustainable, Healthy and Inclusive Food Systems*; Ndongosi, I., Fabbri, K., Eds.; Publications Office: Brussels, Belgium, 2020.
2. Grassini, P.; Eskridge, K.M.; Cassman, K.G. Distinguishing between Yield Advances and Yield Plateaus in Historical Crop Production Trends. *Nat. Commun.* **2013**, *4*, 2918. [CrossRef] [PubMed]
3. The World Bank Food and Energy Price Shocks from Ukraine War Could Last for Years. Available online: https://www.worldbank.org/en/news/press-release/2022/04/26/food-and-energy-price-shocks-from-ukraine-war (accessed on 3 February 2023).
4. FAO. *Climate Change and Food Security: Risks and Responses*; FAO: Rome, Italy, 2015.
5. European Commission. Farm to Fork Strategy for a Fair, Healthy and Environmentally-Friendly Food System. In *Communication from the Commission to the European Parliament, the Council, the European Economic and Social Committee and the Committee of the Regions*; European Commission: Brussels, Belgium, 2020.
6. European Commission. *The European Green Deal*; Communication No. COM/2019/640; European Commission: Brussels, Belgium, 2019.
7. European Commission. *EU Biodiversity Strategy for 2030: Bringing Nature Back into Our Lives*; Publications Office of the European Union: Brussels, Belgium, 2021.
8. Sakellaris, G. Bioeconomy Education. In *Bio#Futures*; Koukios, E., Sacio-Szymańska, A., Eds.; Springer International Publishing: Cham, Switzerland, 2021; pp. 489–506. ISBN 978-3-030-64969-2.

9. Kőmíves, P.M.; Pilishegyi, P.; Novák, N.; Nagy, A.S.; Körösparti, P. The Role of the Higher Education in the Development of the Agriculture. *Int. J. Inf. Educ. Technol.* **2019**, *9*, 607–612. [CrossRef]
10. Akyazi, T.; Goti, A.; Oyarbide, A.; Alberdi, E.; Bayon, F. A Guide for the Food Industry to Meet the Future Skills Requirements Emerging with Industry 4.0. *Foods* **2020**, *9*, 492. [CrossRef] [PubMed]
11. Sorensen, L.B.; Germundsson, L.B.; Hansen, S.R.; Rojas, C.; Kristensen, N.H. What Skills Do Agricultural Professionals Need in the Transition towards a Sustainable Agriculture? A Qualitative Literature Review. *Sustainability* **2021**, *13*, 13556. [CrossRef]
12. Fróna, D.; Szenderák, J.; Harangi-Rákos, M. Economic Effects of Climate Change on Global Agricultural Production. *Nat. Conserv.* **2021**, *44*, 117. [CrossRef]
13. International Labour Organisation; European Centre for the Development of Vocational Training; European Training Foundation; Organisation for Economic Co-Operation and Development. *Skill Needs Anticipation: Systems and Approaches: Analysis of Stakeholder Survey on Skill Needs Assessment and Anticipation*; International Labour Organisation: Geneva, Switzerland, 2017.
14. Gualtieri, V.; Curtarelli, M.; Donlevy, V.; Shater Jannati, M. *ICT for Work: Digital Skills in the Workplace: Final Report*; Publications Office: Brussels, Belgium, 2017.
15. Silva, L.L.; Baptista, F.; Cruz, V.F. Analysis of Skills Needs for Agricultural Workers for a "Sustainable Agriculture". 2017. Available online: http://www.sagriproject.eu/wp-content/uploads/2018/11/D2.1_Analysis-of-skills-needs-for-agricultural-workers_EN_VFinal.pdf (accessed on 3 January 2023).
16. Hamann, K.; Gentile, M.; Loi, A.; Hegyi, A.; Zelenák, L. *New Professions and Career Paths in the Food and Drink Industry: Delivering High-Level Food Industry Skills in the Digital Economy*; European Federation of Food, Agriculture and Tourism Trade Unions, and Food Drink Europe: Brussels, Belgium, 2019.
17. Kelly, E.C.; Brown, G. Who Are We Educating and What Should They Know? An Assessment of Forestry Education in California. *J. For.* **2019**, *117*, 95–103. [CrossRef]
18. Forestry Skills Forum. *Forestry Workforce Research*; RDI Associates Ltd.: Edinburgh, UK, 2021.
19. Jack, C.; Anderson, D.; Connolly, N. Innovation and Skills: Implications for the Agri-Food Sector. *Educ. Train.* **2014**, *56*, 271–286. [CrossRef]
20. Phillips, C. The Role of the Universities in Agriculture Teaching and Research in the Twenty-First Century. *Outlook Agric.* **1999**, *28*, 253–256. [CrossRef]
21. Mayor, L.; Lindner, L.F.; Knöbl, C.F.; Ramalho, A.; Berruto, R.; Sanna, F.; Rossi, D.; Tomao, C.; Goodburn, B.; Avila, C.; et al. Skill Needs for Sustainable Agri-Food and Forestry Sectors (I): Assessment through European and National Focus Groups. *Sustainability* **2022**, *14*, 9607. [CrossRef]
22. Trienekens, J.; Morrenhof, W.; Nanda, A.; Bozou, E.; Lazaro Mojica, J.; Kretschmann, L. D1.8—Trend and Scenario Analysis. 2021. Available online: http://www.erasmus-fields.eu/management/sites/default/files/documents/final/D1.8_T1.5_FIELDS_Trend%20and%20scenario%20analysis_final.pdf (accessed on 5 January 2023).
23. Grbich, C. *Qualitative Research in Health: An Introduction*; Sage: London, UK, 1998; ISBN 1446235017.
24. Elliott, A.C.; Hynan, L.S. A SAS® Macro Implementation of a Multiple Comparison Post Hoc Test for a Kruskal–Wallis Analysis. *Comput. Methods Programs Biomed.* **2011**, *102*, 75–80. [CrossRef]
25. Darnhofer, I.; Bellon, S.; Dedieu, B.; Milestad, R. Adaptiveness to Enhance the Sustainability of Farming Systems. A Review. *Agron. Sustain. Dev.* **2010**, *30*, 545–555. [CrossRef]
26. Aver, B.; Fošner, A.; Alfirević, N. Higher Education Challenges: Developing Skills to Address Contemporary Economic and Sustainability Issues. *Sustainability* **2021**, *13*, 12567. [CrossRef]
27. Duru, M.; Therond, O.; Martin, G.; Martin-Clouaire, R.; Magne, M.-A.; Justes, E.; Journet, E.-P.; Aubertot, J.-N.; Savary, S.; Bergez, J.-E.; et al. How to Implement Biodiversity-Based Agriculture to Enhance Ecosystem Services: A Review. *Agron. Sustain. Dev.* **2015**, *35*, 1259–1281. [CrossRef]
28. Charatsari, C.; Lioutas, E.D. Is Current Agronomy Ready to Promote Sustainable Agriculture? Identifying Key Skills and Competencies Needed. *Int. J. Sustain. Dev. World Ecol.* **2019**, *26*, 232–241. [CrossRef]
29. Bollini, L.; Caccamo, A.; Martino, C. Interfaces of the Agriculture 4.0. In Proceedings of the WEBIST, Vienna, Austria, 18–20 September 2019; pp. 273–280.
30. Romanello, R.; Veglio, V. Industry 4.0 in Food Processing: Drivers, Challenges and Outcomes. *Br. Food J.* **2022**, *124*, 375–390. [CrossRef]
31. Ojo, O.O.; Shah, S.; Coutroubis, A.; Jiménez, M.T.; Ocana, Y.M. Potential Impact of Industry 4.0 in Sustainable Food Supply Chain Environment. In Proceedings of the 2018 IEEE International Conference on Technology Management, Operations and Decisions (ICTMOD), Marrakech, Morocco, 21–23 November 2018; pp. 172–177.
32. Lytos, A.; Lagkas, T.; Sarigiannidis, P.; Zervakis, M.; Livanos, G. Towards Smart Farming: Systems, Frameworks and Exploitation of Multiple Sources. *Comput. Netw.* **2020**, *172*, 107147. [CrossRef]
33. Payne, J.; Willis, M. *Digital Solutions Used by Agriculture Market System Actors in Response to COVID-19*; United States Agency for International Development: Washington, DC, USA, 2021.
34. Fountas, S.; Mylonas, N.; Malounas, I.; Rodias, E.; Hellmann Santos, C.; Pekkeriet, E. Agricultural Robotics for Field Operations. *Sensors* **2020**, *20*, 2672. [CrossRef] [PubMed]
35. Tummers, J.; Kassahun, A.; Tekinerdogan, B. Obstacles and Features of Farm Management Information Systems: A Systematic Literature Review. *Comput. Electron. Agric.* **2019**, *157*, 189–204. [CrossRef]

36. Zeng, Y.; Jia, F.; Wan, L.; Guo, H. E-Commerce in Agri-Food Sector: A Systematic Literature Review. *Int. Food Agribus. Manag. Rev.* **2017**, *20*, 439–460. [CrossRef]
37. Cai, Y.; Lang, Y.; Zheng, S.; Zhang, Y. Research on the Influence of E-Commerce Platform to Agricultural Logistics: An Empirical Analysis Based on Agricultural Product Marketing. *Int. J. Secur. Its Appl.* **2015**, *9*, 287–296. [CrossRef]
38. Bojkić, V.; Vrbančić, M.; Žibrin, D.; Čut, M. Digital Marketing in Agricultural Sector. *Entren. Enterp. Res. Innov.* **2016**, *2*, 419–424.
39. Jouanjean, M.-A. *Digital Opportunities for Trade in the Agriculture and Food Sectors*; Organisation for Economic Co-operation and Development: Paris, France, 2019.
40. Fecke, W.; Danne, M.; Musshoff, O. E-Commerce in Agriculture—The Case of Crop Protection Product Purchases in a Discrete Choice Experiment. *Comput. Electron. Agric.* **2018**, *151*, 126–135. [CrossRef]
41. Khan, N.; Ray, R.L.; Kassem, H.S.; Hussain, S.; Zhang, S.; Khayyam, M.; Ihtisham, M.; Asongu, S.A. Potential Role of Technology Innovation in Transformation of Sustainable Food Systems: A Review. *Agriculture* **2021**, *11*, 984. [CrossRef]
42. Barth, H.; Melin, M. A Green Lean Approach to Global Competition and Climate Change in the Agricultural Sector—A Swedish Case Study. *J. Clean. Prod.* **2018**, *204*, 183–192. [CrossRef]
43. Fanzo, J. Ethical Issues for Human Nutrition in the Context of Global Food Security and Sustainable Development. *Glob. Food Secur.* **2015**, *7*, 15–23. [CrossRef]
44. Mbow, C.; Rosenzweig, C.; Barioni, L.G.; Benton, T.G.; Herrero, M.; Krishnapillai, M.; Liwenga, E.; Pradhan, P.; Rivera-Ferre, M.-G.; Sapkota, T.; et al. Food Security. In *Climate Change and Land: An IPCC Special Report on Climate Change, Desertification, Land Degradation, Sustainable Land Management, Food Security, and Greenhouse Gas Fluxes in Terrestrial Ecosystems*; Shukla, P., Skea, J., Buendia, E.C., Mas-son-Delmotte, V., Pörtner, H.-O., Roberts, D., Zhai, P., Slade, R., Connors, S., van Diemen, R., Eds.; Intergovernmental Panel on Climate Change: Geneva, Switzerland, 2019.
45. Wernaart, B.F.W. *Applied Food Science*; Wernaart, B., van der Meulen, B., Eds.; Wageningen Academic Publishers: Wageningen, The Netherlands, 2022; ISBN 978-90-8686-381-5.
46. McClements, D.J.; Barrangou, R.; Hill, C.; Kokini, J.L.; Lila, M.A.; Meyer, A.S.; Yu, L. Building a Resilient, Sustainable, and Healthier Food Supply Through Innovation and Technology. *Annu. Rev. Food Sci. Technol.* **2021**, *12*, 1–28. [CrossRef] [PubMed]
47. WEF (World Economic Forum). *Innovation with a Purpose: The Role of Technology Innovation in Accelerating Food Systems Transformation*; WEF: Geneva, Switzerland, 2019.
48. Mayor, L.; Flynn, K.; Dermesonluoglu, E.; Pittia, P.; Baderstedt, E.; Ruiz-Bejarano, B.; Geicu, M.; Quintas, M.A.C.; Lakner, Z.; Costa, R. Skill Development in Food Professionals: A European Study. *Eur. Food Res. Technol.* **2015**, *240*, 871–884. [CrossRef]
49. Robles, M.M. Executive Perceptions of the Top 10 Soft Skills Needed in Today's Workplace. *Bus. Commun. Q.* **2012**, *75*, 453–465. [CrossRef]
50. Sample, V.A.; Ringgold, P.C.; Block, N.E.; Giltmier, J.W. Forestry Education: Adapting to the Changing Demands on Professionals. *J. For.* **1999**, *97*, 4–10. [CrossRef]
51. Easterly III, R.G.; Warner, A.J.; Myers, B.E.; Lamm, A.J.; Telg, R.W. Skills Students Need in the Real World: Competencies Desired by Agricultural and Natural Resources Industry Leaders. *J. Agric. Educ.* **2017**, *58*, 225–239. [CrossRef]
52. Dench, S.; Hillage, J.; Reilly, P.; Kodz, J. *Employers Skill Survey: Case Study Food Manufacturing Sector*; DFEE: Sheffield, UK, 2000.
53. Bailey, N.E.; Arnold, S.K.; Igo, C.G. Educating the Future of Agriculture: A Focus Group Analysis of the Programming Needs and Preferences of Montana Young and Beginning Farmers and Ranchers. *J. Agric. Educ.* **2014**, *55*, 167–183. [CrossRef]
54. Bullard, S.H. Forestry Curricula for the 21st Century—Maintaining Rigor, Communicating Relevance, Building Relationships. *J. For.* **2015**, *113*, 552–556. [CrossRef]
55. Sample, V.A.; Bixler, R.P.; McDonough, M.H.; Bullard, S.H.; Snieckus, M.M. The Promise and Performance of Forestry Education in the United States: Results of a Survey of Forestry Employers, Graduates, and Educators. *J. For.* **2015**, *113*, 528–537. [CrossRef]
56. Juhász, T.; Horváth-Csikós, G. The Emergence of Soft Skills in Agricultural Education. *Probl. Perspect. Manag.* **2021**, *19*, 453–466. [CrossRef]
57. Flynn, K.; Wahnström, E.; Popa, M.; Ruiz-Bejarano, B.; Quintas, M.A.C. Ideal Skills for European Food Scientists and Technologists: Identifying the Most Desired Knowledge, Skills and Competencies. *Innov. Food Sci. Emerg. Technol.* **2013**, *18*, 246–255. [CrossRef]
58. Howieson, J.; Lawley, M.; Hastings, K. Value Chain Analysis: An Iterative and Relational Approach for Agri-Food Chains. *Supply Chain. Manag. Int. J.* **2016**, *21*, 352–362. [CrossRef]
59. Reynolds, N.; Fischer, C.; Hartmann, M. Determinants of Sustainable Business Relationships in Selected German Agri-food Chains. *Br. Food J.* **2009**, *111*, 776–793. [CrossRef]
60. Schmiemann, M. Inter-Enterprise Relations in Selected Economic Activities. *Eurostat Stat. Focus* **2007**, *57*, 1–8.
61. Miller, N.J.; Mcleod, H.; Young Ob, K. Managing Family Businesses in Small Communities. *J. Small Bus. Manag.* **2001**, *39*, 73–87. [CrossRef]
62. Stanford-Billington, C.; Cannon, A. Do Farmers Adopt a Strategic Planning Approach to the Management of Their Businesses. *J. Farm Manag.* **2010**, *14*, 3–40.
63. Tell, J.; Hoveskog, M.; Ulvenblad, P.; Ulvenblad, P.-O.; Barth, H.; Ståhl, J. Business Model Innovation in the Agri-Food Sector: A Literature Review. *Br. Food J.* **2016**, *118*, 1462–1476. [CrossRef]
64. McElwee, G. A Taxonomy of Entrepreneurial Farmers. *Int. J. Entrep. Small Bus.* **2008**, *6*, 465–478. [CrossRef]
65. Anderson, A.R.; Jack, S.L. Role Typologies for Enterprising Education: The Professional Artisan? *J. Small Bus. Enterp. Dev.* **2008**, *15*, 259–273. [CrossRef]

66. Alvarez, A.; García-Cornejo, B.; Pérez-Méndez, J.A.; Roibás, D. Value-Creating Strategies in Dairy Farm Entrepreneurship: A Case Study in Northern Spain. *Animals* **2021**, *11*, 1396. [CrossRef]
67. Pindado, E.; Sánchez, M. Researching the Entrepreneurial Behaviour of New and Existing Ventures in European Agriculture. *Small Bus. Econ.* **2017**, *49*, 421–444. [CrossRef]
68. Cf, O. *Transforming Our World: The 2030 Agenda for Sustainable Development*; United Nations: New York, NY, USA, 2015.
69. Augère-Granier, M.-L. Agricultural Education and Lifelong Training in the EU. European Parliamentary Research Service. 2017. Available online: https://www.europarl.europa.eu/RegData/etudes/BRIE/2017/608788/EPRS_BRI(2017)608788_EN.pdf (accessed on 9 January 2023).
70. Erickson, B.; Fausti, S.; Clay, D.; Clay, S. Knowledge, Skills, and Abilities in the Precision Agriculture Workforce: An Industry Survey. *Nat. Sci. Educ.* **2018**, *47*, 1–11. [CrossRef]
71. Laforge, J.M.L.; McLachlan, S.M. Learning Communities and New Farmer Knowledge in Canada. *Geoforum* **2018**, *96*, 256–267. [CrossRef]
72. Lankester, A.J. Conceptual and Operational Understanding of Learning for Sustainability: A Case Study of the Beef Industry in North-Eastern Australia. *J. Environ. Manag.* **2013**, *119*, 182–193. [CrossRef]
73. Shepherd, M.; Turner, J.A.; Small, B.; Wheeler, D. Priorities for Science to Overcome Hurdles Thwarting the Full Promise of the 'Digital Agriculture' Revolution. *J. Sci. Food Agric.* **2020**, *100*, 5083–5092. [CrossRef]
74. Vlachopoulou, M.; Ziakis, C.; Vergidis, K.; Madas, M. Analyzing AgriFood-Tech e-Business Models. *Sustainability* **2021**, *13*, 5516. [CrossRef]

MDPI
St. Alban-Anlage 66
4052 Basel
Switzerland
Tel. +41 61 683 77 34
Fax +41 61 302 89 18
www.mdpi.com

Sustainability Editorial Office
E-mail: sustainability@mdpi.com
www.mdpi.com/journal/sustainability

www.ingramcontent.com/pod-product-compliance
Lightning Source LLC
LaVergne TN
LVHW070127170726
843515LV00007B/1764
* 9 7 8 3 0 3 6 5 8 3 1 8 1 *